Delivery and Mixing in the Subsurface: Processes and Design Principles for *In Situ* Remediation

---

SERDP and ESTCP Remediation Technology Monograph Series

Series Editor: C. Herb Ward, Rice University

Delivery and Mixing in the Subsurface: Processes and Design Principles for *In Situ* Remediation

---

Edited by

Peter K. Kitanidis

Stanford University, Stanford, CA

Perry L. McCarty

Stanford University, Stanford, CA

Authors

Linda M. Abriola
John A. Christ
Olaf A. Cirpka
Craig S. Criddle
Robert W. Gillham
Mark N. Goltz
Paul C. Johnson
Richard L. Johnson
Peter K. Kitanidis
Jian Luo
Perry L. McCarty
Kurt D. Pennell
C. Andrew Ramsburg
Albert J. Valocchi
John Vogan

Editors
Peter K. Kitanidis
Department of Civil and Environmental Engineering
Yang and Yamazaki Environment & Energy Building
Stanford University
473 Via Ortega
Stanford CA 94305
USA
peterk@stanford.edu

Perry L. McCarty
Department of Civil and Environmental Engineering
Yang and Yamazaki Environment & Energy Building
Stanford University
473 Via Ortega
Stanford, CA 94305
USA
pmccarty@stanford.edu

ISSN 1869-6864 ISSN 1869-6856 (electronic)
ISBN 978-1-4614-2238-9 ISBN 978-1-4614-2239-6 (eBook)
DOI 10.1007/978-1-4614-2239-6
Springer New York Heidelberg Dordrecht London

Library of Congress Control Number: 2012933155

Cover illustration: Graphical depiction of a remediated plume (blue) emerging downgradient from an injection-extraction well pair used for treatment of a contaminated aquifer (red) (see Chapter 6).

Cover design: The image on the cover of this volume was created through mathematical simulation using COMSOL Multiphysics® by Peter K. Kitanidis. Cover layout designed by Kenneth C. Arevalo, Noblis Inc., Falls Church, VA.

Printed on acid-free paper

Springer is part of Springer Science+Business Media (www.springer.com)

SERDP and ESTCP Remediation Technology Monograph Series
Series Editor: C. Herb Ward, Rice University

SERDP and ESTCP have joined to facilitate the development of a series of monographs on remediation technology written by leading experts in each subject area. This volume provides a review of the state-of-the-art on processes and design principles for subsurface delivery and mixing for *in situ* remediation. Volumes previously published in this series include:

- *In Situ* Bioremediation of Perchlorate in Groundwater
- *In Situ* Remediation of Chlorinated Solvent Plumes
- *In Situ* Chemical Oxidation for Remediation of Contaminated Groundwater

Additional volumes planned for publication in the near future include:

- Bioaugmentation for Groundwater Remediation
- Processes, Assessment and Remediation of Contaminated Sediment
- Chlorinated Solvent Source Zone Remediation

U.S. Department of Defense Strategic Environmental Research & Development Program (SERDP)
901 North Stuart Street, Suite 303
Arlington, VA 22203

U.S. Department of Defense Environmental Security Technology Certification Program (ESTCP)
901 North Stuart Street, Suite 303
Arlington, VA 22203

Preface

In the late 1970s and early 1980s, our nation began to grapple with the legacy of past disposal practices for toxic chemicals. With the passage in 1980 of the Comprehensive Environmental Response, Compensation, and Liability Act (CERCLA), commonly known as Superfund, it became the law of the land to remediate these sites. The U.S. Department of Defense (DoD), the nation's largest industrial organization, also recognized that it too had a legacy of contaminated sites. Historic operations at Army, Navy, Air Force, and Marine Corps facilities, ranges, manufacturing sites, shipyards, and depots had resulted in widespread contamination of soil, groundwater, and sediment. While Superfund began in 1980 to focus on remediation of heavily contaminated sites largely abandoned or neglected by the private sector, the DoD had already initiated its Installation Restoration Program in the mid-1970s. In 1984, the DoD began the Defense Environmental Restoration Program (DERP) for contaminated site assessment and remediation. Two years later, the U.S. Congress codified the DERP and directed the Secretary of Defense to carry out a concurrent program of research, development, and demonstration of innovative remediation technologies.

As chronicled in the 1994 National Research Council report, "Ranking Hazardous-Waste Sites for Remedial Action," our early estimates on the cost and suitability of existing technologies for cleaning up contaminated sites were wildly optimistic. Original estimates, in 1980, projected an average Superfund cleanup cost of a mere $3.6 million per site and assumed only around 400 sites would require remediation. The DoD's early estimates of the cost to clean up its contaminated sites were also optimistic. In 1985, the DoD estimated the cleanup of its contaminated sites would cost from $5 billion to $10 billion, assuming 400–800 potential sites. A decade later, after an investment of over $12 billion on environmental restoration, the cost-to-complete estimates had grown to over $20 billion and the number of sites had increased to over 20,000. By 2007, after spending over $20 billion in the previous decade, the estimated cost to address the DoD's known liability for traditional cleanup (not including the munitions response program for unexploded ordnance) was still over $13 billion. Why did we underestimate the costs of cleaning up contaminated sites? All of these estimates were made with the tacit assumption that existing, off-the-shelf remedial technology was adequate to accomplish the task, that we had the scientific and engineering knowledge and tools to remediate these sites, and that we knew the full scope of chemicals of concern.

However, it was soon and painfully realized that the technology needed to address the more recalcitrant environmental contamination problems, such as fuels and chlorinated solvents in groundwater and dense nonaqueous phase liquids (DNAPLs) in the subsurface, was seriously lacking. In 1994, in the "Alternatives for Ground Water Cleanup" document, the National Research Council clearly showed that as a nation we had been conducting a failed 15-year experiment to clean up our nation's groundwater and that the default technology, pump-and-treat, was often ineffective at remediating contaminated aquifers. The answer for the DoD was clear. The DoD needed better technologies to clean up its contaminated sites and better technologies could only arise through a better scientific and engineering understanding of the subsurface and the associated chemical, physical, and biological processes. Two DoD organizations were given responsibility for initiating new research, development, and demonstrations to obtain the technologies needed for cost-effective remediation of facilities across the DoD: the Strategic Environmental Research and Development Program (SERDP) and the Environmental Security Technology Certification Program (ESTCP).

SERDP was established by the Defense Authorization Act of 1991 as a partnership of the DoD, the U.S. Department of Energy, and the U.S. Environmental Protection Agency; its mission is "to address environmental matters of concern to the Department of Defense and the Department of Energy through support of basic and applied research and development of technologies that can enhance the capabilities of the departments to meet their environmental obligations." SERDP was created with a vision of bringing the capabilities and assets of the nation to bear on the environmental challenges faced by the DoD. As such, SERDP is the DoD's environmental research and development program. To address the highest-priority issues confronting the Army, Navy, Air Force, and Marine Corps, SERDP focuses on cross-service requirements and pursues high-risk and high-payoff solutions to the DoD's most intractable environmental problems. SERDP's charter permits investment across the broad spectrum of research and development, from basic research through applied research and exploratory development. SERDP invests with a philosophy that all research, whether basic or applied, when focused on the critical technical issues, can impact environmental operations in the near term.

A DoD partner organization, ESTCP, was established in 1995 as the DoD's environmental technology demonstration and validation program. ESTCP's goal is to identify, demonstrate, and transfer technologies that address the DoD's highest priority environmental requirements. The program promotes innovative, cost-effective environmental technologies through demonstrations at DoD facilities and sites. These technologies provide a large return on investment through improved efficiency, reduced liability, and direct cost savings. The current cost and impact on DoD operations of environmental compliance is significant. Innovative technologies are reducing both the cost of environmental remediation and compliance and the impact of DoD operations on the environment, while enhancing military readiness. ESTCP's strategy is to select laboratory-proven technologies with potential broad DoD application and use DoD facilities as test beds. By supporting rigorous test and evaluation of innovative environmental technologies, ESTCP provides validated cost and performance information. Through these tests, new technologies gain end-user and regulatory acceptance.

In the 15–19 years since SERDP and ESTCP were formed, much progress has been made in the development of innovative and more cost-effective environmental remediation technology. Since then, recalcitrant environmental contamination problems for which little or no effective technology had been available are now tractable. However, we understand that newly developed technologies will not be broadly used in government or industry unless the consulting engineering community has the knowledge and experience needed to design, cost, market, and apply them.

To help accomplish the needed technology transfer, SERDP and ESTCP have facilitated the development of a series of monographs on remediation technology written by leading experts in each subject area. Each volume will be designed to provide the background in process design and engineering needed by professionals who have advanced training and 5 or more years of experience. The first volume in this series, *In Situ Bioremediation of Perchlorate in Groundwater*, met a critical need for state-of-the-technology guidance on perchlorate remediation. The second volume, *In Situ Remediation of Chlorinated Solvent Plumes*, addresses the diverse physical, chemical, and biological technologies currently in use to treat what has become one of the most recalcitrant contamination problems in the developed world. The third volume, *In Situ Chemical Oxidation for Remediation of Contaminated Groundwater*, describes the principles and practices of this emerging technology. This volume, *Delivery and Mixing in the Subsurface: Processes and Design Principles for In Situ Remediation*, describes the principles of chemical delivery and mixing systems and their design and implementation for effective *in situ* remediation. Other volumes will follow, including additional volumes that will be written as new remediation technologies are developed and proven to be effective.

This volume has ten chapters including the introduction. The two chapters that follow the introduction are tutorials on reactions and processes of importance in groundwater remediation. In Chapter 2, the focus is on chemistry and biology and Chapter 3 focuses on transport. The synthesis of these concepts is addressed in the remaining chapters. This division should make it easier for readers who have sufficient background and may not wish to study either or both of these two chapters.

The next two chapters present conceptualization of the problem and modeling approaches. Review of available hydrogeochemical models is given in Chapter 4 together with their application. This chapter addresses issues of site characterization and model calibration for practical implementation. Chapter 5 emphasizes the travel-time approach to modeling – a specialized approach well suited for engineered remediation problems. Included are protocols for practical implementation for design and monitoring of progress.

The next chapters deal with design issues. Chapter 6 covers an important method for introducing and mixing chemicals with groundwater contaminants, the recirculation system, where mixing is controlled through and takes place mostly in wells. Design principles are addressed as are the challenges of operating and maintaining injection-extraction wells, the non-uniform distribution of biomass, plugging issues, and strategies such as pulsing. Case studies are provided. Chapter 7 has similar objectives, covering the design and use of reactive barrier walls. Contaminants are brought through normal groundwater advection to the reacting chemicals contained within permeable barrier walls through which the groundwater passes.

Technologies such as air sparging used for introducing and mixing gaseous reactants as well as for removing volatile contaminants are discussed in Chapter 8. Starting with a general background on processes for mixing of gases, the discussion proceeds to design issues of air sparging. Case studies are included. Chapter 9 covers the special case of intrinsic remediation in natural-gradient systems, where the kinetics are controlled through mixing of contaminant plumes with ambient species, such as dissolved oxygen, over long periods of time. Issues covered include monitoring and estimation of long-term reaction rates, lengths of plumes, and other important parameters. Case studies again are provided. Chapter 10 addresses the very special problem of source remediation, as opposed to plume remediation, and its challenges.

Each chapter in this volume has been thoroughly reviewed for technical content by one or more experts in the subject area covered. The editors and chapter authors have produced a state-of-the-art volume that we hope will prove to be a useful reference for those making decisions on remediation of contaminated groundwater and for those involved in research and development of advanced technology for the *in situ* remediation of groundwater.

SERDP and ESTCP are committed to the development of new and innovative technologies to reduce the cost of remediation of soil, groundwater, and sediment contamination as a result of past operational and industrial practices. We are also firmly committed to the widest dissemination of these technologies to ensure that our investments continue to yield savings for not only the DoD, but also the nation. In facilitating this monograph series, we hope to provide the broader remediation community with the most current knowledge and tools available in order to encourage full and effective use of these technologies.

Jeffrey A. Marqusee, PhD, Executive Director, SERDP and ESTCP
Andrea Leeson, PhD, Environmental Restoration Program Manager, SERDP and ESTCP

About the Editors

Peter K. Kitanidis

Dr. Kitanidis is Professor in the Department of Civil and Environmental Engineering and in the Institute for Computational and Mathematical Engineering at Stanford University. Before joining Stanford University in 1987, he held faculty appointments at the University of Iowa and the University of Minnesota.

Dr. Kitanidis has a diploma in Civil Engineering from the National Technical University of Athens, Greece (1974), an M.S. in Civil Engineering (1976) and a Ph.D. in Water Resources (1978) from the Massachusetts Institute of Technology.

Dr. Kitanidis specializes in the analysis of data and the development and implementation of mathematical models that describe and predict flow and transport rates in the environment. He has devised methods for the analysis of spatially distributed hydrologic and water-quality data, the calibration of groundwater models, and the optimization of sampling and control strategies when the available information is incomplete. He is active in the study of mixing processes in groundwater, the development of cost-effective enhanced *in situ* remediation methods and on how the heterogeneous nature of hydrogeochemical variables affects the net rates of flow and transport at hydrologic scales. He was the recipient of the L. G. Straub Award in 1979, the W. L. Huber Civil Engineering Research Prize in 1994, and he was recognized as an ISI Highly Cited Researcher in 2001. He delivered the Pioneers in Groundwater Lecture (ASCE [American Society of Civil Engineers]) in 2011.

Perry L. McCarty

Dr. McCarty is the Silas H. Palmer Professor Emeritus at Stanford University where he joined the faculty in 1962. From 1989 to 2002 he served as Director of the Western Region Hazardous Substance Research Center. He has a B.S. degree in Civil Engineering from Wayne State University, and M.S. and Sc.D. degrees in Sanitary Engineering from the Massachusetts Institute of Technology.

His research focus has been on biological processes for the control of environmental contaminants. He was elected to the National Academy of Engineering in 1977, the American Academy of Arts and Sciences in 1996, and to Honorary Membership in the American Academy of Environmental Engineers in 2009. He received the John and Alice Tyler Prize for Environmental Achievement in 1992, the Athalie Richardson Irvine Clarke Prize for Outstanding Achievements in Water Science and Technology in 1997, the Stockholm Water Prize in 2007, and was inducted into the Water Industry Hall of Fame in 2009. Dr. McCarty has over 350 publications, and is coauthor of the textbooks, *Chemistry for Environmental Engineering and Science*, and *Environmental Biotechnology – Principles and Applications.*

About the Authors

Linda M. Abriola

Dr. Abriola is Dean of Engineering and Professor of Civil and Environmental Engineering at Tufts University. Dr. Abriola received her Ph.D. from Princeton University in Civil Engineering. Prior to joining Tufts, she was the Horace Williams King Collegiate Professor of Environmental Engineering at the University of Michigan, where she also directed the Environmental and Water Resources Engineering Program. Dr. Abriola's research relates to the integration of mathematical modeling and laboratory experiments to investigate multiphase and reactive transport of contaminants in the subsurface.

An author of more than 130 refereed publications, she has been the recipient of a number of awards, including the Association for Women Geoscientist's Outstanding Educator Award, the National Ground Water Association's Distinguished Darcy Lectureship, designation as an ISI Highly Cited Author in Ecology/Environment, and the Strategic Environmental Research and Development Program Project of the Year Award in Remediation. She is a Fellow of the American Geophysical Union and a member of both the American Academy of Arts and Sciences and the National Academy of Engineering.

John A. Christ

Lt. Col. Christ is the Deputy Department Head for Curriculum and Faculty Development, Environmental Engineering Division Chief, and Associate Professor in the Department of Civil and Environmental Engineering at the U.S. Air Force Academy. He received his Ph.D. from the University of Michigan in 2005 and has research interests in mathematical modeling of multiphase flow and transport in heterogeneous porous media, the influence of bioactivity on nonaqueous phase liquid (NAPL) dissolution, upscaled modeling of dense nonaqueous phase liquid (DNAPL) source zones, vapor intrusion, engineering education and socially responsible engineering. Lt. Col. Christ is a registered professional engineer (PE).

Olaf A. Cirpka

Olaf A. Cirpka is Professor of Hydrogeology at the Center of Applied Geoscience at Eberhard Karls Universität Tübingen, Germany. He holds a Diploma in Geoecology from the University of Karlsruhe, Germany, and a Ph.D. in Civil Engineering from the University of Stuttgart, Germany. He was consulting faculty member at the Department of Civil and Environmental Engineering at Stanford University for 9 years and worked at the Swiss Federal Institute of Aquatic Science and Technology, Eawag, from 2004 to 2008.

Dr. Cirpka is a groundwater hydrologist who has specialized in modeling and analyzing mixing-controlled reactive transport in heterogeneous aquifers using numerical simulations and stochastic analysis. His other research interests focus on the identification of aquifer properties by geostatistical inverse modeling. He has authored more than 80 scientific papers and is associate editor of several journals.

Craig S. Criddle

Dr. Criddle is a Professor of Civil and Environmental Engineering at Stanford University and Senior Fellow in the Woods Institute for the Environment. He received his B.S. in Civil and Environmental Engineering and B.A. in Spanish from Utah State University in 1982 followed by an M.S. in Environmental Engineering in 1984. In 1990, he completed his Ph.D. at Stanford

University in Civil Engineering (Environmental Engineering and Science). His research focus is environmental biotechnology.

He began his academic career in 1989 as a faculty member at Michigan State University (MSU). While at MSU, he served as Project Director for the Schoolcraft project, a field-scale test of bioaugmentation that involved faculty from diverse disciplines and institutions, staff scientists, students, consultants, and outreach education to members of the public and officials in Michigan. The National Ground Water Association named it one of two Outstanding Remediation Projects for 2002.

Since 1998, Dr. Criddle has been a member of the Stanford faculty, serving as Associate Chair in 2003, and as Director of a bioremediation field project at the U.S. Department of Energy (USDOE) Field Research Center in Oak Ridge, Tennessee (2000–2006). This team project entailed a multi-year series of experiments focused on *in situ* sequestration of uranium in groundwater, and involved faculty from several universities, USDOE scientists, and many students. Dr. Criddle has over 100 peer refereed publications, and is coauthor with award-winning artist Larry Gonick of *The Cartoon Guide to Chemistry* (2006), a widely used supplementary text for high school and freshman chemistry.

Robert W. Gillham

Dr. Robert Gillham was a Professor in Hydrogeology in the Department of Earth and Environmental Sciences at the University of Waterloo for more than 30 years. His main research interests included physical hydrogeology, contaminant transport processes and groundwater remediation. Dr. Gillham has over 130 refereed publications, a similar number of non-refereed contributions, and 66 M.Sc. and 23 Ph.D. students have graduated under his supervision. Currently, Dr. Gillham is a Distinguished Professor Emeritus of the University of Waterloo.

Mark N. Goltz

Dr. Goltz is Professor of Engineering and Environmental Management at the Air Force Institute of Technology, Wright Patterson AFB, Ohio. He has a B.S. in Electrical Engineering from Cornell University, an M.S. in Sanitary Engineering from the University of California, Berkeley, and a Ph.D. in Environmental Engineering and Science from Stanford University. He is a retired Air Force officer with 20 years of service as an Air Force civil engineer. He has a number of publications reporting the results of field demonstrations of innovative contaminated groundwater treatment technologies. A particular interest of Dr. Goltz is development of methods to more effectively transfer these technologies to commercial use. He also has conducted modeling and field studies of the fate and transport of subsurface contaminants, as well as developed methods to characterize groundwater transport of contaminants.

Paul C. Johnson

Dr. Johnson is a Professor of Civil, Environmental, and Sustainable Engineering and is also the Dean of the Ira A. Fulton School of Engineering at Arizona State University (ASU). Prior to joining the faculty at ASU, he was a Senior Research Engineer for Shell Development in Houston, Texas. He received his B.S. degree in Chemical Engineering from the University of California, Davis and his Ph.D. from Princeton University. For over 20 years he has been active in the development and evaluation of technologies and strategies for the management and cleanup of contaminated soil and groundwater sites. He has also developed approaches for assessing the risks posed by chemicals in the subsurface. His work in the areas of soil vapor extraction, *in situ* air sparging, aerobic biobarriers, thermal treatment, vapor intrusion to buildings, and risk-based corrective action is reflected today in practice and in many guidance documents related to these topics. Dr. Johnson also served as the editor of the National Ground Water Association's journal *Ground Water Monitoring and Remediation* from 1993 to 2011 and serves as a consultant to regulatory agencies, government agencies and industry.

Richard L. Johnson

Dr. Johnson is a Professor of Environmental and Biomolecular Systems in the Institute for Environmental Health at Oregon Health & Science University (OHSU). He received his B.S. degree in chemistry from the University of Washington and M.S. and Ph.D. degrees from the Oregon Graduate Institute (now part of OHSU). His research interests include the transport and fate of organic chemicals in the environment, the development and evaluation of technologies for restoration of contaminated soil and groundwater, and physical and numerical modeling of subsurface and watershed-scale processes.

Jian Luo

Dr. Luo is an Assistant Professor in the School of Civil and Environmental Engineering at the Georgia Institute of Technology. He received his B.S., with Summa Cum Laude, and M.S. degrees in Environmental Engineering from Tsinghua University, Beijing, China and a Ph.D. degree in Civil and Environmental Engineering from Stanford University. The research Dr. Luo completed and is conducting with coworkers involves numerical and experimental investigations of contaminant fate and reactive transport in the subsurface; flow and transport in coastal aquifers; parameter estimation and inverse modeling; and the use of computational methods and models to assess subsurface contamination and to aid in the optimal design of bioremediation operations.

Kurt D. Pennell

Dr. Pennell is Professor and Chair of the Department of Civil and Environmental Engineering at Tufts University. He received his B.S. from the University of Maine, M.S. from North Carolina State University, Ph.D. from the University of Florida, and completed a postdoctoral fellowship at the University of Michigan. His current research focuses on the environmental fate and neurotoxicity of engineered nanomaterials, groundwater remediation technologies, and the role of persistent organic pollutants in Parkinson's disease. In 2006, Dr. Pennell received an NIH Career Award to study the "Quantitative Metabolomic Analysis of Chronic Exposures to Environmental Toxicants." He is a member of the Integrated Multiphase Environmental Systems (IMPES) laboratory at Tufts, and an investigator in the NIH-funded Parkinson's Disease Collaborative Environmental Research Center (PD-CERC) at Emory University. Dr. Pennell has published over 100 refereed journal articles and book chapters, and he currently serves as an associate editor for the Journal of Contaminant Hydrology. Dr. Pennell is a registered professional engineer (PE) and a board certified environmental engineer (BCEE).

C. Andrew Ramsburg

Dr. Ramsburg is an Assistant Professor in the Department of Civil and Environmental Engineering at Tufts University. His research combines fundamental and applied projects which focus on understanding and/or engineering of chemical, biological, and physical processes occurring on multiple scales within the subsurface environment. He is interested in innovative subsurface remediation, emulsion-based delivery of remedial amendments, implications of contaminant partitioning in DNAPL source zones, and fate and transport of pharmaceuticals in water reuse scenarios. Dr. Ramsburg has a Bachelor of Chemical Engineering, Master of Science in Environmental Engineering, and Doctor of Philosophy (in Environmental

Engineering) from the Georgia Institute of Technology. His postdoctoral training was completed at the University of Michigan.

Albert J. Valocchi

Dr. Valocchi is a Professor in the Department of Civil and Environmental Engineering at the University of Illinois. He received his B.S. in Environmental Systems Engineering from Cornell University, and M.S. and Ph.D. degrees in Civil Engineering from Stanford University.

Dr. Valocchi has nearly 30 years of experience teaching and conducting research in water resources engineering, groundwater hydrology and contaminant transport, groundwater modeling, and numerical methods. His research has focused on mathematical modeling of flow and transport in porous media, with applications to groundwater contamination and remediation. He has over 80 refereed journal papers, and from 2003 to 2010 he was Editor-in-Chief of the *Journal of Contaminant Hydrology*. In 2009 he became a Fellow of the American Geophysical Union. He was recently named an Abel Bliss Professor by the University of Illinois, College of Engineering.

John Vogan

Mr. Vogan is a senior hydrogeologist and national operations manager with Arcadis Canada. Mr. Vogan earned an M.S. (Earth Science-Hydrogeology) degree from the University of Waterloo in 1993. He has over 20 years of experience in site investigation and remediation, including more than 15 years in the application of permeable reactive barriers (PRBs) and related technologies. Mr. Vogan has over 25 publications concerning PRB technology and has contributed to numerous book chapters, U.S. Environmental Protection Agency (USEPA), Interstate Technology & Regulatory Council (ITRC) and international guidance documents. He served as an instructor for a USEPA short course entitled *In Situ* Permeable Reactive Barriers: Application and Deployment.

External Reviewers

Wilson S. Clayton
Trihydro Corp
Evergreen, CO USA
Email: wclayton@trihydro.com

Jeffrey A. Cunningham
Civil and Environmental Engineering
University of South Florida
Tampa, FL USA
Email: cunning@usf.edu

John C. Fountain
Marine Earth and Atmospheric Sciences
North Carolina State University
Raleigh, NC USA
Email: fountain@ncsu.edu

Tissa H. Illangasekare
Environmental Science and Engineering
Colorado School of Mines
Golden, CO USA
Email: tillanga@mines.edu

Shu-Guang Li
Department of Civil and Environmental Engineering
Michigan State University
East Lansing, MI USA
Email: lishug@egr.msu.edu

Douglas M. Mackay
Land, Air and Water Resources
University of California Davis
Davis, CA USA
Email: dmmackay@ucdavis.edu

David Major
Geosyntec Consultants
Guelph, ON Canada
Email: dmajor@geosyntec.com

John E. McCray
Environmental Science and Engineering
Colorado School of Mines
Golden, CO USA
Email: jmccray@mines.edu

James W. Mercer
GeoTrans, Inc. (a Tetra Tech Company)
Sterling, VA USA
Email: jim.mercer@geotransinc.com

Jack Parker
Civil and Environmental Engineering and Institute for a Secure and Sustainable Environment
University of Tennessee
Knoxville, TN USA
Email: jparker@utk.edu

Bruce A. Robinson
Earth and Environmental Sciences
Los Alamos National Laboratory
Los Alamos, NM USA
Email: robinson@lanl.gov

Lewis Semprini
School of Chemical, Biological and Environmental Engineering
Oregon State University
Corvallis, OR USA
Email: lewis.semprini@oregonstate.edu

Hans F. Stroo
HydroGeoLogic, Inc
Ashland, OR USA
Email: hstroo@hgl.com

Paul G. Tratnyek
Department of Science and Engineering
Oregon Health and Science University
Beaverton, OR USA
Email: tratnyek@ebs.ogi.edu

Michael J. Truex
Pacific Northwest National Laboratory
Richland, WA USA
Email: mj.truex@pnnl.gov

Albert J. Valoccchi
Department of Civil and Environmental Engineering
University of Illinois at Urbana-Champaign
Urbana, IL USA
Email: valocchi@mx.uillinois.edu

Brian D. Wood
School of Chemical, Biological and Environmental Engineering
Oregon State University
Corvallis, OR USA
Email: brian.wood@oregonstate.edu

Irene Xagoraraki
Department of Civil and Environmental Engineering
Michigan State University
East Lansing, MI USA
Email: xagorara@egr.msu.edu

Contents

List of Figures

List of Tables

CHAPTER 1

INTRODUCTION

Peter K. Kitanidis[1] and Perry L. McCarty[1]

[1]Department of Civil and Environmental Engineering, Stanford University, Stanford, CA, USA

1.1 BACKGROUND

The remediation of a site contaminated with one or more hazardous chemicals normally involves removing, destroying, or stabilizing in place the chemicals of concern. Chemical destruction or stabilization generally involves chemical or biological reactions that require the bringing together of the contaminant with one or more chemical or biological ingredients necessary for the reaction to proceed. For example, the aerobic biological destruction of the gasoline derived contaminant benzene in groundwater would require that oxygen be present or introduced in some manner and mixed with the benzene so that naturally occurring microorganisms residing within the aquifer could bring about benzene oxidation to carbon dioxide and water. Nutrients for growth of the microorganisms, such as nitrogen and phosphorus, may also need to be added to sustain the reaction. Additionally in some cases, acidic or basic chemicals such as hydrochloric acid or sodium bicarbonate may need to be mixed in with the water to achieve a pH condition that is satisfactory for biological growth. In some cases, such as in the anaerobic biological destruction of vinyl chloride, special strains of microorganisms may also need to be added as they may not be present naturally in the aquifer. In all such cases, the important processes of mass transfer and mixing are involved. Because of the complexity of aquifer systems, such mass transfer and mixing often becomes one of the most difficult and expensive aspects of site remediation. Thus, it behooves the designer or operator of a remediation system to be well versed in the fundamentals and applications of mass transfer and mixing processes as they apply to aquifer systems. The purpose of this volume is to aid in that understanding.

Numerous basic and applied studies have improved our scientific understanding of the underlying physical, chemical, and biological processes that take place in completely mixed batch reactors or in laboratory columns of porous media. However, before one can describe or make predictions of *field-scale processes*, one must properly account for mass-transfer processes, in addition to the biological and abiotic reactions in aqueous and non-aqueous phases. In particular, a grasp of the rates of hydrologic transport and mixing of the species involved in the reactions is required in order to predict the effective rate and extent of biochemical transformations in the field.

When comparing laboratory and field data, the most attention-grabbing discrepancy is in the kinetics, i.e., the observed reaction rates. Reactions in the field appear to occur much slower than in the laboratory, often by orders of magnitude. The literature is replete with evidence of such discrepancies, and occasionally with explanations for the discrepancies. Reaction rates seem to decrease exponentially with increase in the scale of the field application being studied. Selected representative references for this include Roberts et al., 1989; Rifai and Bedient, 1990; MacQuarrie and Sudicky, 1990; Anderson and McCarty, 1994; Dykaar and Kitanidis, 1996; Steefel and Lichtner, 1998; Malmstrom et al., 2000, 2004; Bryant and Thompson, 2001; Lichtner and Tartakovsky, 2003; Maher et al., 2006; Li et al., 2006; and Cirpka and Valocchi, 2007.

P.K. Kitanidis and P.L. McCarty (eds.), *Delivery and Mixing in the Subsurface: Processes and Design Principles for In Situ Remediation*, doi: 10.1007/978-1-4614-2239-6_1, © Springer Science+Business Media New York 2012

In a review, Bryant and Thompson (2001) noted that "On the one hand, modeling, theory, and experiment continue to provide useful insights into the behavior of natural and engineered systems. On the other hand, real systems continue to reveal instances of non-classical behavior that is not explainable by traditional approaches." The term non-classical behavior is often understood to mean either behavior inconsistent with that of textbook models, which may presume homogeneity, or that from laboratory studies.

While several mechanisms may be responsible for the discrepancies between laboratory and field observations, a consensus is gradually developing that the differences often can be attributed to the fact that processes in the field are *mass-transfer limited* or *mixing controlled* while processes in the laboratory tend to be *kinetically controlled*. These terms signify that the mechanism that limits the rate of reaction in the field is physical: the mixing of reactants. This point has been made time and again. For example, Simoni et al. (2001) point out that "Microbial degradation rates in the subsurface are not only limited by the physiological capacity of the organisms, but also by inefficient supply of nutrients to the microbes." Also, to maintain reaction rates, it is often important to remove products that may inhibit the rate of reaction. These mass transfer limitations become more crucial as the spatial extent of the reactor increases and, thus, must be taken into account in the analysis of all engineered or intrinsic *in situ* remediation projects.

Mass transfer limitations have profound effects on the effectiveness of enhanced (i.e., engineered) or intrinsic (i.e., natural) methods of *in situ* remediation. Consider enhanced *in situ* bioremediation as a case in point. Its main advantage is that it does not require the extraction of pollutants from below the ground surface and their subsequent treatment in an on-site facility or disposal in a landfill. Another important benefit of *in situ* treatment is that one can often take advantage of native microbial populations and natural processes that are likely to occur. Last but not least, *in situ* remediation can be relatively inexpensive. To speed up *in situ* bioremediation, for example, one only needs to modify the conditions in the subsurface by adding oxidants or electron donors, amendments to control pH, or micronutrients that may be required to promote microbial growth. But here one is then faced with the challenge of devising an efficacious and cost-effective system for delivery of the chemical or biological amendments into the subsurface and their mixing with polluted water. Simple injection may not be adequate since the injected water containing the additives may simply displace the polluted water such that microbial populations would be stimulated and transformations would take place only in the mixing zone between the two waters. However, such mixing at the boundaries tends to be quite slow, largely controlled by the slow process of molecular diffusion within the aqueous phase. This may be accompanied by several undesirable effects, including the biofouling or clogging of injection wells as excess biomass builds up in the narrow mixing zone so that additives fail to penetrate further.

The process of designing an effective *in situ* bioremediation project can be seen as consisting of the following successive stages, as illustrated by the pyramid in Figure 1.1. At the base of the pyramid are bench-scale studies using completely mixed batch reactors or small packed columns that are often used to provide the proof of concept in terms of the biochemical reactions that hopefully will affect the transformation of the pollutants into innocuous substances. These studies help us to learn what microbial populations, under what conditions, and with what byproducts can bring about the breakdown of the contaminants. Chemical delivery and mixing become issues at the next stages – implementation at the pilot- and full-scale. The major engineering challenge here is to devise a chemical delivery and mixing scheme that takes into account the opportunities and limitations of site geology in order to achieve satisfactory rates of remediation. In the process, one may have to make special efforts to manipulate the flow to prevent contaminated water from escaping untreated, to prevent

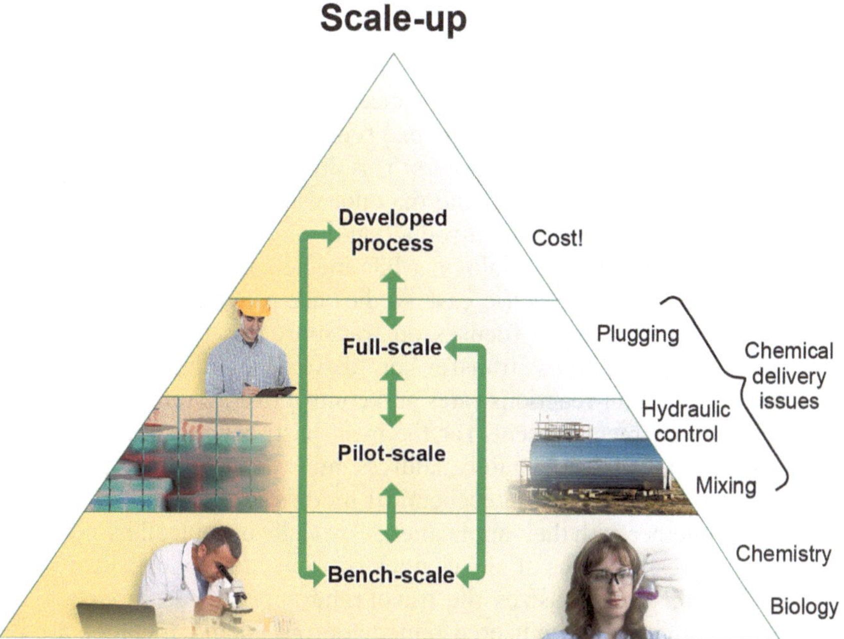

Figure 1.1. Stages in designing an *in situ* engineered bioremediation system. Illustration developed by Peter Kitanidis and Craig Criddle, Stanford University, Stanford, CA, USA.

biofouling, and to prevent unwanted reaction products from accumulating. All these objectives must be achieved at reasonable cost.

Geologic formations can be represented as biochemical reactors, like the ones studied by chemical engineers (see for example Nauman, 2001). However, there are important differences between manmade reactors used in chemical plants or treatment facilities and natural geologic formations. Manmade reactors are designed to maximize performance, have a known structure, and are readily amenable to monitoring and control. The performance of such reactors is contingent on the provision of effective chemical delivery and mixing systems, which often involve volatilization, taking advantage of fast mixing in gas phases, or intense stirring of liquid phases (Tatterson, 2003). By contrast, natural porous or fractured formations are non-ideal reactors that are hard to characterize, monitor, or control. They also tend to be large, which makes it even more difficult to overcome mass transfer limitations. It is virtually impossible to utilize the mechanical stirring systems than can be so crucial in speeding up mixing in manmade reactors. Turbulence, which can be quite highly efficient for inducing mixing, is practically absent in subsurface systems because flow takes place at low velocities through very small passages.

There is no question that basic understanding of processes and forethought on principles of design are essential for developing successful chemical delivery and mixing systems. This volume is meant to provide the practitioner with information on the natural mixing processes occurring in aquifers as well as to describe basic strategies that can be implemented to enhance mixing in particular cases. For example, when it comes to mixing miscible liquids, one can speed up mixing in the formation by manipulating the flow such as through the use of recirculation wells. Furthermore, much of the mixing can be achieved partially within recirculation wells themselves, where contaminated water is admixed with additives, volatile products may be removed through a vapor mass exchanger, etc. Thus, adding mixing wells can significantly increase the performance of the delivery and mixing system and speed up the process of remediation.

1.2 OVERVIEW OF THE CONTENTS OF THIS VOLUME

This volume has ten chapters including the introduction. The two chapters that follow are tutorials on reactions and processes of importance in groundwater remediation. In Chap. 2, the focus is on chemistry and biology and Chap. 3 focuses on transport. The synthesis of these concepts is addressed in the remaining chapters. This division should make it easier for readers who already have sufficient background and may not wish to study either or both of the two tutorial chapters. Chapter 2 provides an introduction/tutorial on *in situ* remediation processes. The focus here is on chemistry and biology, while incidentally providing the motivation for addressing mixing issues. Some specific case studies are provided to illustrate the problems and challenges involved. Chapter 3 focuses on transport and mixing processes, emphasizing advection, dispersion, and "mass transfer" or "diffusional" limitations. Discussed is how transport and mixing can limit reactions rates and control overall remediation progress.

The following two chapters present, for the most part, conceptualization of the problem and modeling approaches. To adequately address mass transfer and mixing processes in complex groundwater systems, hydrogeochemical models are needed. Models available are reviewed in Chap. 4 together with their application, especially in controlling mixing and reaction rates. This chapter addresses issues of site characterization and model calibration for practical implementation. Chapter 5 emphasizes the travel-time approach to modeling, a specialized approach that is well suited for engineered remediation problems. Included are protocols for practical implementation for design and monitoring of progress.

The next few chapters deal with design issues. Chapter 6 covers an important method for introducing and mixing chemicals with groundwater contaminants, that is, the recirculation system, a system where mixing is controlled through and takes place mostly in wells. Here, design principles are addressed, including how to estimate (mostly with simpler models or "screening tools") travel times, reactor volumes, flow-through rates, and other useful design parameters. Covered also are the challenges of operating and maintaining injection-extraction wells, the non-uniform distribution of biomass, plugging issues, and strategies such as pulsing. Case studies are provided. Chapter 7 has objectives similar to Chap. 6, but this time the use of reactive barrier walls is covered. In this case, the contaminants move by normal groundwater advection to the reacting chemicals that are contained within permeable barrier walls through which the groundwater passes.

Technologies such as air sparging that are used for introducing and mixing gaseous reactants such as oxygen as well as for removing volatile contaminants are discussed in Chap. 8. Starting with a general background on processes for mixing of gases, the discussion then proceeds to design issues of air sparging. Case studies are included. Chapter 9 covers the special case of intrinsic remediation in natural-gradient systems. These are systems where the kinetics are controlled through mixing of contaminant plumes with ambient species, such as dissolved oxygen, over long periods of times. Issues covered include monitoring and estimation of long-term reaction rates, lengths of plumes, and other important parameters. Case studies again are provided. Finally, Chap. 10 addresses the special problem of source remediation, as opposed to plume remediation, and its challenges.

1.3 ONGOING RESEARCH AND OUTSTANDING CHALLENGES

Topics related to chemical delivery and mixing remain current and continue to be studied in applied and theoretical research through modeling studies and field experiments. Following is a brief review of some of the remaining challenges and ongoing research, without any pretension of completeness.

There is considerable interest in understanding and improving the performance of injection galleries and wells as well as permeable reactive walls. Recirculating wells can be used to serve as active barriers to contaminant transport (using the wells to mix reactive chemicals into contaminated water as it moves downgradient). The challenge is to quantify reliably the efficiency of the process. Several field experiments have been conducted where a portion of a plume was intercepted, and upgradient and downgradient concentrations compared. In actual practice, the technology can be used to intercept an entire plume, and the true measure of efficiency will be reduction of overall contaminant flux. An evaluation of a full-scale remediation system using recirculating wells to establish a barrier to contaminant transport, using mass flux as a measure of efficiency, has not been accomplished. And, it is especially important that either flux (contaminant mass per area per time), or total mass transport (contaminant mass per time), be used as the metric to quantify the efficiency of the circulating well technology. Even if there are large concentration reductions, it is possible that the flow field established by the recirculating wells will allow a significant fraction of the total natural gradient flow to bypass the treatment system, so that flux reduction may not be sufficient to meet remediation goals.

From an applied standpoint it is important to delineate the utility and the pros and cons of strategies for enhanced chemical delivery. For example, pulsing and intermittent substrate delivery generally reduce biomass growth near injection points and prevent or delay the plugging (or biofouling) of injection wells and galleries. A possible approach would be the addition of inhibitory levels of substrates to facilitate deeper or more extensive penetration of aquifer formations. Subsequent dilution would decrease the concentrations to non-inhibitory levels and degradation would then proceed. Research is needed on substrates, such as alcohols, that might be appropriate for this case. A similar strategy would be to deliver degradable substrates together with additives or co-substrates that stop or minimize substrate degradation during delivery periods. Degradation would proceed as the concentration of the inhibitor or co-substrate falls due to dilution or degradation. Research is needed to identify and test useful additives and co-substrates that might be appropriate for this case.

Another innovative idea is "trap-and-treat" – temporarily retarding the movement of contaminants through reactive zones to enhance their contact time with the agents of degradation/detoxification. New sorptive materials might be developed for that purpose, or surface modification methods employed to change subsurface biogeochemistry. Thinking further down the road, the value of emerging nanotechnology and its potential for enhancing chemical delivery and mixing should be evaluated. For example, can electrically conductive "wires" or gels be formed *in situ* to "wire-up" the subsurface for more efficient electron delivery? Can microbial nanowire formation be used for electron delivery?

In most cases, mathematical models play a crucial role in designing *in situ* remediation projects and in assimilating data to monitor performance. The modeler is faced with the challenge of assigning appropriate parameters, especially those that control rates of mass transfer and biochemical parameters. The challenge is greatest when the geologic environment includes fractured media, separate gas phases, and multiple liquid phases.

REFERENCES

Anderson JE, McCarty PL. 1994. Model for treatment of trichloroethylene by methanotrophic biofilms. J Environ Eng 120:379–400.

Bryant L, Thompson KE. 2001. Theory, modeling and experiment in reactive transport in porous media. Curr Opin Colloid Interface Sci 6:217–222.

Cirpka OA, Valocchi AJ. 2007. Two-dimensional concentration distribution for mixing-controlled bioreactive transport in steady state. Adv Water Resour 30:1668–1679.

Dykaar BB, Kitanidis PK. 1996. Macrotransport of a biologically reacting solute through porous media. Water Resour Res 32:307–320.

Li L, Peters CA, Celia MA. 2006. Upscaling geochemical reaction rates using pore-scale network modeling. Adv Water Resour 29:1351–1370.

Lichtner PC, Tartakovsky DM. 2003. Stochastic analysis of effective rate constant for heterogeneous reactions. Stochastic Environ Res Risk Assess 17:419–429.

MacQuarrie KTB, Sudicky EA. 1990. Simulation of biodegradable organic contaminants in groundwater, 2. Plume behavior in uniform and random flow fields. Water Resour Res 26:223–239.

Maher K, Steefel CI, DePaolo DJ, Viani BE. 2006. The mineral dissolution rate conundrum: Insights from reactive transport modeling of U isotopes and pore fluid chemistry in marine sediments. Geochimica et Cosmochimica Acta 70:337–363.

Malmstrom ME, Destouni G, Banwart SA, Stromberg BHE. 2000. Resolving the scale dependence of mineral weathering rates. Environ Sci Technol 34:1375–1378.

Malmstrom ME, Destouni G, Martinet P. 2004. Modeling expected solute concentration in randomly heterogeneous flow systems with multicomponent reactions. Environ Sci Technol 38:2673–2679.

Nauman B. 2001. Handbook of Chemical Reactor Design, Optimization, and Scaleup. McGraw-Hill, New York, NY, USA. 600 p.

Rifai HS, Bedient PB. 1990. Comparison of biodegradation kinetics with an instantaneous reaction model for groundwater. Water Resour Res 26:637–645.

Roberts PV, Semprini L, Hopkins GD, Grbic-Galic D, McCarty PL, Reinhard M. 1989. In-situ aquifer restoration of chlorinated aliphatics by methanotrophic bacteria. EPA/600/S2-89/033. Robert S. Kerr Environmental Research Laboratory, Ada, OK, USA. August.

Simoni SF, Schaefer A, Harms H, Zehnder AJB. 2001. Factors affecting mass transfer limited biodegradation in saturated porous media. J Contam Hydrol 50:99–120.

Steefel CI, Lichtner PC. 1998. Multicomponent reactive transport in discrete fractures, 1. Controls on reaction front geometry. J Hydrol 209:186–199.

Tatterson GB. 2003. Scaleup and Design of Industrial Mixing Processes. McGraw-Hill, New York, NY, USA. 392 p.

CHAPTER 2

CHEMICAL AND BIOLOGICAL PROCESSES: THE NEED FOR MIXING

Perry L. McCarty[1] and Craig S. Criddle[1]

[1]Department of Civil and Environmental Engineering, Stanford University, Stanford, CA, USA

2.1 INTRODUCTION

Except for spontaneous reactions such as radioactive decay, chemical transformations often require that two or more substances be brought together for the transformation to occur. Examples of particular interest in groundwater are oxidations of inorganic or organic species, which require the presence of some oxidant, such as diatomic oxygen (O_2), nitrate, sulfate, or ferric iron (Fe(III)). In biological reactions, three entities generally are required, the compound being oxidized (electron donor), the oxidant (electron acceptor), and the microorganism carrying out the transformation. At times, the required entities are already present together, and then transformation occurs based simply on normal reaction kinetics. However, this is often not the case in groundwater remediation, and then the missing reactants must be supplied through some means and mixed with the substance or substances targeted for removal. The speed of the reaction is then likely to be governed primarily by the rate at which the required substances can be brought together. Natural attenuation for transformation of materials may require mixing brought about by the diffusion of oxygen into an aquifer from the vadose zone above, or from an adjacent groundwater flow stream. The process of adding and mixing needed substances for desired transformation is one of the most challenging and costly aspects of *in situ* remediation of contaminated groundwater and soil. This is a much more difficult process than with an aboveground reactor because of complex and often undefined hydrogeology and the general uncertainty of the exact location of the contaminants.

Some form of mixing may also be required for processes other than chemical oxidations. Included are the addition of reducing compounds for chemical reductions; acids or bases for pH control; chemicals that promote precipitation for in-place stabilization; detergents, solvents, or other chemicals that promote solubilization of the compound of interest for easier removal; addition of a separate phase such as air; use of thermal treatment to enhance vaporization; as well as chemical changes resulting from groundwater-surface water interactions that are driven by variability in rates of precipitation, extraction, and aquifer recharge. All such processes involve mixing in one form or another. The emphasis in this chapter is not on the mixing processes themselves, but on the chemical and biological requirements for contaminant transformation, destruction, or removal. A few examples of field studies where mixing has been used to bring the reactants together are provided for illustration, and many others are provided in other chapters of this volume.

2.2 GROUNDWATER CONTAMINANTS

The most frequently found chemicals in groundwater at hazardous waste sites are listed in Table 2.1 (NRC, 1994). Among organic contaminants, the chlorinated solvents, trichloroethene (TCE), perchloroethene (PCE), methylene chloride (dichloromethane or MC), and

P.K. Kitanidis and P.L. McCarty (eds.), *Delivery and Mixing in the Subsurface: Processes and Design Principles for In Situ Remediation*, doi: 10.1007/978-1-4614-2239-6_2,

Table 2.1. Most Frequently Detected Groundwater Contaminants at Hazardous Waste Sites (after NRC, 1994)

Organic contaminants		Inorganic contaminants	
Rank	**Chemical**	**Rank**	**Chemical**
1	Trichloroethene (TCE)	1	Lead
2	Tetrachloroethene (PCE)	2	Chromium
3	Benzene	3	Zinc
4	Toluene	4	Arsenic
5	Methylene chloride (MC)	5	Cadmium
6	1,1,1-Trichloroethane (TCA)	6	Manganese
7	Chloroform	7	Copper
8	1,1-Dichloroethane (1,1-DCA)	8	Barium
9	1,2-Dichloroethene (1,2-DCE)	9	Nickel
10	1,1-Dichloroethene (1,1-DCE)		
11	Vinyl chloride (VC)		
12	1,2-Dichloroethane (1,2-DCA)		
13	Ethylbenzene		
14	Di(2-ethylhexyl)phthalate		
15	Xylenes		
16	Phenol		

1,1,1-trichloroethane (TCA) are among the six most frequently found organic chemicals. These chemicals are denser than water such that when spills of the liquid solvents reach groundwater, they continue downward under the force of gravity, often penetrating deeply into a groundwater aquifer. They are poorly biodegradable and represent the most difficult and costly chemicals for remediation. It is for this reason that so much attention has been paid to them. Others among the list of frequently found organic chemicals are degradation products of these four chlorinated solvents, including 1,1-dichloroethane (1,1-DCA), 1,2-dichloroethene (1,2-DCE), 1,1-dichloroethene (1,1-DCE), and vinyl chloride (VC). Thus, 8 of the 11 most frequently found organic chemicals are chlorinated solvents themselves and their degradation products.

The second group of organic chemicals includes benzene and toluene, the third and fourth most frequently found on the list. These aromatic hydrocarbons are the more soluble components of gasoline that partition into groundwater from gasoline spills. Gasoline itself is lighter than water and so tends to spread out over the surface of the groundwater, rather than penetrating into it. Two other aromatic hydrocarbon components of gasoline are also on the list, ethylbenzene and xylenes (of which there are three different isomers). These four aromatic hydrocarbons are collectively known as the BTEX compounds (benzene, toluene, ethylbenzene, xylene).

Only 4 of the top 16 organic chemicals are not among the chlorinated solvent or BTEX groups. These include chloroform, generally formed from the chlorination of water through its interaction with humic materials; 1,2-dichloroethane (1,2-DCA), a chlorinated compound used widely in chemical synthesis and as a solvent; di(2-ethylhexyl)phthalate, a chemical used in plastics manufacture; and phenol and its derivatives, including the chlorinated phenols used in treating wood.

Other organic chemicals of importance as groundwater contaminants but not included on this list are carbon tetrachloride (CT), another widely used solvent in the past; methyl tertiary-butyl ether (MTBE), an oxygenate additive of gasoline; 1,4-dioxane, an industrial chemical commonly used as a solvent stabilizer; and chlorinated benzenes and benzoates, which have a wide variety of industrial and commercial uses. These chemicals are all persistent organic pollutants (POPs) that need to be addressed in groundwater remediation.

Table 2.1 includes nine inorganic chemicals. These substances are not destroyed chemically or biologically. Consequently, their remediation is through removal from the groundwater by extraction or immobilization. Five of the inorganic chemicals in Table 2.1 (lead, zinc, cadmium, barium and nickel) are metals that exist primarily as stable cations and so are not susceptible to oxidation and reduction, but can be removed from water by adsorption or chemical precipitation. The other metals in Table 2.1 are directly susceptible to oxidation-reduction reactions that alter their solubility and thus mobility in groundwater. As shown in Table 2.2, chromium (Cr), arsenic (As), selenium (Se), and uranium (U) are present as either cations or oxyanions depending upon oxidation state and pH. Under neutral to basic conditions, hexavalent chromium exists as the highly soluble and toxic chromate oxyanion (CrO_4^{2-}). Under acidic conditions, however, it exists as dichromate ($Cr_2O_7^{2-}$). It is also readily reduced to a trivalent state that is nontoxic and precipitates as $Cr(OH)_3(s)$, a solid with low solubility in water and low toxicity. Arsenic can be found in the soluble trivalent (AsO_3^{3-}) or pentavalent (AsO_4^{3-}) states. The relative solubility and mobility of soluble arsenic species depends on interactions with the solid phase. Selenium is a metalloid that is naturally present in some groundwaters where it may be present as the soluble oxyanions selenite (SeO_3^{2-}) or selenate (SeO_4^{2-}). These species can be

Table 2.2. Regulated Metals and Metalloids That Are Susceptible to Changes in Solubility Through Microbial or Chemically Mediated Redox Reactions (adapted from Nyman et al., 2005)

Metal or metalloid	Oxidation state	Oxidized species	Reduced species (often less soluble)	Common sources
As[a]	-II		AsS	Erosion of natural deposits; runoff from orchards; runoff from glass & electronics production
	0		FeAsS, As	
	III	H_2AsO_3, $H_2AsO_3^-$, $HAsO_3^{2-}$, AsO_3^{3-}	As_2O_3	
	V	H_3AsO_4, $H_2AsO_4^-$, $HAsO_4^{2-}$, AsO_4^{3-}		
Cr	III		Cr_2O_3	Steel and pulp mills; erosion of natural deposits
	VI	H_2CrO_4, $HCrO_4^-$, CrO_4^{2-}, $Cr_2O_7^{2-}$		
Se	-II	H_2Se, HSe^-, Se^{2-}		Refineries; natural deposits; mines
	0		Se	
	IV	H_2SeO_3, $HSeO_3^-$, SeO_3^{2-}	SeO_2	
	VI	H_2SeO_4, $HSeO_4^-$, SeO_4^{2-}		
	VII	SeO_4^-		
U	IV		UO_2, $USiO_4$	Mine tailings; atomic bomb fabrication sites; weapons use; erosion of natural sources
	VI	UO_2^{2-}, $UO_2(CO_3)$, $UO_2(CO_3)_2^{2-}$, $UO_2(CO_3)_3^{4-}$		

[a]The normal valence states of arsenic are III and V. As(III) can be more mobile and toxic than As(V)

biologically reduced to zero-valent selenium (low solubility) or to selenium hydride (H_2Se). Finally, uranium is a radionuclide that is often present in nature in the +IV oxidation state as uraninite UO_2, a sparingly soluble mineral. During extraction and refining operations, the U(IV) is oxidized to toxic, soluble, and mobile complexes. At low pH, the uranyl cation UO_2^{2+} is dominant; at near neutral pH and above, carbonate complexes dominate.

Inorganic chemicals of concern that are not listed in Tables 2.1 or 2.2 include nitrate, perchlorate (ClO_4^-), and ferrous iron. Nitrate is a common contaminant from agricultural operations and from the use of nitric acid for mineral extraction. It is also a common electron acceptor for bacteria, and can be removed from water by denitrification. Perchlorate is used in rocket fuel, fireworks, and road flares. Like nitrate, it can serve as an electron acceptor for microbial growth, and as such can be biologically reduced to harmless chloride. Although iron and manganese are not listed in Table 2.1 as prevalent contaminants, they can be present at high levels in solution, often formed from natural aquifer minerals through biological reduction.

Knowledge of the physical properties of contaminants (Tables 2.3, 2.4) is of interest to help better understand processes that affect their movement and fate in groundwater. As already indicated and as Table 2.3 illustrates, the chlorinated solvents, which are liquid at room temperature, have densities greater than water (1.0 gram per cubic centimeter [g/cm^3]) and thus tend to penetrate deeply into groundwater. BTEX compounds have densities lower than water and so will not penetrate downward into groundwater, but will remain in the capillary fringe above. Water solubility of chemicals indicates the extent to which the free phase liquid of the solvent can dissolve in water. Solubilities of most chemicals listed in Table 2.3 are in the gram per liter (g/L) range or less, and are thus called "sparingly soluble."

Table 2.3. Physical and Chemical Properties of Chlorinated Solvents and Their Transformation Products at 25 Degrees Celsius (°C) (after Yaws, 1999)

Compound	Density (g/cm^3)	Henry's law constant, *H* (atm/M)	Water solubility (mg/L)	Octanol-water partition coefficient (log K_{ow})
Methanes:				
Carbon tetrachloride (CT)	1.59	29	790	2.83
Trichloromethane	1.48	4.1	7,500	1.97
Methylene chloride (MC)	1.33	2.5	19,400	1.25
Chloromethane	0.92	8.2	5,900	0.91
Ethanes:				
1,1,1-Trichloroethane (1,1,1-TCA)	1.34	22	1,000	2.49
1,1-Dichloroethane (1,1-DCA)	1.18	5.8	5,000	1.79
1,2-Dichloroethane (1,2-DCA)	1.24	1.2	8,700	1.48
Chloroethane	0.90	6.9	9,000	1.43
Ethenes:				
Tetrachloroethene (PCE)	1.62	27	150	3.4
Trichloroethene (TCE)	1.46	12	1,100	2.42
cis-1,2-Dichloroethene (*cis*-DCE)	1.28	7.4	3,500	1.85
trans-1,2-Dichloroethene (*trans*-DCE)	1.26	6.7	6,300	2.09
1,1-Dichloroethene (1,1-DCE)	1.22	23	3,400	2.13
Vinyl chloride (VC)	0.91	22	2,700	1.62

(continued)

Table 2.3. (continued)

Compound	Density (g/cm^3)	Henry's law constant, H (atm/M)	Water solubility (mg/L)	Octanol-water partition coefficient ($\log K_{ow}$)
Aromatic compounds:				
Benzene	0.88	5.6	1,760	2.13
Toluene	0.87	6.4	540	2.73
Ethylbenzene	0.86	8.1	165	3.15
o-xylene	0.88	4.2	221	3.12
m-xylene	0.86	6.8	174	3.20
p-xylene	0.86	6.2	200	3.15
Methyl tertiary-butyl ether (MTBE)	0.74	0.54	51,000	0.94
Chlorobenzene	1.10	4.5	300	2.84
1,2-Dichlorobenzene	1.30	2.8	92	3.43
Phenol[a]	#	0.00076	80,000	1.46

Note: *atm/M* atmosphere liters per mole, *mg/L* milligrams per liter
[a]Solid at room temperature

Table 2.4. Mineral Solubility Products (from Nyman et al., 2005)

Compound	Formula	pK_{sp}	K_{sp}	Reference
Arsenic(III) sulfide	As_2S_3	21.68	2.1×10^{-22}	Dean, 1999
Cadmium sulfide	CdS	26.10	8.0×10^{-27}	Dean, 1999
Chromium(III) hydroxide	$Cr(OH)_3$	30.20	6.3×10^{-31}	Dean, 1999
Cobalt sulfide	CoS	20.40	4.0×10^{-21}	Dean, 1999
	CoS	24.70	2.0×10^{-25}	
Copper(I) sulfide	Cu_2S	47.60	2.5×10^{-48}	Dean, 1999
Copper(II) sulfide	CuS	35.20	6.3×10^{-36}	Dean, 1999
Ferrihydrite	$Fe(OH)_3$	39.5	3.16×10^{-40}	Cornell and Schwertmann, 1996
Goethite	FeOOH	40.7	2.00×10^{-41}	Cornell and Schwertmann, 1996
Hematite	Fe_2O_3	42.75	1.78×10^{-43}	Cornell and Schwertmann, 1996
Iron(II) sulfide	FeS	17.20	6.3×10^{-18}	Dean, 1999
Lead sulfide	PbS	27.10	8.0×10^{-28}	Dean, 1999
Manganese hydroxide	$Mn(OH)_2$	12.72	1.9×10^{-13}	Dean, 1999
Mercury(II) sulfide	HgS red	52.4	4×10^{-53}	Dean, 1999
	HgS black	51.8	1.6×10^{-52}	
Nickel α-sulfide	NiS	18.5	3.2×10^{-19}	Dean, 1999
β-sulfide	β-NiS	24.0	1.0×10^{-24}	
γ-sulfide	NiS	25.70	2.0×10^{-26}	
Technicium	TcO_2	8	10^{-8}	Rard et al., 1999
Uraninite	UO_2	60.6	2.5×10^{-61}	Langmuir, 1978
Zinc sulfide: sphaelerite	ZnS	23.8	1.6×10^{-24}	Dean, 1999
wurtzite	ZnS	21.6	2.5×10^{-22}	

Note: K_{sp} solubility product constant; $pK_{sp} = -\log K_{sp}$

The Henry's Law constant (H) indicates the potential of a compound to partition between water and air and, therefore, the tendency of a compound to be removed from water by air stripping, the higher the value the easier it is to be removed such as by air sparging. Ionic (i.e., charged) compounds and compounds with H less than about 0.2 atm/M are not likely to be removed readily by air stripping. The octanol-water partition coefficient (K_{ow}) indicates the potential of a compound to partition from water onto aquifer solids, and particularly into the organic portion of aquifer solids. This partitioning impacts on the compound's rate of movement through an aquifer and on the ease with which a chemical injected into the aquifer can move and interact with a contaminant. Compounds with log K_{ow} in the range of 2 or above will partition moderately onto aquifer solids, depending upon the organic content, and this applies to most of the chemicals listed in Table 2.3. Compounds such as DDT (dichlorodiphenyltrichloroethane) and PCBs (polychlorinated biphenyls) have log K_{ow} values around 6 or higher and thus sorb very strongly to aquifer solids. It is for this reason that they are not major groundwater contaminants as they sorb so strongly to soils that they rarely penetrate sufficiently downward to contaminate groundwater.

2.3 REACTION AND MASS TRANSFER PROCESSES

2.3.1 Overview

The emphasis in this chapter is on chemical or biological transformations or reactions that require the bringing together of two or more chemical or biological species for the reaction to occur. Mass transfer refers to the process or processes by which they come together. The discussion of these processes is rather brief, more detailed information can be found in environmental chemistry textbooks (Benjamin, 2002; Morel and Hering, 1993; Sawyer et al., 2003; Stumm and Morgan, 1996). Reaction stoichiometry, i.e., the relative amounts of chemicals needed for transformations to go to completion, is an important component of reaction and mass transfer analyses that is needed for the design of a delivery system or for analysis of natural attenuation. Stoichiometry also makes it possible to quantify the products of a transformation, which sometimes are also contaminants of concern. Examples of products include methane, sulfide, the soluble reduced forms of iron and manganese, and partially reduced or oxidized contaminant species. Reaction stoichiometry depends to some extent on the type of reaction involved, so this requires some consideration. Next comes understanding of mass transfer and reaction kinetics to determine when the rate of a reaction will be controlled primarily by the intrinsic kinetics of the reaction itself and when it will be controlled by the rate at which reactants are brought into contact with one another.

2.3.2 Stoichiometry

In the design of a system involving chemical transformation, the making of a mass balance is critical for determining how much chemical must be added to bring about a given amount of change and what will be the products of the reaction. These quantities can be provided through use of a stoichiometric equation that describes the overall reaction of interest (Sawyer et al., 2003). If a stoichiometric equation for the reaction of interest cannot be written because of inadequate information, then knowledge of the reaction is insufficient to make a good judgment on chemical requirements. In such a case, more study is needed before *in situ* remediation is attempted, or else costly mistakes may be made, either in adding too much of a needed substance or too little. In order to address stoichiometry, knowledge of reaction and mass-transfer processes is useful.

2.3.3 Reaction and Mass-Transfer Processes

Table 2.5 summarizes important reaction and mass-transfer processes involved in contaminant movement and fate in water. The significance of acid–base reactions is that they change the active species of a chemical under given chemical conditions in water. They also dominate the acid–base buffering of a system.

Table 2.5. Examples of Reaction and Mass-Transfer Processes of Interest in Groundwater Remediation

Reaction process	Description	Examples
Acid–base	Change in an element in solution from one chemical form to another without a change in the valance state – generally in response to pH conditions.	$H^+ + OH^- = H_2O$
		$HCO_3^- = H^+ + CO_3^{2-}$
		$CO_2 + H_2O = H^+ + HCO_3^-$
		$H_2S = H^+ + HS^-$
		$Zn_2^+ + OH^- = ZnOH^+$
		$Cr_2O_7^{2-} + H_2O = 2CrO_4^{2-} + 2H^+$
Oxidation-reduction	Change in the oxidation state of an element in a chemical, generally requires change in oxidation state of two elements, one is oxidized, the electron acceptor, and the other reduced, the electron donor.	$4Fe(OH)_2 + O_2 + 2H_2O = 4Fe(OH)_3$
		$4Cr^{3+} + 3O_2 + 8H_2O = 2Cr_2O_7^{2-} + 16H^+$
		$CH_3COOH + 2O_2 = 2CO_2 + 2H_2O$
		$CH_3COOH + SO_4^{2-} = 2CO_2 + H_2S + 2OH^-$
Precipitation	Formation of a solid phase from reaction between chemicals in solution.	$Ca^{2+} + CO_3^{2-} = CaCO_3$ (*s*)
		$Cr^{3+} + 3OH^- = Cr(OH)_3$ (*s*)
		$2Fe^{3+} + 6OH^- = Fe_2O_3$ (*s*) + $3H_2O$
		$Fe^{2+} + S^{2-} = FeS$ (*s*)
		$Zn^{2+} + S^{2-} = ZnS$ (*s*)
Mass transfer process	**Description**	**Examples**
Solubilization	May represent dissolution of a chemical from a solid phase into a soluble form, the reverse of precipitation. It may also represent the partitioning of a chemical from a non-miscible liquid phase into the aqueous phase.	$CaCO_3$ (*s*) = $CaCO_3$ (*aq*)
		Fe_2O_3 (*s*) = Fe_2O_3 (*aq*)
		TCE (*l*) = TCE (*aq*)
		benzene (*l*) = benzene (*aq*)
Volatilization	The movement of a chemical from an aqueous phase to a gaseous phase.	TCE (*aq*) = TCE (*g*)
		benzene (*aq*) = benzene (*g*)
Sorption	The partitioning of a chemical from the aqueous phase onto or into a solid phase.	TCE (*aq*) = TCE (*sorbed*)
		$Fe^{3+} = Fe^{3+}$ (*sorbed*)
Advection	The transport of a chemical by being carried along in a moving fluid such as water or air.	
Diffusion-dispersion	Diffusion is the net transport of molecules from a region of higher concentration to one of lower concentration by random molecular motion. Dispersion is similar but is a faster process brought about in addition by dynamic mixing of the fluid in which the chemical is contained.	

Note: (*aq*) aqueous phase, (*g*) gas phase, (*l*) liquid phase, (*s*) solid phase

Oxidation-reduction reactions are perhaps the most important reactions used in groundwater remediation. Here, the chemical being oxidized is termed the electron donor as electrons are removed from it in the process. The chemical being reduced is the electron acceptor because it accepts the electrons. This electron exchange is illustrated by the half reactions shown in Table 2.6 for typical electron donors and Table 2.7 for typical electron acceptors. Stoichiometric equations for oxidation-reduction reactions can be written by adding a given electron donor half-reaction to that of an electron acceptor half reaction. For example, the oxidation of the electron donor ethanol with the electron acceptor carbon dioxide (CO_2) results in the following stoichiometric equation for the conversion of ethanol into methane:

$$\frac{1}{12}CH_3CH_2OH = \frac{1}{8}CH_4 + \frac{1}{24}CO_2 \quad \text{(Eq. 2.1)}$$

Table 2.6. Electron Donor Half Reactions

Electron donor		End product	Half reaction
Hydrogen	H_2	H^+	$\frac{1}{2}H_2 = H^+ + e^-$
Zero-valent iron	Fe^0	Fe^{2+}	$\frac{1}{2}Fe(s) = \frac{1}{2}Fe^{2+} + e^-$
Acetate	CH_3COO^-	CO_2	$\frac{1}{8}CH_3COO^- + \frac{3}{8}H_2O = \frac{1}{8}CO_2 + \frac{1}{8}HCO_3^- + H^+ + e^-$
Lactate	$C_3H_5O_2^-$	CO_2	$\frac{1}{12}CH_3CHOHCOO^- + \frac{1}{3}H_2O = \frac{1}{6}CO_2 + \frac{1}{12}HCO_3^- + H^+ + e^-$
Fatty acid	$C_{18}H_{31}O_2^-$	CO_2	$\frac{1}{100}C_{18}H_{31}O_2^- + \frac{7}{20}H_2O = \frac{17}{100}CO_2 + \frac{1}{100}HCO_3^- + H^+ + e^-$
Methanol	CH_3OH	CO_2	$\frac{1}{6}CH_3OH + \frac{1}{6}H_2O = \frac{1}{6}CO_2 + H^+ + e^-$
Ethanol	CH_3CH_2OH	CO_2	$\frac{1}{12}CH_3CH_2OH + \frac{3}{12}H_2O = \frac{1}{6}CO_2 + H^+ + e^-$
Carbohydrate	$C_6H_{12}O_6$	CO_2	$\frac{1}{24}C_6H_{12}O_6 + \frac{1}{4}H_2O = \frac{1}{4}CO_2 + H^+ + e^-$
Benzene	C_6H_6	CO_2	$\frac{1}{30}C_6H_6 + \frac{2}{5}H_2O = \frac{1}{5}CO_2 + H^+ + e^-$
Toluene	$C_6H_5CH_3$	CO_2	$\frac{1}{36}C_6H_5CH_3 + \frac{7}{18}H_2O = \frac{7}{36}CO_2 + H^+ + e^-$
Ethylbenzene	$C_6H_5C_2H_5$	CO_2	$\frac{1}{42}C_6H_5C_2H_5 + \frac{8}{21}H_2O = \frac{4}{21}CO_2 + H^+ + e^-$
Xylene	$C_6H_4(CH_3)_2$	CO_2	$\frac{1}{42}C_6H_4(CH_3)_2 + \frac{8}{21}H_2O = \frac{4}{21}CO_2 + H^+ + e^-$
TCE	$CHCl{=}CCl_2$	$CO_2 + Cl^-$	$\frac{1}{6}CHCl{=}CCl_2 + \frac{2}{3}H_2O = \frac{1}{3}CO_2 + \frac{1}{2}Cl^- + \frac{3}{2}H^+ + e^-$
DCE	$CHCl{=}CHCl$	$CO_2 + Cl^-$	$\frac{1}{8}CHCl{=}CHCl + \frac{1}{2}H_2O = \frac{1}{4}CO_2 + \frac{1}{4}Cl^- + \frac{5}{4}H^+ + e^-$
VC	$CH_2{=}CHCl$	$CO_2 + Cl^-$	$\frac{1}{10}CH_2{=}CHCl + \frac{2}{5}H_2O = \frac{1}{5}CO_2 + \frac{1}{10}Cl^- + \frac{11}{10}H^+ + e^-$
Chlorobenzene	C_6H_5Cl	$CO_2 + Cl^-$	$\frac{1}{28}C_6H_5Cl + \frac{3}{7}H_2O = \frac{3}{14}CO_2 + \frac{1}{28}Cl^- + \frac{29}{28}H^+ + e^-$
Dichlorobenzene	$C_6H_4Cl_2$	$CO_2 + Cl^-$	$\frac{1}{26}C_6H_4Cl_2 + \frac{6}{13}H_2O = \frac{3}{13}CO_2 + \frac{1}{13}Cl^- + \frac{14}{13}H^+ + e^-$

Table 2.7. Electron Acceptor Half Reactions

Electron acceptor		End product	Half reaction
Oxygen	O_2	H_2O	$\frac{1}{8}O_2 + H^+ + e^- = \frac{1}{2}H_2O$
Nitrate	NO_3^-	N_2	$\frac{1}{5}NO_3^- + \frac{6}{5}H^+ + e^- = \frac{1}{10}N_2 + \frac{3}{5}H_2O$
Manganate	MnO_2	Mn^{2+}	$\frac{1}{2}MnO_2 + 2H^+ + e^- = \frac{1}{2}Mn^{2+} + H_2O$
Ferric iron	Fe_2O_3	Fe^{2+}	$\frac{1}{2}Fe_2O_3 + 3H^+ + e^- = Fe^{2+} + \frac{3}{2}H_2O$
Sulfate	SO_4^{2-}	$H_2S + HS^-$	$\frac{1}{8}SO_4^{2-} + \frac{19}{16}H^+ + e^- = \frac{1}{16}H_2S + \frac{1}{16}HS^- + \frac{1}{2}H_2O$
Carbon dioxide	CO_2	CH_4	$\frac{1}{8}CO_2 + H^+ + e^- = \frac{1}{8}CH_4$
Perchlorate	ClO_4^-	Cl^-	$\frac{1}{8}ClO_4^- + H^+ + e^- = \frac{1}{8}Cl^- + \frac{1}{2}H_2O$
PCE	$CCl_2{=}CCl_2$	$CH_2{=}CH_2$	$\frac{1}{8}CCl_2{=}CCl_2 + \frac{1}{2}H^+ + e^- = \frac{1}{8}CH_2{=}CH_2 + \frac{1}{2}Cl^-$
TCE	$CHCl{=}CCl_2$	$CH_2{=}CH_2$	$\frac{1}{6}CHCl{=}CCl_2 + \frac{1}{2}H^+ + e^- = \frac{1}{6}CH_2{=}CH_2 + \frac{1}{2}Cl^-$
Chromate	CrO_4^{2-}	$Cr(OH)_2$ (*s*)	$\frac{1}{3}CrO_4^{2-} + \frac{5}{3}H^+ + e^- = \frac{1}{3}Cr(OH)_3 + \frac{1}{3}H_2O$
Permanganate	MnO_4^-	MnO_2 (*s*)	$\frac{1}{3}MnO_4^- + \frac{4}{3}H^+ + e^- = \frac{1}{3}MnO_2 + \frac{2}{3}H_2O$
Peroxide	H_2O_2	H_2O	$\frac{1}{2}H_2O_2 + H^+ + e^- = H_2O$

Multiplying by the least common denominator of 24 yields the typical reaction:

$$2CH_3CH_2OH = 3CH_4 + CO_2 \quad \text{(Eq. 2.2)}$$

This balanced equation indicates that in this anaerobic reaction, 2 moles (mol) (92 g) ethanol is converted to 3 moles methane and 1 mole carbon dioxide.

Oxidation-reduction reactions of interest may be purely chemical (abiotic) or biological. An abiotic example is permanganate oxidation of an organic contaminant to carbon dioxide and water. A biological example is microbial oxidation of an organic contaminant to carbon dioxide and water when oxygen is available. Permanganate and oxygen are just two of the many different oxidants or electron acceptors that are used to enhance oxidations of interest. At times, rather than adding an oxidant to transform a contaminant, a reductant might be added. Hexavalent chromium (CrO_4^{2-}) is very soluble, but it can be reduced chemically or biologically by adding a suitable electron donor to form the insoluble trivalent chromium form ($Cr(OH)_3(s)$) which precipitates and is thus removed from the aqueous phase. The trivalent form is also less toxic than the hexavalent form, so reduction reduces both the solution concentration and the toxicity. For chemical reduction, sulfur dioxide might be added, or for biological reduction, hydrogen (H_2) or an organic electron donor might be added. Biological reduction is also commonly used for bioremediation of chlorinated solvents. Here, H_2 or an organic electron donor is added for the reduction of chlorinated solvents, a process in which the chlorines on the compound are biologically replaced with hydrogen atoms. Thus, tetrachloroethene ($CCl_2{=}CCl_2$) might be converted to the less harmful ethene ($CH_2{=}CH_2$). In this case, the chlorine removed enters solution as hydrochloric acid, thus tending to lower pH. Thus, pH control may be necessary in order to maintain the near neutral range generally desired for biological reactions.

The first eight electron donors in Table 2.6 (hydrogen through carbohydrates) are often added to groundwater for chemical or biological remediation of some of the hazardous electron acceptors such as nitrate and perchlorate through chromate listed in Table 2.7. Fatty acids are often added in the form of emulsified vegetable oil and carbohydrates in the form of compounds such as sugar or molasses. Also listed as electron donors in Table 2.6 are several organic compounds from benzene through dichlorobenzene. These are at times oxidized by the addition of electron acceptors listed in Table 2.7 such as oxygen, nitrate, or through the action of an electron acceptor commonly present in groundwater or formed in the reaction itself, carbon dioxide. Sulfate and ferric iron are also often present naturally in groundwater and may serve as electron acceptors for oxidation. When the electron acceptors required for oxidation of an electron donor are already present in the aquifer, then natural attenuation is possible, but may require a mixing process to bring the reactants together.

Precipitation reactions (the precipitate is indicated by (*s*) following the chemical) are of importance when stabilization of a chemical is desired, such as by its removal from the water phase and formation of a solid phase that does not contaminate or move with groundwater. For example, formation of the precipitate $Cr(OH)_3(s)$ removes chromium from water. Some other important low-solubility metal complexes are listed in Table 2.4. The low solubility product of many sulfide species suggests that they would be good candidates for removal from groundwater. Sulfides for this purpose might be formed from sulfate reduction under anaerobic conditions.

Precipitation, while often beneficial, can also cause serious problems, such as clogging by calcium carbonate ($CaCO_3(s)$) which is often encountered in groundwater remediation. Clogging may be undesirable because it can re-route the direction of groundwater flow leading to migration of contaminated water into previously uncontaminated regions and/or delivery of added chemicals to regions that are uncontaminated. The outcome may be an inefficient and wasteful use of added chemicals and the creation of regions left untreated or poorly treated.

Solubilization is a mass-transfer process related to the movement of a chemical between a solid phase and the aqueous phase. Solubilization may also occur through the dissolution of a non-miscible liquid into water, such as benzene or trichloroethene. Mixing often enhances solubilization by enhancing mass transfer. Additionally, chemicals can be added that enhance solubilization. For example, detergents may be used to increase the solubility of liquid-phase chlorinated solvents so that they can be extracted more readily from groundwater. Solutions containing high concentrations of water-soluble solvents such as ethanol may be used for this purpose as well. Detergent and solvent enhanced solubilization are major remediation processes that require the introduction and mixing of chemicals for groundwater remediation.

Sorption is another mass-transfer process that results in the movement of a chemical species from one phase to another, i.e. from an aqueous phase to a solid phase. At times this process also may not involve addition of a different chemical species, but instead may be aided by mixing to enhance mass transfer rates. However, it should be noted that different forms of a chemical differ in their susceptibility to volatilization or sorption. For example, CO_2 is a volatile gas, while HCO_3^- (bicarbonate) is not, just as H_2S (hydrogen sulfide) is a volatile gas, while HS^- (bisulfide) is not. The sorption characteristics of Zn^{2+} are different from those of $ZnOH^+$. The pH affects the relative proportions of these different species, and thus by implementing pH control, the potential for volatilization or sorption can be made to vary considerably. This again illustrates the importance that pH control can have on the movement and fate of chemicals in groundwater.

Advection and diffusion or dispersion are transport processes associated with the fluid in which the chemicals are contained. For example a chemical discharged into a flowing river is carried downstream with the flowing water by advection. As it moves downstream, the chemical spreads out and becomes more dilute through mixing caused by the turbulent action of water, a process called dispersion. In very still waters or in water moving by laminar flow, mixing may be more limited. Advection, dispersion, and diffusion are major processes of importance in bringing chemicals together for reaction in groundwater, and are addressed in more detail in Chapter 3, as well as later in this chapter and elsewhere in this volume.

2.3.4 Reaction Kinetics

Reaction rate processes are discussed in detail in general textbooks (Bailey and Ollis, 1986; Levenspiel, 1999; Weber and DiGiano, 1996) and will only briefly be summarized here. There are two basic classifications of reactions, *homogeneous* and *heterogeneous*. A homogeneous reaction is one that takes place in one phase only, such as in water. A heterogeneous reaction occurs in two phases, or at an interphase, such as between groundwater and aquifer solids, or between groundwater and microorganisms. Thus, in groundwater systems both homogeneous and heterogeneous reactions are likely to occur. Many variables may affect reaction rates such as temperature and pressure. Heterogeneous reactions are much more complex; here mass transfer effects are likely to play a key role in overall observed reaction rates. Mass transfer effects such as diffusion of a chemical to and into aquifer solids are likely to be involved. When a reaction consists of a number of steps in series, it is the slowest step in that series that controls the overall rate of the reaction. If one knows what step that is, whether mass transfer or reaction rate, then the rate can be modeled by consideration of that step alone. The transformation of a contaminant in a biofilm is just one case where both mass transfer rate and reaction rate are involved, an example of such a case is discussed in Section 2.4.4, while mass transfer effects overall are discussed in detail in Chapter 3. The following discussion concentrates only on the reaction term portion of a reaction rate series.

Let us first consider the rate of change, r_i, in one component i in a reaction, we indicate this rate by the change with time in its molar concentration N_i to be dN_i/dt. The reaction rate may be expressed in different ways, depending upon the basis of the reaction:

Basis for reaction	Reaction form	
Unit volume of reacting fluid	$r_i^0 = \frac{1}{V}\frac{dN_i}{dt} = \frac{\text{moles } i \text{ formed}}{(\text{volume of fluid})(\text{time})}$	(Eq. 2.3)
Unit mass of solid in fluid	$r_i^1 = \frac{1}{W}\frac{dN_i}{dt} = \frac{\text{moles } i \text{ formed}}{(\text{mass of solid})(\text{time})}$	(Eq. 2.4)
Unit interfacial surface of solid in fluid	$r_i^2 = \frac{1}{S}\frac{dN_i}{dt} = \frac{\text{moles } i \text{ formed}}{(\text{unit suface})(\text{time})}$	(Eq. 2.5)

Equation 2.3 is generally the form used in homogenous groundwater reactions when all the reactants are in the aqueous phase. Equations 2.4 and 2.5 are used primarily with heterogeneous reactions. Equation 2.5 may be the more accurate equation of the two, but frequently surface area is not readily determined because of the greatly differing characteristics and sizes of aquifer solid particles, so Equation 2.4 is frequently used as a more convenient substitute.

Beginning with a homogeneous reaction and Equation 2.3, let us first consider a simple reaction involving two reactants in aqueous phase that form two aqueous phase products:

$$\mathrm{aA + bB = cC + dD}$$

We may then become interested in the rate of loss of component A, having a molar concentration C_A. This may be expressed in many different ways depending upon the factors affecting the reaction. Some example reaction expressions are:

Reaction type	Reaction equation	Units for *k*	
First-order	$-r_A = kC_A$	T^{-1}	(Eq. 2.6)
Second-order	$-r_A = kC_A^2$	$L^3M^{-1}T^{-1}$	(Eq. 2.7)
Second-order	$-r_A = kC_AC_B$	$L^3M^{-1}T^{-1}$	(Eq. 2.8)
Zero-order	$-r_A = k$	$ML^{-3}T^{-1}$	(Eq. 2.9)
Complex reaction	$-r_A = \frac{kC_AC_X}{K+C_A}$	T^{-1}	(Eq. 2.10)
Complex reaction	$-r_A = \frac{kC_AC_X}{K_A+C_A}\frac{C_B}{K_B+C_B}$	T^{-1}	(Eq. 2.11)

Where the symbols M, L and T refer to standard units – M is mass (generally expressed in milligrams [mg] or micrograms [μg]), L is length (usually expressed in meters [m] or centimeters [cm]), and T is time (generally expressed in days [d] or seconds [s]).

The order of the reaction is generally given by the sum of the exponents on the concentration terms in the reaction. Thus, Equations 2.7 and 2.8 are both second order reactions, the first depending upon the square of component A's concentration and the second on the product of the concentration of two different components. In the zero-order reaction, the rate is independent of the concentration of any of the reactants.

Complex reactions, however, cannot be described by the order concept. The complex reactions shown are just two of many possibilities. These are non-linear equations that are difficult to use when an analytical solution for a groundwater model is sought, their use generally requires some form of numerical solution. These two particular equations are similar to variations used in the Monod expression for biological processes. Here, C_X would represent the concentration of the acting microorganisms. Equation 2.10 is the form generally used when component A is in limiting supply and controls the overall reaction. Equation 2.11 is used when either reactant component, A or B, may be limiting at times, so both need consideration in a numerical model. An example where Equation 2.11 might be useful is in modeling the biological oxidation of toluene by organisms using nitrate as an electron acceptor. At the point where toluene first comes in contact with an aquifer that contains nitrate, the nitrate concentration may be high and non-limiting compared with toluene. But as the groundwater moves through the toluene spill, nitrate concentration decreases – the nitrate concentration then may become rate limiting. If nitrate is taken to be component B in Equation 2.11, we see that in the first case of high nitrate (this means high with respect to the constant K_B), then the expression $C_B/(K_B + C_B)$ approaches 1. When C_B decreases to the point where it equals K_B, then the expression equals one-half, meaning the overall rate is halved. This is the reason K_B is often called the half-velocity coefficient.

In selecting the most appropriate rate expression, the modeler should chose one that is complex enough to describe the situation adequately for the purpose intended, but not so complex that the model solution becomes overly difficult. At times, one may wish to use a more appropriate rate expression, but the information required for input to the model is not available. Perhaps too often, simple models that are inadequate for predictive purposes are used simply because they are simpler to use, often leading to grossly erroneous predictions. However, simpler models are sometimes justified for use when the field situation deems it appropriate. For example, which model might be most appropriate for conversion of acetate to methane (methanogenesis)?

High concentrations in the thousands of mg/L range of acetate often result from fermentation of organic electron donor added to aquifers for biological remediation of chlorinated solvents. The acetate emerging in resulting anaerobic plumes can be converted to methane gas by methanogens. One may wish to model this process and might first consider using Equation 2.11. Here, no electron acceptor is needed, so that C_B in Equation 2.11 is zero, thus, Equation 2.10 would be adequate instead. Also, the K_A for acetate is on the order of 100 mg/L, so if acetate concentration is 1,000 mg/L or above, the term $C_A/(K_A + C_A)$ essentially equals 1. Eliminating that element means that the first order Equation 2.6 is adequate with C_X being substituted for C_A. However, measuring C_X is very difficult as the organisms it represents are mostly attached to aquifer solids and not adequately determined from analysis of extracted groundwater. The organism concentration also changes with growth through acetate utilization. Because of this difficulty, modelers often then tend to assume C_X is constant, essentially meaning that the zero-order Equation 2.9 is sufficient. Others just assume Equation 2.6 is adequate. Neither really fits the case. It would be better here to develop a model that includes changes in C_X with time and acetate utilization. A typical model for change in C_X through normal biological growth and decay is as follows:

$$\frac{dC_X}{dt} = Yr_A - bC_X \qquad \text{(Eq. 2.12)}$$

Here, Y equals the yield of organisms per mole of acetate consumed, r_A is the rate of acetate utilization, and b is a first-order decay rate coefficient (T^{-1}) for the microorganisms.

We see here that one could obtain appropriate results using the more complex Equation 2.11 or the simplified Equation 2.9 as long as an appropriate value as derived from Equation 2.12 were included in the overall model. Modeling thus sometimes becomes as much of an art as it is a science.

In the above example for biological transformation, it is seen that microorganisms were considered to be part of a homogeneous reaction. Microorganisms actually act as a catalyst to bring about the reaction, extracting energy for growth from the process. Thus, Equations 2.10 and 2.11 may be used as well to describe rates resulting from catalyst addition to an aquifer for chemically enhancing a reaction rate. Current interest is in using nanoparticles for this purpose. However, like microorganisms, catalyst or reactants may be attached to aquifer material, so treatment as if it were a homogeneous reaction may not be appropriate. Reaction rate instead may be a function of surface area exposed rather than solution concentration. A good example here is a permeable reactive barrier wall, such as one containing zero-valent iron, as described in Chapter 7. Chemicals, such as a chlorinated solvent contained in groundwater passing through the barrier wall must then be mass transported such as by diffusion from the water to the iron surface, where the dechlorination reaction takes place, oxidizing the iron in the process. Equation 2.5 then becomes the appropriate reaction term for use, and the reaction rate for solutes is then expressed in mass per time per unit surface area. The difficulty here is that diffusive mass transport to the reacting surface becomes of importance as does knowledge of the surface area of the material with which it is reacting. These may be difficult to determine. Simplifications such as use of Equation 2.4 are then often resorted to in zero-valent barrier walls as the mass quantity of iron added is generally known, if not its surface area. In other cases, modelers simply resort to first- or zero-order reaction rates as determined from empirical field measurements. Such models generally do not involve sufficient knowledge of system characteristics to be useful for sound predictions. Great care thus needs to be taken in their use.

Temperature is an important factor affecting reaction rates as is pH and reaction inhibitors. There are several different theoretical models that indicate how reaction rate varies with temperature. In general, most result in a logarithmic expression that modifies the rate coefficient as in the following:

$$k_T = k_{T^0} e^{K_T(T - T^0)} \quad \text{(Eq. 2.13)}$$

where k_T is the rate constant at temperature T, k_{T^0} is the rate at some standard temperature T^0 such as 20°C, and K_T is a temperature constant. In the normal groundwater temperature range between 10°C and 30°C, rate is commonly considered to double with each 10°C rise in temperature. This corresponds with a value for K_T of 0.069/°C.

Inhibiting the reaction rate are such things as high concentration of the substrate being consumed, high concentration of a reaction product, or competition for key enzymes by different substrates. Typical models for each are listed below, illustrating how they might be incorporated to modify Equation 2.10.

Inhibition factor	Example incorporation into Equation 2.10	
Substrate inhibition	$-r_A = \frac{kC_AC_X}{K + C_A\left(1 + \frac{C_A}{K_I}\right)}$	(Eq. 2.14)
Product inhibition	$-r_A = \frac{kC_AC_X}{K + C_A}\frac{K_I}{K_I + C_P}$	(Eq. 2.15)
Competitive inhibition	$-r_A = \frac{kC_AC_X}{K\left(1 + \frac{C_c}{K_I}\right) + C_A}$	(Eq. 2.16)
Non-competitive inhibition	$-r_A = \frac{kC_AC_X}{\left(1 + \frac{C_c}{K_I}\right)(K + C_A)}$	(Eq. 2.17)

Here, K_I is the relevant inhibition constant, C_P is concentration of a product of a reaction, and C_c is the concentration of a reactant C that is competing for a key enzyme involved in transforming reactant A. Substrate inhibition may be experienced in the reductive dehalogenation of a chlorinated solvent such as TCE by high TCE concentrations that exist near a dense nonaqueous phase liquid (DNAPL) or of benzene near a gasoline-spill produced light nonaqueous phase liquid (LNAPL). Product inhibition may result during TCE reductive dehalogenation from a large increase in the concentration of *cis*-DCE, the product of TCE reduction. Competitive inhibition in chlorinated solvent biodegradation can occur during the reductive dehalogenation of *cis*-DCE and VC when these two electron acceptors compete for the same electron transfer train in a single organism. Generally the organism, which can use either, will select to use that electron acceptor in highest relative concentration. Non-competitive inhibition represents the adverse impact of one compound on the transformation of another. The similarity between Equations 2.17 and 2.15 should be readily apparent, they are mathematically the equivalent of each other.

2.3.5 Summary

In summary, there are many different reactions and phase changes that might be brought about through the delivery and mixing of chemicals for *in situ* remediation of groundwater. Selecting the correct chemical and correct amount is one part of the challenge. Reaction stoichiometry helps in this selection. The other is in the delivery and mixing of the chemical where needed in order to bring about the desired change. These are rate processes that also need to be understood. Both are challenges, but the latter is perhaps the bigger of the two, and the major emphasis given in this volume. This chapter, however, emphasizes the first challenge, the selection of the right chemicals and amounts for *in situ* remediation, although some brief discussion of mass transfer and reaction rates is also provided.

2.4 BIOLOGICAL PROCESSES

Most naturally occurring organics that percolate down through the soil are degraded by naturally occurring bacteria, thus rendering them harmless so that they pose no serious threat to groundwater quality. Many anthropogenic chemicals can be destroyed readily by microorganisms. It is only some of the anthropogenic organic chemicals that pose a significant threat, and these, for the most part are the ones that are difficult to biodegrade, compounds that are termed "persistent organic pollutants," or POPs. Included here are many halogenated compounds such as pesticides, chlorinated solvents, chlorinated benzenes and phenols, and dioxin, many of which are listed as frequently detected contaminants in Table 2.1. Also included are some with difficult to degrade structures such as complex ethers (e.g., MTBE). There are many inorganic chemicals of concern in groundwater as well that can be transformed biologically to less harmful forms, such as nitrate, perchlorate, chromate, and uraninite. Most biological reactions of interest in remediation are oxidation-reduction reactions, and in these reactions, the target contaminant may be rendered less harmful either through its oxidation or its reduction as already indicated. More detailed information about biological processes can be obtained from textbooks (Rittmann and McCarty, 2001).

2.4.1 Biological Processes

Microorganisms bring about oxidation-reduction reactions in order to obtain energy for growth, thus organism growth must be considered as part of the reaction. In order to grow, microorganisms also need certain mineral nutrients to form necessary cellular components such as nucleic acids, enzymes, proteins, carbohydrates, and fats. Of major importance here are the elements carbon, nitrogen, phosphorus, sulfur, and iron. Certain trace chemicals such as nickel and manganese may also be required for enzyme activity. These may or may not be present in excess in the aquifer solids surrounding groundwater—often they are, and so such nutrient additions may not be needed. As an example, a balanced stoichiometric equation of the overall reaction for transformation of an organic contaminant (benzoate) through reduction of an inorganic contaminant (nitrate) is as follows:

$$C_6H_5COO^- + 3.29NO_3^- + 3.29H^+ = 0.588C_5H_7O_2N + 1.35N_2 + HCO_3^- + 3.05CO_2 + 1.58H_2O \qquad \text{(Eq. 2.18)}$$

Here, $C_5H_7O_2N$ is used as an empirical formula for cells and indicates the relative proportion of various elements in the cells. Nitrogen represents about 12% of the weight of the cell. Phosphorus, another major element required is not shown in this formulation, but represents about 2% of the cell weight.

Equation 2.18 indicates that for oxidation of 1 mole of benzoate (121 g) 3.29 moles of nitrate (46 g nitrate-N) would be reduced, with most being converted to N_2 gas. In this process, 0.588 mole of cells (66 g) would be formed. The reaction is a basic one as indicated by consumption of 3.29 moles H+ on the left side and formation of 1.0 mole of the basic bicarbonate anion on the right side. This balanced equation is thus useful for indicating how much of one chemical is required in order to bring about the destruction of the other. This is the kind of information needed in order to properly design a chemical feed system. Interesting here is that according to this reaction, benzoate could be added to destroy nitrate contamination, or nitrate could be added to treat benzoate contamination. However, benzoate itself can be toxic, so if the goal is to remove nitrate, a different electron donor would generally be added, such as acetate, ethanol, or lactate.

Reaction 2.18 can be divided into two components (Rittmann and McCarty, 2001), the energy component and the synthesis component:

Energy component:

$$0.45C_6H_5COO^- + 2.7NO_3^- + 2.7H^+ = 1.35N_2 + 0.45HCO_3^- + 2.7CO_2 + 2.25H_2O \quad \text{(Eq. 2.19)}$$

Synthesis component:

$$0.55C_6H_5COO^- + 0.59NO_3^- + 0.59H^+ + 0.67H_2O = 0.588C_5H_7O_2N + 0.55HCO_3^- + 0.35CO_2 \quad \text{(Eq. 2.20)}$$

Adding Equation 2.19 to Equation 2.20 results in Equation 2.18. From this it can be seen that here 45% of the benzoate is consumed in denitrification, or the conversion of nitrate into N_2, while 55% is used for synthesis of cells. Most of the nitrate is destroyed by denitrification, but about 18% is used in cell synthesis. In considering demand for electron donor, that portion associated with both energy production and synthesis needs evaluation.

The synthesis component of the biological reaction can be obtained by adding the synthesis half reaction to the electron donor half reaction. The synthesis half reaction is:

$$\frac{1}{5}CO_2 + \frac{1}{20}NH_4^+ + \frac{1}{20}HCO_3^- + H^+ + e^- = \frac{1}{20}C_5H_7O_2N + \frac{9}{20}H_2O \quad \text{(Eq. 2.21)}$$

While the stoichiometry of a biological reaction is given by a balanced overall reaction, such as Equation 2.18, the quantity of electron donor required for the reaction can also be estimated by considering just the energy portion of the reaction and then including in the calculations sufficient excess donor to satisfy the need for biological synthesis. The fraction of donor used for synthesis is highest for aerobic reactions and denitrification, with as much as 50% then being used for synthesis during active bacterial growth. Thus, about twice the electron donor required for the energy reaction would need to be present to also satisfy the need for biological growth. In groundwater remediation, growth rate is usually not maximal, and perhaps only about 50% excess donor is needed to satisfy the synthesis demand in the above cases. However, with anaerobic reactions (those not involving O_2), the amount of donor associated with synthesis is generally much less. When methane production or sulfate reduction are the dominant reactions, the excess amount of donor needed for synthesis varies between about 5% when fatty acids are used as donors up to about 20% with carbohydrates. In reductive dehalogenation, the additional amount needed for synthesis may be closer to 10–15%.

Some discussion is justified concerning the energy reactions involved in anaerobic processes. For this example, acetate will be used. Several possible energy reactions with acetate are listed in Table 2.8. The first four reactions represent the typical ones for which microorganisms are common and ubiquitous in the environment. The first is the aerobic reaction with oxygen as electron acceptor. The next three are anoxic reactions, the first, denitrification with nitrate, next, sulfate reduction or sulfidogenesis, and the fourth, methanogenesis. The energy derived from each reaction is noted on the right side of Table 2.8. Aerobic oxidation of organic substances yields the highest energy and so growth on a given amount of acetate is higher here, that is the portion of electron donor used for synthesis is higher as already noted. Nitrate energy yield is not far behind. However, the energy from sulfate reduction and methanogenesis are much less, with that from methanogenesis the smallest. Methanogens and sulfate-reducing

Table 2.8. Energy Reactions Involving Acetate

Electron acceptor	Energy reaction	$\Delta G^{0\prime}$ (kJ)
O_2	$CH_3COO^- + 2O_2 \rightarrow CO_2 + HCO_3^- + H_2O$	−849
NO_3^-	$CH_3COO^- + 1.6NO_3^- + 1.6H^+ \rightarrow CO_2 + HCO_3^- + 0.8N_2 + 1.8H_2O$	−797
SO_4^{2-}	$CH_3COO^- + SO_4^{2-} + 1.5H^+ \rightarrow CO_2 + HCO_3^- + 0.5H_2S + 0.5HS^- + H_2O$	−52
CO_2	$CH_3COO^- + H^+ \rightarrow CO_2 + CH_4$	−36
Fe(III)	$CH_3COO^- + 4Fe_2O_3 + 16H^+ \rightarrow CO_2 + HCO_3^- + 8Fe^{2+} + 9H_2O$	
ClO_4^-	$CH_3COO^- + ClO_4^- \rightarrow CO_2 + HCO_3^- + Cl^- + H_2O$	−972
PCE	$CH_3COO^- + 2CCl_2{=}CCl_2 + 3H_2O \rightarrow CO_2 + HCO_3^- + 2CHCl{=}CHCl + 4Cl^- + 4H^+$	−463

Note: *kJ* kilojoules

bacteria (SRB) must therefore oxidize a larger fraction of the electron donor so as to have sufficient energy for cell synthesis. This is why the cell yield from these reactions is so low, and why the organisms grow so slowly under sulfidogenic and methanogenic conditions. Doubling times for aerobic organisms are on the order of hours, while that for sulfidogenic and methanogenic conditions are on the order of days.

Significant rates of conversion of substrates by microorganisms require an organism concentration on the order of one million per milliliter (mL) of water. When the doubling time for the organism is 1 h – as in the case of aerobic growth on organics – the concentration of organisms can increase from one to one million per milliliter in less than 1 day. The same job requires 60 days when the doubling time is 3 days, as is the case for methanogens. The slow doubling time of anaerobic microorganisms is why it often takes months to begin to see significant degradation of hazardous compounds once the remediation process is initiated, even if the needed microorganisms may already be present in small concentrations.

Another factor of importance when considering the first four reactions in Table 2.8 is that the fourth reaction, methanogenesis, occurs in the absence of an external electron acceptor. In other words, if a compound is amenable to decomposition under methanogenic conditions, it can be degraded in groundwater without an added electron acceptor. All that is needed are sufficient microorganisms capable of degrading the target contaminants and the trace nutrients necessary for their growth. Necessary trace nutrients are commonly present in aquifer minerals, so they may not need to be added either. Most commonly, natural attenuation of hazardous organic compounds occurs because the compounds are amenable to methanogenesis, which generally requires a consortium of different species working together to process the organic through the steps of fermentation, acidogenesis, and then methanogenesis. Potential for conversion through methanogenesis is the case with most naturally occurring organic compounds. Included are many hazardous compounds, such as phenol, styrene, and the BETX compounds. Some numerical models of natural attenuation assume that external electron acceptors are required for anaerobic degradation of these compounds in groundwater, but this is not actually necessary through methanogenesis as well demonstrated in the landmark publication by Gribić-Galić and Vogel (1987) and numerous subsequent articles. While the consortia of anaerobic microorganisms required for the conversion of these compounds to methane are not always present in groundwaters, they are sufficiently common that natural attenuation often can be counted upon to rid groundwater of such chemicals. When the required organisms are not present, then bioaugmentation with suitable microorganisms might be considered. The process used for introduction and mixing of the microorganisms then becomes an issue.

2.4.2 Chlorinated Solvents

Because of their importance as major groundwater contaminants and the variety of ways by which they may be transformed in groundwater (Vogel et al., 1987), some specific comments about them are included here. Methylene chloride can be biodegraded under either aerobic or anaerobic conditions while supplying energy to the microorganisms and using typical electron acceptors as listed in Table 2.7 just as is the case with many other common organic non-halogenated compounds. However, this is not the case with the other four main chlorinated solvents, PCE, TCE, TCA, and CT. There is little evidence that any of them can be degraded aerobically or through denitrification in a manner that is beneficial to microorganisms. TCE and TCA, however, can be aerobically transformed through cometabolism, primarily by organisms that contain an oxygenase used for initiating oxidation of hydrocarbons or ammonia. Anaerobically, when neither oxygen nor nitrate is present, PCE, TCE, and TCA, but not CT, can be used by certain microorganisms as electron acceptors in energy metabolism. Here, the reaction is stepwise, one chlorine at a time is removed and replaced with hydrogen, a process termed reductive dehalogenation. In this process, several intermediate chlorinated species result as illustrated in Figure 2.1. Generally, compounds with more chlorine atoms tend to be transformed faster than those with fewer chlorine atoms, often resulting in the buildup of the intermediate compounds. Frequently, specific dechlorinating microorganisms can remove only some of the chlorine atoms from some of the compounds of concern so that complete removal of all chlorine atoms from a chlorinated compound may require the action of more than one dehalogenating organism. The electron donor that appears to be most generally preferred by dehalogenating organisms is H_2, and this is the only electron donor found so far to be acceptable by organisms that reductively dehalogenate *cis*-DCE and VC. Some organisms can use other electron donors, such as acetate or lactate, for at least partial dehalogenation of some compounds, such as TCE and PCE. Additionally, TCA can be transformed partially abiotically to form other chemicals of concern.

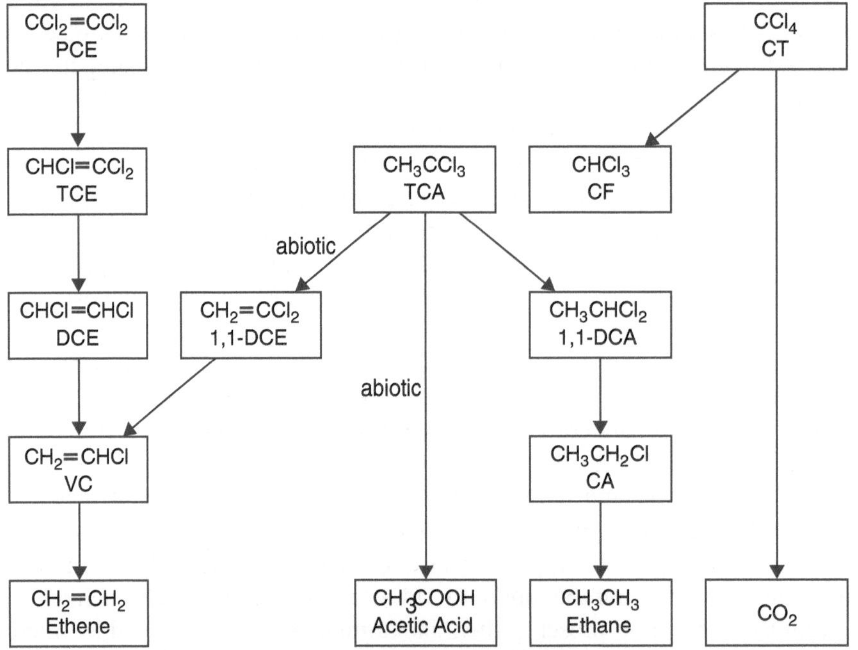

Figure 2.1. General scheme for anaerobic biological transformations of chlorinated aliphatic compounds (some spontaneous abiotic steps also indicated).

Table 2.9. Abiotic and Biotic Reactions for PCE, TCE, TCA, and CT

Reaction	Reactant	Product	Other electron donors possible?
ANAEROBIC – METABOLIC ENERGY YIELDING			
Tetrachloroethene (PCE)			
$CCl_2{=}CCl_2 + H_2 \rightarrow CHCl{=}CCl_2 + H^+ + Cl^-$	PCE	TCE	Yes
Trichloroethene (TCE)			
$CHCl{=}CCl_2 + H_2 \rightarrow CHCl{=}HCl + H^+ + Cl^-$	TCE	*cis*-DCE	Yes
$CHCl{=}CHCl + H_2 \rightarrow CH_2{=}CHCl + H^+ + Cl^-$	*cis*-DCE	VC	–
$CH_2{=}CHCl + H_2 \rightarrow CH_2{=}CH_2 + H^+ + Cl^-$	VC	Ethene	–
1,1,1-Trichloroethene (TCA)			
$CH_3CCl_3 + H_2 \rightarrow CH_3CHCl_2 + H^+ + Cl^-$	TCA	1,1-DCA	–
$CH_3CHCl_2 + H_2 \rightarrow CH_3CH_2Cl + H^+ + Cl^-$	1,1-DCA	CA	–
ABIOTIC			
1,1,1-Trichloroethene (TCA)			
$CH_3CCl_3 \rightarrow CH_2{=}CCl_2 + H^+ + Cl^-$	TCA	1,1-DCE	–
$CH_3CCl_3{+}2H_2O \rightarrow CH_3COOH + 3H^+ + 3Cl^-$	TCA	Acetic acid	–
COMETABOLIC			
Trichloroethene (TCE)			
$CHCl{=}CCl_2 + NADH + H^+ + O_2 \rightarrow CHClOCHCl + NAD^+ + H_2O$	TCE	TCE Epoxide	–
Carbon Tetrachloride (CT)			
$aCCl_4 + \text{cofactors} \rightarrow bCHCl_3 + cCO_2 + d\text{Other}$	CT	$CHCl_3$[a]	–

[a]Chloroform generally is one of the products formed from CT transformation, but not always depending upon the organism involved

Table 2.9 provides a listing of chemical (abiotic) and biological (biotic) transformations commonly observed in groundwater. In the examples provided where oxidation-reduction is involved, H_2 is indicated as the electron donor for simplicity with a note indicating when other electron donors might also be used.

The anaerobic transformation of organic compounds is fairly complex and often relies on a variety of microorganisms to complete the transformation. A general scheme for anaerobic transformation is illustrated in Figure 2.2. Here, complex organics such as carbohydrates, proteins, and fats, are first hydrolyzed to form simple sugars, amino acids, and fatty acids, which are then fermented and partially oxidized by a variety of microorganisms to produce hydrogen and acetic acid. Generally, about 2 moles H_2 will be produced per mole of acetate that is formed, but this ratio varies somewhat depending upon the starting electron donor. The hydrogen and acetic acid formed can then be used by methanogens and converted into methane, or by other organisms that compete for hydrogen, such as sulfate reducers, iron reducers, or dehalogenators (Table 2.10). In order to supply hydrogen as needed by *cis*-DCE and VC dehalogenators, any of a variety of organic donors might be used, as the anaerobic degradation of most will produce the needed hydrogen. Elemental hydrogen itself might be added to satisfy

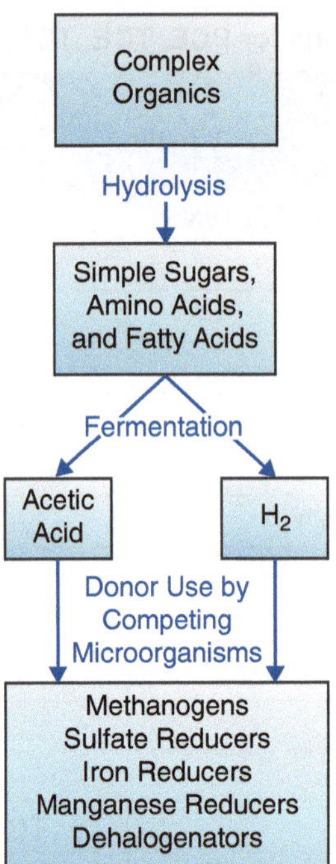

Figure 2.2. The mixed-culture anaerobic transformation of organic compounds.

Table 2.10. Energy Reactions Using H_2 as Electron Donor with Various Electron Acceptors

Electron acceptor	Energy reaction	Effect on pH
O_2	$H_2 + \frac{1}{2}O_2 \rightarrow H_2O$	Neutral
NO_3^-	$H_2 + \frac{1}{5}NO_3^- + \frac{1}{5}H^+ \rightarrow \frac{1}{10}N_2 + \frac{3}{10}H_2O$	Basic
SO_4^{2-}	$H_2 + \frac{1}{4}SO_4^{2-} + \frac{3}{8}H^+ \rightarrow \frac{1}{8}H_2S + \frac{1}{8}HS^- + \frac{1}{2}H_2O$	Basic
CO_2	$H_2 + \frac{1}{4}HCO_3^- + \frac{1}{4}H^+ \rightarrow \frac{1}{4}CH_4 + \frac{3}{4}H_2O$	Basic
Fe(III)	$H_2 + Fe_2O_3 + 4H^+ \rightarrow 2Fe^{2+} + 3H_2O$	Basic
ClO_4^-	$H_2 + \frac{1}{4}ClO_4^- \rightarrow \frac{1}{4}Cl^- + H_2O$	Neutral
PCE	$H_2 + \frac{1}{4}CCl_2{=}CCl_2 \rightarrow \frac{1}{4}CH_2{=}CH_2 + Cl^- + H^+$	Acidic

this need, but at the higher concentrations that result, homoacetogenic microorganisms can grow from the energy produced by reducing carbon dioxide with hydrogen to produce acetic acid. This is generally not considered a desirable outcome, because it results in some unwanted loss of the hydrogen and produces an acid that may adversely impact solution pH. Generally for reductive dehalogenation, organic electron donors that release hydrogen slowly and only when the concentration is brought below a threshold for the homoacetogens of about 300 nanomolar (nM) are desired. This is generally the case with fatty acids containing three or more carbon atoms such as propionic or butyric acids. Another example is the vegetable oils commonly added as electron donors and consisting primarily of 16- to 18-carbon fatty acids such as palmitic, oleic and linoleic acids.

2.4.3 Biological Reaction Kinetics

The reaction rate for a biological reaction is often characterized by Monod kinetics, which can be formulated as follows (Rittmann and McCarty, 2001):

Rate of substrate utilization:

$$-\frac{dS}{dt} = qX\frac{S}{K+S} \qquad \text{(Eq. 2.22)}$$

where,

S = rate-limiting substrate concentration, mg/L
t = time, days
q = maximum substrate utilization rate, mg substrate per mg cells per day (d)
X = cell concentration, mg/L
K = half-velocity coefficient, mg/L

This equation assumes that only a single substrate is rate limiting, all other nutrients needed by the organisms for growth are in excess concentration and so do not affect the rate of the reaction. The relationship between substrate utilization rate and substrate concentration is illustrated in Figure 2.3. At low substrate concentration, the rate is directly proportional to substrate concentration, but at high substrate concentration, the rate reaches a maximum with a value of q.

Rate of organism growth:

$$\frac{dX}{dt} = Y\left(-\frac{dS}{dt}\right) - bX \qquad \text{(Eq. 2.23)}$$

where,

Y = organism yield, mg organism produced per mg substrate consumed
b = organism decay rate, day^{-1}

Combining Equations 2.22 and 2.23 yields:

$$\frac{dX/X}{dt} = Yq\frac{S}{K+S} - b \qquad \text{(Eq. 2.24)}$$

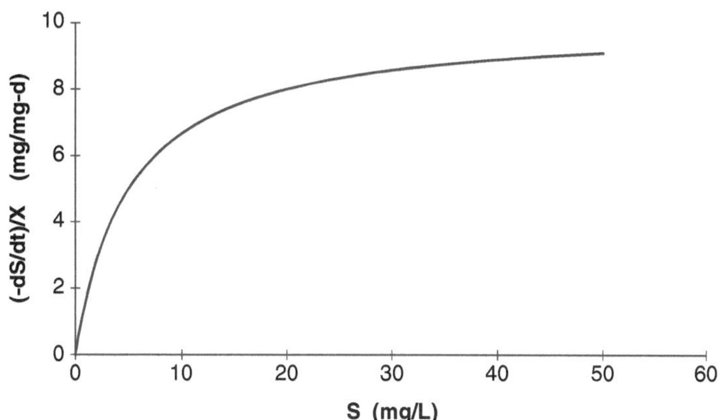

Figure 2.3. Relationship between the concentration of a rate-limiting substrate and biological reaction rate (Y = 0.6 mg cells/mg substrate, K = 5 mg/L, q = 10 g substrate/g cells/day, b = 0.2 day^{-1}).

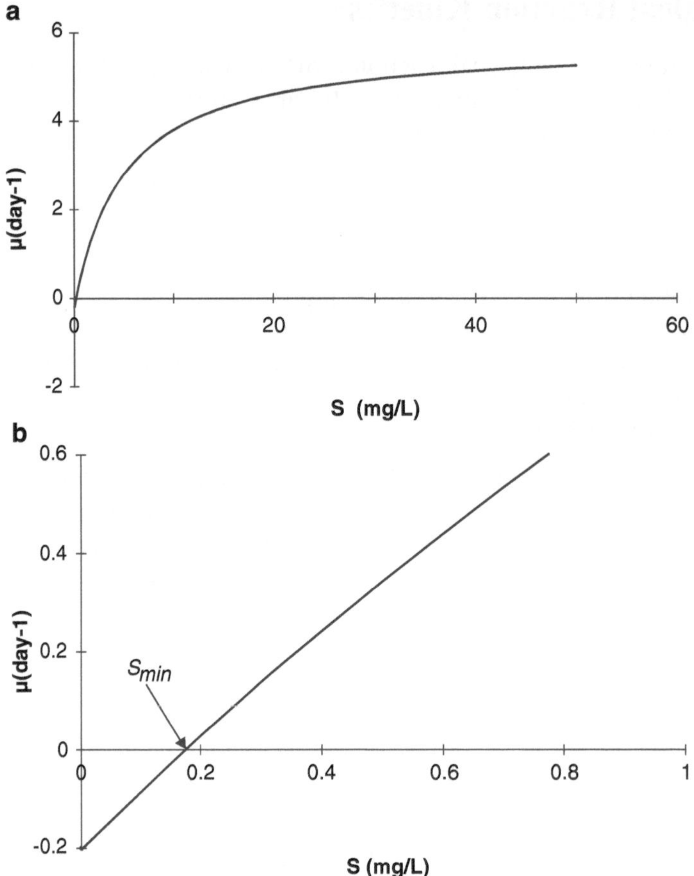

Figure 2.4. Relationship between the concentration of a rate-limiting substrate and microorganism growth rate (same conditions as in Figure 2.3). Expanded Figure 2.4b illustrates negative growth rate when S is below S_{min} of 0.17 mg/L).

The net specific growth rate of microorganisms (*dX/Xdt*) is generally represented by the symbol μ with units of day^{-1}, and the product Yq equals the maximum growth rate μ_m so that,

$$\mu = \mu_m \frac{S}{K + S} - b \qquad \text{(Eq. 2.25)}$$

The relationship between organism growth rate and substrate concentration is illustrated in Figure 2.4. Of interest to note is that there is a substrate concentration S_m, below which the net growth rate of organisms is less than zero, in other words, the organisms are in net decay because they decay away faster than they grow. The relationship between S_m and other variables of interest can be found by setting the net growth rate to zero in Equations 2.24 and 2.25. This results in the following:

$$S_m = K \frac{b}{Yk - b} = K \frac{b}{\mu_m - b} \qquad \text{(Eq. 2.26)}$$

Equation 2.26 indicates that S_m is a function of K. The relationship between the two is given by the ratio $(\mu_m - b)/b$. Typical values for this ratio or maximum growth rate to decay rate are 20–100, suggesting that S_m typically is in the range of perhaps 10–500 micrograms per liter (μg/L) when K is in the range of 1–10 mg/L. S_m represents the lowest concentration for a substrate under steady-state conditions when that substrate is the only substrate for organism growth and all other growth requirements are in excess supply. At times then, the minimum concentration to which a contaminant can be biodegraded in an aquifer can be limited by S_m.

Often it is observed that compounds in groundwater are being degraded to concentrations below S_m. This can occur when they are used as secondary substrates or through cometabolism. Degradation as a secondary substrate occurs when an organism is provided with a sufficient amount of a primary substrate in order to maintain itself and produce the enzymes necessary for the simultaneous consumption of the secondary substrate. For example, an organism might be able to aerobically consume and grow on either acetate or benzene. Benzene by itself at a concentration of 10 μg/L might not be able to support net biological growth, but if the organism were at the same time given 1,000 μg/L of acetate, which is above its S_m level, it could grow on the acetate and simultaneously degrade the benzene down to 1 μg/L of benzene or less.

Cometabolism is the degradation of a compound by an organism using enzymes that serve some purpose for the cell other than degradation of that compound. The organism obtains no benefit from the transformation, indeed it may harm them. For example, some organisms that aerobically oxidize toluene initiate the oxidation using an enzyme called an oxygenase that adds elemental oxygen to toluene forming cresol. Commonly, the oxygenase also fortuitously adds elemental oxygen across the double bond in TCE to form TCE epoxide, which is chemically unstable and degrades to a series of simpler compounds that are used by other organisms for food. In this manner, TCE is destroyed by an organism that obtains no benefit from the transformation, indeed the epoxide formed may not only sap some energy away from the organism, but the epoxide itself can also be quite lethal to them. Nevertheless, cometabolism has been demonstrated to be useful as a method for aerobic destruction of TCE in groundwater (McCarty et al., 1998a). Another example is the cometabolic transformation of CT by the denitrifying bacterium *Pseudomonas stutzeri* KC. Strain KC secretes a biomolecule – pyridine-2,6-bis-thiocarboxylate (PDTC) – that has a primary role in trace metal acquisition, but also fortuitously degrades CT to harmless end products when it is chelated to copper (Dybas et al., 1995b; Lee et al., 1999).

S_m as a concept in groundwater is important in setting a lower substrate bound below which organisms cannot be in net growth. However, when the concentration is above S_m, as it usually is at the point of injection when substrates are added to aquifers to stimulate microbial growth, growth rate will be positive. Indeed, it will remain positive as long as a rate limiting substrate is above S_m. When this occurs at the point of chemical injection into an aquifer, organisms can continue to grow until the pore spaces between aquifer minerals are filled with them, clogging the aquifer. This is a problem that needs to be prevented at points of continuous substrate injection into aquifers, such as in wells. Methods to address this potential problem are outlined in detail in ESTCP (2005). These include pulsing of substrates instead of continuous injection so that periods of organism starvation and population decrease will occur, or periodic or continuous injection of a bacterial toxicant such as hydrogen peroxide to reduce clogging by organism growth near the injection well.

2.4.4 Mass Transfer Limitations

Frequently in groundwater, reaction rates are limited by the rate of transport of a needed substance to the point of reaction. Transport processes include advection, dispersion, sorption, and diffusion. Advection, dispersion, and sorption are covered adequately in other parts of this volume. Diffusion controlled reactions are as well, but will be mentioned briefly here to compliment the discussion of biological kinetics. The rate of a chemical or biological reaction

Table 2.11. Spreading Time for Chemicals as Function of Distance. Chemicals must move by diffusion to site of reaction. Assumed coefficient of molecular diffusion $D = 10^{-9}$ m²/s. Spreading in time *t* is $t = l^2/2D$, where *l* is the diffusion distance.

Diffusion distance	Time required
1 μm (scale of a bacterium)	10^{-3} s
1 mm (scale of a grain of sand)	8 min
1 cm	1 day
10 cm	5 months
1 m	16 years
10 m	4,000 years

at some specific location may be controlled mainly by the intrinsic rate of a reaction or by the rate of diffusion of a needed substance to that location. The two rate processes involved are diffusion and biotransformation. At times one may be more limiting than the other. Which is limiting in a given case affects how best to operate a chemical delivery system.

As shown in Table 2.11, spreading by molecular diffusion is fast over the distance scale of a bacterium or a grain of sand, occurring in seconds to minutes. But the time required for spreading is proportional to the distance squared. So over longer distances, much more time is needed. If reactants can only be delivered to a location within 10 cm of a target contaminant, 5 months are required. Clearly, patience is needed when contaminants and/or other reactants must diffuse through micro-fractures or small channels before becoming accessible for degradation.

Even when chemicals can be effectively distributed or delivered close to the contaminants, diffusion remains important. Microorganisms in aquifers for the most part are attached to aquifer material or exist as large immobile bundles of organisms living in the interstitial spaces between aquifer mineral particles. As such, they act as biofilms. Here as groundwater moves past, substrates must be conveyed from the water to and into the biofilm for biodegradation. Mass transfer from the water to the biofilm, and diffusion within the biofilm is required to bring substrate to the microorganisms. Mass transfer rather than intrinsic biodegradation rate may limit the rate of the biological reaction. This is often the case with natural attenuation.

Consider a simple case of steady-state diffusion of a rate-limiting substrate from the aqueous phase to a biofilm attached to the surface of some aquifer minerals. This is illustrated in Figure 2.5. The rate of mass diffusion ($\mathrm{d}M/\mathrm{d}t$) across a unit area of the boundary layer to the biofilm is proportional to the concentration gradient (Rittmann and McCarty, 2001):

$$-\frac{\mathrm{d}M}{\mathrm{d}t} = k_d(S_x - S_s)A \qquad \text{(Eq. 2.27)}$$

where $\mathrm{d}M/\mathrm{d}t$ represents the mass of substance moving across the boundary layer into an area A of biofilm per unit time, k_d is the rate of mass transport (length over time), and S_x and S_s are the concentration of the substance in the bulk water and at the biofilm surface at the given location within the aquifer. Biodegradation of the limiting substrate within the biofilm itself is a function of the concentration as given by Equation 2.27, but the substrate concentration decreases with distance within the biofilm, making the relationship somewhat complicated. A general solution for this case is (Rittmann and McCarty, 2001):

$$-\frac{dM}{dt} = \left[2qXD\left((S_s - S_w) + K\ln\left(\frac{K+S_w}{K+S_s}\right)\right)\right]^{1/2} \qquad \text{(Eq. 2.28)}$$

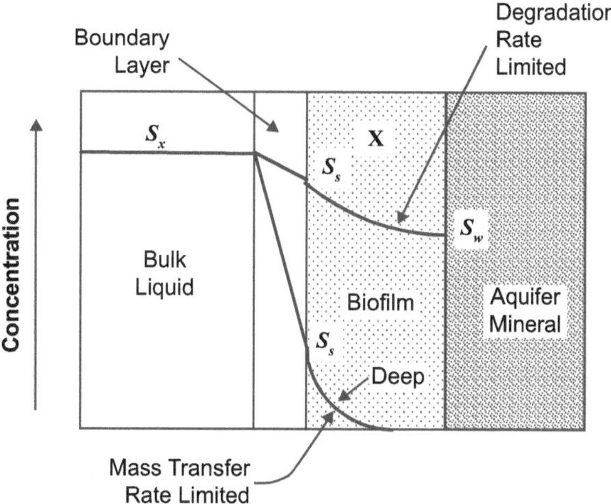

Figure 2.5. Substrate diffusion across a boundary layer and into a biofilm, illustrating cases of reaction rate limited by mass transfer and by biological reaction kinetics.

where D is the rate of molecular diffusion of the substrate through the biofilm and S_w is the substrate concentration at the point of biofilm attachment to the aquifer minerals. Under near steady-state conditions, which are typical in aquifers, the rate of diffusion of substrate into the biofilm just equals the rate of biological transformation, thus,

$$k_d(S_x - S_s) = \left[2qXD\left((S_s - S_w) + K\ln\left(\frac{K+S_w}{K+S_s}\right)\right)\right]^{1/2} \qquad \text{(Eq. 2.29)}$$

Figure 2.5 illustrates various likely outcomes from the solution of this equation, depending upon the relative values of the various rate constants involved. If the process is diffusion limited (relatively low k_d) then S_s will be much less than S_x, but if biodegradation is rate-limiting (relatively high k_d), then S_s will be similar in value to S_x. With a small starting seed of microorganisms, the depth of the biofilm may be very small so that S_w is almost equal to S_s. The rate of biotransformation then is very low and thus rate limiting. Changing the rate of mass transfer to the biofilm will make little difference. However, as microorganisms grow and the biofilm thickens, S_w decreases and may eventually approach zero, in which case, biotransformation becomes maximum. The limiting factor may then be the rate of mass transfer to the biofilm surface. In this case, changing the rate of mass transfer may increase reaction speed. This might be accomplished by increasing the fluid velocity passed the biofilm, such as by artificially increasing mixing speed through groundwater recirculation. This indicates the importance of understanding what factors are affecting reaction rate within an aquifer.

2.4.5 Bioaugmentation

Frequently, the microorganisms required for biodegradation of contaminants are naturally present in an aquifer and to bring about contaminant destruction only requires a non-toxic environment and that the microorganisms be brought into contact with an adequate mixture of electron donor and electron acceptor for energy production, and necessary nutrients for growth. In some cases, however, the needed organisms may not be present so that bioaugmentation may be desirable. External production of the degrading organisms and introduction into the aquifer with adequate mixing is then required. Bioaugmentation at times has not been successful, but if conditions are correct, bioaugmentation can be very successful. For bioaugmentation to succeed well, the bioaugmented microorganisms must be filling a niche not already being filled by other microbes. Examples of unsuccessful, partially successful, and highly successful bioaugmentation are provided in the following.

Bioaugmentation was attempted to enhance the aerobic cometabolism of chlorinated solvents in two separate studies at the Moffett Field experimental field site in Mountain View, California. The first was unsuccessful and the second was partially successful. In previous studies at Moffett Field, the use of toluene as an electron donor was found to be quite successful for stimulating microorganisms that cometabolize TCE. However, use of toluene for this purpose is sometimes of concern because it is a regulated compound, and so a more generally acceptable donor is desired. Toward this end, an organism that produces the toluene ortho monooxygenase enzyme that cometabolizes TCE efficiently was genetically modified to grow on lactate, a generally acceptable donor, while still maintaining a high concentration of toluene ortho monooxygenase. Laboratory studies indicated that the organism, *Burkholderia* G4, performed well in pure culture, but column studies with natural aquifer materials indicated that maintaining good activity over time might be difficult (Munakata-Marr et al., 1996). This was demonstrated in subsequent field studies in which high concentrations of the microorganisms were continually injected into the aquifer in hopes of allowing it to compete well with native lactate using microorganisms (McCarty et al., 1998b). Good TCE cometabolism was achieved for about 10 days, and then it declined as competing lactate users that did not cometabolise toluene came into dominance. It was also noted that continuous addition of high organism concentrations resulted in growth of a predatory population of protozoa that consumed the bioaugmented organisms. This study indicated the difficulty of trying to out compete native organisms with the same or better ability at electron donor and acceptor utilization.

In another study at Moffett Field, an organism that cometabolises TCA well while growing on butane was used for bioaugmentation (Semprini et al., 2007). Here, bioaugmentation resulted in a rapid increase in the ability to grow on injected butane and to cometabolize TCA, but after about 1 month, TCA degradation decreased. In a control system without bioaugmentation, a native population of butane-oxidizing bacteria that lacked the ability to cometabolize TCA eventually became established. Cometabolism could be continued if the native population was controlled by periodic additions of high hydrogen peroxide concentration. In this case, bioaugmentation was partially successful, but its maintenance was difficult in the presence of non-TCA utilizing organisms that could compete effectively for the added butane.

Bioaugmentation has been successful when competition by other organisms for electron donor or electron acceptor is either not an issue or is suppressed. This has been the case with bioaugmentation for anaerobic PCE, TCE, DCE, and VC reductive dehalogenation. Since these electron acceptors are not used under anaerobic conditions for processes other than reductive dehalogenation, the organisms carrying out these reactions have no competition for their respective electron acceptors, even though competition is quite strong for the electron donors that they use. As long as the dehalogenators can compete successfully with other hydrogen-using microorganisms, they survive well in anaerobic groundwater environments. This has been demonstrated in field-scale demonstrations at Dover Air Force Base (AFB), Deleware (Ellis et al., 2000) and Kelly AFB, Texas (Major et al., 2002).

Suppression of competition was demonstrated in pilot- and demonstration-scale bioaugmentation studies conducted at Schoolcraft, Michigan, in an aquifer contaminated with CT and nitrate. The added organism was *Pseudomonas stutzeri* KC, the denitrifying, CT-degrading bacterium described earlier (Dybas et al., 1998, 2002; Hyndman et al., 2000). The concept was to introduce strain KC into the aquifer ahead of the CT plume and to maintain it as a biofilm through weekly additions of acetate. A challenge was how to prevent indigenous denitrifying bacteria from outcompeting strain KC for the added acetate. When stimulated by acetate addition, the other denitrifying organisms at the site could also convert CT to chloroform, an unwanted and persistent product. Failure to selectively stimulate strain KC would result in the formation of chloroform and failure of the bioaugmentation effort.

A laboratory comparison of the specific growth rates of strain KC and the indigenous microflora at different pH levels revealed a solution to the problem of competition (Dybas et al., 1995a). At Schoolcraft, the native denitrifying bacteria had adapted to the background groundwater at a pH of 7.2, and as expected, their maximum specific growth rate was highest at that pH level. Increasing the pH to 8.0–8.2 caused precipitation of Fe(III) and created conditions unfavorable for the indigenous microflora, but favorable for the growth of strain KC, an effective iron scavenger. Thus, adjusting the aquifer pH to 8 prior to introduction of strain KC conferred a colonization advantage on the strain KC and enabled long-term control of the CT degradation pathway.

Other challenges at Schoolcraft included: how to introduce alkalinity, strain KC, and acetate across a large and deep aquifer in a uniform fashion, and how to maintain sufficient concentrations of strain KC within the biocurtain to insure reliable CT degradation to levels below the U.S. Environmental Protection Agency (USEPA) drinking water standard (five parts per billion [ppb]) over a period of years, as the CT plume slowly passed through. These challenges were overcome by weekly 6-h chemical delivery periods in which groundwater amended with acetate and adjusted to pH 8 was recirculated through a "picket fence" of closely spaced (1 m apart) extraction/injection wells screened over the entire depth of contamination and positioned normal to the direction of groundwater flow. The resulting recirculation patterns between these wells allowed for pH adjustment, introduction of strain KC, formation of a well-colonized biocurtain, and maintenance of the biocurtain for a period of years.

Delivery of strain KC into the subsurface was not problematic. This was because wells for chemical and organism delivery were spaced close together (1 m apart) and because delivery of chemicals occurred in the same wells used to deliver the organism. In general, organism delivery has not been a problem for bioaugmented systems when the added organisms are introduced at the same wells or near wells where donor or acceptor are later added. Growth near the well is rapid, and the added organisms tend to spread rapidly through the aquifer as they multiply in response to the presence of growth factors. Only a small fraction of the organisms need to be carried through the aquifer to act as seed throughout the system. This was clearly shown at the Dover AFB demonstration (Ellis et al., 2000).

The general strategy used at Schoolcraft to control competition – chemical conditioning a region of the subsurface to prepare for the introduction of a new organism – is useful when specific organisms or groups of organisms need to be encouraged or discouraged. For bioaugmentation, the native microflora will typically be adapted to the pH of their environment, and that pH is likely to be different from the optimum for the added organism. A pH shift can thus encourage survival and growth of the introduced organism while selecting against the indigenous competitors. Table 2.12 lists strain KC along with other major microbial groups and some optimal pH ranges for each. It is important to keep in mind, however, that most of the listed groups in Table 2.12 also contain highly specialized representatives capable of growth under extreme acidic conditions (acidophiles) and extreme alkaline conditions (alkalophiles), so the ranges indicated represent "non-extreme" values for each group.

Table 2.12. Optimum pH Ranges for Different Microorganisms and Functions

Organism type	Function	Optimal pH range
Heterotrophs	Oxidize ammonia	6–9
Nitrifiers	Oxidize ammonia	6–9
Denitrifiers	Reduce nitrate to N_2	6–9
Acidogens	Convert complex organic matter to weak acids	3–6
Acetogens	Convert propionic acid and butyric acid to acetic acid	6–7
H_2-utilizing methanogens	Convert H_2 + CO_2 to methane	6.2–7.2
Acetoclastic methanogens	Convert acetic acid to methane	6.6–7.2
Sulfate-reducing bacteria (SRB)	Reduce sulfate to hydrogen sulfide, remove metals as sulfide precipitates	4–10
Specialized cultures, example: *Pseudomonas stutzeri* KC	Function depends on the organism. In the case of strain KC, denitrification and dechlorination of CT are important	>7 (8.2 optimal)

2.4.6 Organic Bioremediation Example: Edwards AFB, California

A full-scale evaluation for *in situ* aerobic cometabolic biodegradation of TCE at Edwards AFB in southern California serves as an example to illustrate how chemicals needed for biodegradation can be successfully introduced and mixed to enhance biodegradation (McCarty et al., 1998a). Cometabolism is the fortuitous biodegradation of a compound by enzymes that are used by organisms to carry out some other essential function in the organism. There have been several field demonstrations of successful use of cometabolism for biodegradation of TCE and other halogenated aliphatic compounds.

At Edwards AFB a TCE contaminated plume emanated from a location where TCE contaminated wastewater was discharged onto the ground surface in the 1950s and 1960s. At the downgradient location where *in situ* cometabolism was applied, the groundwater was divided between two aquifers separated by a 2 m thick clay aquitard (Figure 2.6). The upper unconfined aquifer was 9 m below ground surface (bgs) and was 8 m in depth. The lower aquifer was 5 m deep. The substrate selected here for cometabolism was toluene, which was shown from earlier pilot studies at Moffet Federal Air Field to be a good substrate for efficient cometabolism of TCE (Hopkins and McCarty, 1995). Studies with aquifer material from Edwards AFB indicated that the necessary microorganisms for toluene consumption and efficient TCE cometabolism were naturally present throughout the aquifer (Jenal-Wanner and McCarty, 1997). Bioaugmentation was not necessary. In order to achieve cometabolism, both toluene and oxygen for its oxidation were added to the aquifers, and both were mixed with the TCE contaminated water and brought together for consumption by toluene-using microorganisms in order to enhance biodegradation of TCE. Two potential problems had to be considered in designing the delivery system.

The first potential problem was how to bring toluene, oxygen, TCE, and the toluene-consuming microorganisms together at the same location within the aquifer. Oxygen must be present in order for microorganisms to oxidize the toluene and grow, producing the toluene monooxygenase enzyme needed for TCE cometabolism. TCE had to be present when the enzyme was induced so that it would be biodegraded. However, toluene and TCE compete

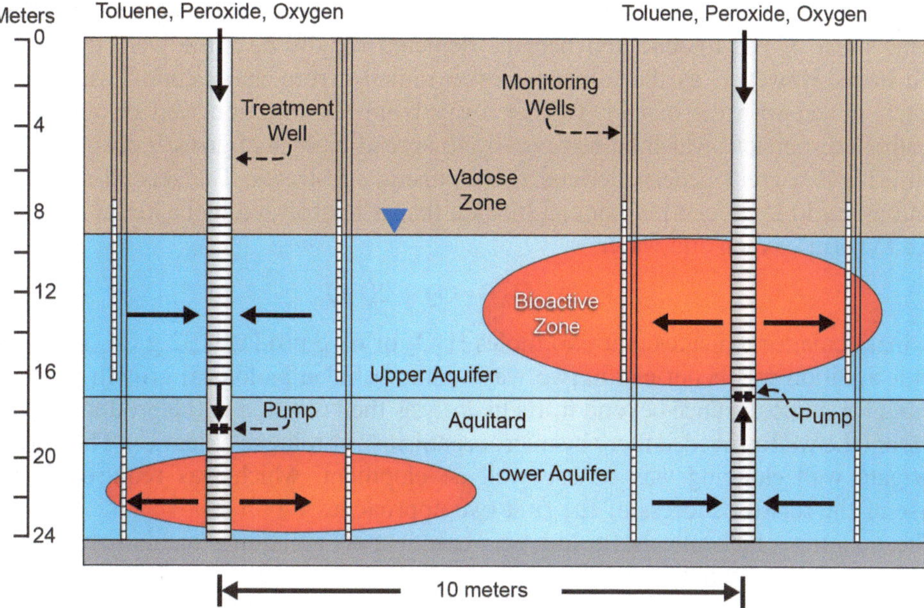

Figure 2.6. Treatment scheme used for adding and mixing chemicals for *in situ* aerobic cometabolic biodegradation of TCE at Edwards AFB, California. Reprinted from McCarty et al., 1998a with permission by American Chemical Society.

for the same enzyme, and so it was desirable to have the toluene present in low concentration, at least occasionally, so that TCE could be biodegraded efficiently. Also, TCE cometabolism as well as toluene oxidation could take place only under aerobic conditions. Thus, aerobic conditions in the aquifer had to be maintained.

The second major potential problem was how to prevent clogging of well screens in wells where toluene and oxygen were to be added to the aquifer. The two potential problems were addressed at Edwards AFB through use of tandem recirculating wells as illustrated in Figure 2.6. The two tandem wells were established for mixing and for adding toluene and oxygen. Each well had two screens, one in the upper aquifer and one in the lower aquifer. Each well contained a pump for water circulation, water in one was pumped downward from the upper aquifer to the lower aquifer, and in the other water was pumped upward from the lower aquifer to the upper aquifer. This created a circular pattern of water movement in the aquifer. The circulating water contained the TCE contaminant. A static mixer was placed in the exit line from the pump so that toluene and oxygen could be added to the circulating water as it passed through the well. Neat toluene was pumped and gaseous oxygen was allowed to flow from a pressure cylinder into the static mixer for mixing with the TCE contaminated water. In this manner, mixing between the toluene and oxygen and the TCE could be achieved within the well. The mixture then flowed into the aquifer to come in contact with the biodegrading microorganisms. Thus, all four necessary ingredients for organism growth, enzyme induction, and TCE cometabolism were brought together as needed.

The other problem to solve was that of well clogging through excessive biological growth near the well screens, a problem that was likely to exist as the injection concentrations of toluene and oxygen were well above their respective S_m values so that microorganism growth would continue until the pores between aquifer minerals were completely filled. Three strategies were used here to reduce this problem. The first was toluene pulsing. The needed oxygen was added continuously, but toluene was added in high-concentration pulses only three times per day. During the toluene pulse, the oxygen concentration at that time was insufficient for oxidizing most of the toluene and became depleted rapidly near the well, thus minimizing growth there. However, as the toluene moved radially from the injection well and into the aquifer it mixed with the oxygen further away from the injection well through lateral and longitudinal dispersion, which helped greatly to spread biological growth out into the aquifer system. This is a great benefit derived from pulsing substrates. The second strategy was to continuously add hydrogen peroxide, a biocide that is hydrolyzed by bacterial enzymes in the aquifer to form oxygen and water:

$$2H_2O_2 \rightarrow O_2 + 2H_2O \qquad \text{(Eq. 2.30)}$$

A mass balance indicates that two moles H_2O_2 or 68 g produces 32 g O_2. Thus, hydrogen peroxide addition was beneficial in two ways. First, it killed biological growth within the well itself and for some distance beyond it. Its hydrolysis then resulted in the production of oxygen away from the well as needed there by microorganisms for toluene oxidation. The third strategy to mitigate well clogging was to use well development, which was required infrequently because of the relative success of the first two approaches.

The quantity of chemicals to add was based upon reaction stoichiometry. The energy reaction for toluene oxidation is:

$$C_6H_5CH_3 + 9O_2 \rightarrow 7CO_2 + 4H_2O \qquad \text{(Eq. 2.31)}$$

This equation indicates that 9 moles oxygen is required for each mole of toluene. However, some of the toluene is converted to cells and so the actual oxygen requirement is less than this. A laboratory study to simulate field conditions was conducted in order to determine reaction stoichiometry for aerobic oxidation of toluene when organism growth was included, showing that the actual need was for only 6 moles oxygen (192 g) per mole toluene (92 g), or 2.1 g oxygen per g toluene (Jenal-Wanner and McCarty, 1997).

The pulsing of toluene to give a time-averaged concentration of 9.0 mg/L, and continuous addition of pure oxygen (44 mg/L) and hydrogen peroxide (47 mg/L) resulted in good steady-state operation with little problems from aquifer clogging. With 9 mg/L toluene, only about 20 mg/L of oxygen was actually needed based upon reaction stoichiometry, but the excess added insured that aerobic conditions remained throughout the aquifer, a desirable condition. TCE removal during this period varied between 83% and 87% of the TCE passing through the treatment well. However, toluene removal was much higher. Within 2.5 m of the injection wells, toluene concentration was mostly consumed, and by the time the water reached the four 15-m sampling locations surrounding the treatment system, the concentration of toluene had been reduced to an average of 1.3 μg/L, well below the taste and odor threshold of 20 μg/L and the drinking water maximum contaminant level (MCL) of 1,000 μg/L.

2.5 CHEMICAL PROCESSES

An advantage of biological processes is that they can result in the destruction as well as removal of chlorinated solvents. The same can be said of some chemical processes. As with biological processes, the chemical processes can be divided into oxidative and reductive processes. Oxidative processes would logically result in the oxidation of organic carbon in the chlorinated solvents to carbon dioxide, while releasing organic chlorine as chloride. Reductive processes on the other hand, reduce the organic carbon in chlorinated solvents to a lower oxidation state such as ethane, while again releasing the organic chlorine as chloride. Some chemical processes for chlorinated solvent transformation occur under the ambient environmental conditions associated with aquifers (Rheinhard et al., 1997). Often natural chemical transformations do not result in complete conversion to harmless end products. Nevertheless, an understanding of these natural processes is important for assessing the source of contaminants that may be found at a site and in selecting processes and strategies for remediation. Engineered remediation using chemical processes for *in situ* contaminant destruction have been broadly studied, and some have been frequently applied.

2.5.1 Oxidative Chemical Processes

Chemical oxidants have been used in the water treatment industry for decades for the destruction of unwanted organic chemicals. Most frequently used have been ozone, permanganate, and Fenton's reagent. However, possible use of chemical oxidants for *in situ* destruction of chlorinated solvents has been explored in detail only in recent years, and that has been for addressing the difficult problem of DNAPL destruction. This has become known as ISCO or *in situ* chemical oxidation. Perhaps the first to explore the use of permanganate for this purpose was Schnarr et al. (1998). They reported on both laboratory and field experiments for PCE and TCE destruction in which 10 g/L permanganate was found to completely oxidize the compounds to carbon dioxide and chloride. Two field experiments were conducted. In the first, 1 L PCE that was added to a confined area was completely removed within 120 days by flushing through 100 L/day of the 10 g/L $KMnO_4$ solution. For the second, 8 L of a mixed PCE/TCE DNAPL was added to a test cell, and after 290 days of flushing with 10 g/L permanganate, 62% of the initial source had been oxidized. In this oxidation process, the MnO_4^- oxidant is reduced to form the insoluble MnO_2. Subsequently, many studies by a wide range of researchers have been conducted to further evaluate the use of permanganate.

Fenton's reagent is a mixture of hydrogen peroxide and ferrous iron, which serves as a catalyst, forming hydroxyl radicals, the main oxidizing species in Fenton's reagent. An earlier experiment using Fenton's reagent for oxidation of PCE was conducted by Leung et al. (1992), who reported mineralization of 1 g PCE per kilogram (kg) aquifer solids within 3 h with a solution containing 2.1 molar (M) H_2O_2 and 5 millimolar (mM) $FeSO_4$. TCE appears to be oxidized somewhat more slowly than PCE (Teel et al., 2001). Fenton's reagent also degrades CT even though its carbon is already in the fully-oxidized state (Teel and Watts, 2002). This apparently occurs by a reduction mechanism in which a superoxide radical anion is involved (Smith et al., 2006). Many studies using Fenton's reagent for destruction of chlorinated solvents have now been conducted.

2.5.2 Reductive Chemical Processes

Perhaps the first to recognize the potential for abiotic reduction of chlorinated solvents for *in situ* destruction was Gillham and O'Hannesin (1994), who found that 100-mesh zero-valent iron was capable of removing chloride from 14 different chlorinated methanes, ethenes, and ethanes, and replacing the chlorides with hydrogen. In the process Fe(0) is converted to Fe(II). The rates of transformation were sufficiently fast for field application, except perhaps for dichloromethane. Gillham and O'Hannesin proposed that zero-valent iron might be used for either *in situ* or aboveground applications for remediation of contaminated groundwater. A field demonstration of the technology was initiated in 1991 at Canadian Forces Base, Borden, Ontario, to treat a plume containing 268 mg/L TCE and 58 mg/L PCE (O'Hannesin and Gillham, 1998). Here, a mixture of 22% granular iron and 78% sand installed as a permeable "wall" across the path of the plume removed approximately 90% of the TCE and 86% of the PCE. The first full-scale application of granular zero-valent iron was a reactive wall installed in 1996 in North Carolina to treat overlapping plumes of chromate and chlorinated solvents (Puls et al., 1998).

This passive approach to the control of plume migration, while involving a relatively high capital expenditure, has been an attractive alternative to those wishing to avoid an active program of control, which has lower capital but higher maintenance costs.

Experiments with zero-valent iron have been conducted for other than plume-migration control. For example, a demonstration was conducted in which zero-valent iron was mixed with aquifer material contaminated with TCE DNAPL using a large-diameter mixing blade (Wadley and Gillham, 2003). Here, bentonite was added as well to serve as a lubricant to facilitate injection of the iron and to isolate the contaminated zone. PCE was reported to decrease to non-detectable levels within the 13-month monitoring period. Alternatively, Cantrell and Kaplan (1997) proposed using colloidal sized suspensions of zero-valent iron that could be injected directly into an aquifer without the need to build a reactive wall. This has been carried further by Zhang et al. (1998), who have suggested use of nanoscale bimetallic particles in which one metal (Fe or Zn) serves as the reductant, while palladium or platinum serves as a catalyst to speed up the reaction. Much research and field studies have been conducted on this alterative approach. In a further alternative that also uses a palladium catalyst, Schreier and Reinhard (1995) demonstrated that molecular hydrogen could be used instead of iron as the reductant. Here, the reaction is sufficiently faster so that the system lends itself to a down-well or surface reactor. Thus, many alternatives for treatment of chlorinated solvent contaminated plumes as well as DNAPL source areas using reductive chemical processes have emerged in recent years.

2.5.3 Precipitation

Precipitation may be desired for stabilization of hazardous chemicals within an aquifer so that they do not contaminant water passing by. Chemical species for which this may be an option are generally metal cations that have very low solubility in water under given aquifer chemical, redox and pH conditions.

Metals as such cannot be destroyed, and so either stabilization in some manner within the subsurface where contamination exists or removal may be the only viable remediation options. Possible metals for the stabilization option are chromium, cadmium, zinc, lead, mercury, uranium, and plutonium. If stabilization by precipitation is to be an option for remediation, then the aquifer conditions that promote precipitation should not change over time, otherwise the metals may become soluble to contaminate groundwater. Metals most frequently occur as cations, which is the form most susceptible to precipitation. Some metals, such as hexavalent chromium (CrO_4^{2-}), may also exist as oxyanions that generally do not precipitate well. Some metal cations precipitate well in one oxidation state, but not in another. For example, Fe(III) hydroxide is quite insoluble, while Fe(II) hydroxide is not.

Factors involved in precipitation can be quite complex and are not discussed in detail here. Further information can be found in general textbooks on environmental chemistry (Benjamin, 2002; Morel and Hering, 1993; Sawyer et al., 2003; Stumm and Morgan, 1996). The general principle involved, however, is the solubility product (K_{sp}) of the metal with an anion:

$$Cd^{2+} + 2OH^- = Cd(OH)_2(s) \qquad [Cd^{2+}][OH^-]^2 = K_{sp} = 2 \times 10^{-14} \qquad \text{(Eq. 2.32)}$$

$$Fe^{3+} + 3OH^- = Fe(OH)_3(s) \qquad [Fe^{3+}][OH^-]^3 = K_{sp} = 6 \times 10^{-38} \qquad \text{(Eq. 2.33)}$$

$$Pb^{2+} + S^{2-} = PbS(s) \qquad [Pb^{2+}][S^{2-}] = K_{sp} = 1 \times 10^{-28} \qquad \text{(Eq. 2.34)}$$

From the ionization product of water, ($[H^+][OH^-] = 10^{-14}$), the hydroxide concentration at pH 7.0 is found to equal $10^{-14}/10^{-7} = 10^{-7}$. At pH 7, cadmium(II) in the form of its hydroxide is quite soluble while iron (III) is not. Based upon Equations 2.32 and 2.33, the concentrations of

Cd^{2+} and Fe^{3+} are thus 2 M and $6(10^{-14})$M, respectively. Thus, we would not expect Cd^{2+} to be stabilized in groundwater at pH 7 while we would with iron. But these simple calculations are not sufficient. An important regulatory consideration is the total solubility of a metal, including all soluble forms in equilibrium with the solid phase. An estimate for this value requires a knowledge of the solubility product data in Table 2.4 along with equilibrium coefficients for other equilibria that involve the metal of interest.

As discussed previously, chromium can be removed by reduction to Cr(III) hydroxide, Cr $(OH)_3$. This solid will be in equilibrium with Cr^{3+}. At pH 7.5, the concentration of Cr^{3+} can be estimated from the solubility product:

$$\begin{aligned} \mathrm{Cr(OH)_3}(s) &= \mathrm{Cr^{3+}} + 3\mathrm{OH^-} \\ K_{sp} &= 10^{-30.2} \\ K_{sp} &= [Cr^{3+}][OH^-]^3 \qquad \text{(Eq. 2.35)} \\ 10^{-30.22} &= [Cr^{3+}](10^{-6.5})^3 \\ [Cr^{3+}] &= 1.9 \times 10^{-11}\mathrm{M} = 9.9 \times 10^{-4}\mu g/L \end{aligned}$$

The above concentration is much less than the USEPA regulatory standard of 100 μg/L Cr, but it only represents the soluble Cr(III) that is chelated to H_2O ligands. Other dissolved Cr(III) species are present and must be accounted for. The nature of these species will depend on whatever additional ligands are present and their equilibrium binding constants. If the only other ligands are hydroxyl groups from water, the total soluble Cr(III) at pH 7.5 can be estimated from K_{sp} and the relevant equilibrium constants (K_1 through K_4), where:

$$K_1 = 10^{10.0} = [\mathrm{Cr(OH)}^{2+}]/\{[\mathrm{OH}^-][\mathrm{Cr}^{3+}]\}$$

Thus, $[\mathrm{Cr(OH)}^{2+}] = 10^{10.0}[10^{-6.5}][1.9 \times 10^{-11}] = 6.0 \times 10^{-8}\mathrm{M} = 3\ \ \mu g/L\ \mathrm{Cr}$

$$K_2 = 10^{8.3} = [\mathrm{Cr(OH)_2}^+]/\{[\mathrm{OH}^-][\mathrm{Cr(OH)}^{2+}]\}$$

So that, $[\mathrm{Cr(OH)_2}^+] = 10^{8.3}[10^{-6.5}][6.0 \times 10^{-7}] = 3.8 \times 10^{-6}\mathrm{M} = 198\ \ \mu g/L\ \mathrm{Cr}$

$$K_3 = 10^{5.7} = [\mathrm{Cr(OH)_3}(aq)]/\{[\mathrm{OH}^-][\mathrm{Cr(OH)_2}^+]\}$$

Solving gives, $[\mathrm{Cr(OH)_3}(aq)] = 10^{5.7}[10^{-6.5}][1.2 \times 10^{-5}] = 6.0 \times 10^{-7} = 31\ \ \mu g/L\ \mathrm{Cr}$

$$K_4 = 10^{4.6} = [\mathrm{Cr(OH)_4}^-]/\{[\mathrm{OH}^-][\mathrm{Cr(OH)_3}(aq)]$$

This means $[\mathrm{Cr(OH)_4}^-] = 10^{4.6}[10^{-7}][6.0 \times 10^{-6.5}] = 7.6 \times 10^{-9}\mathrm{M} = 0.4\ \ \mu g/L\ \mathrm{Cr}$

The above calculations show that for this pH, most of the dissolved Cr(III) is present as $Cr(OH)_2^+$. The total mass concentration in solution is the sum of the concentrations of all dissolved species: $9.9 \times 10^{-4} + 3 + 198 + 31 + 0.4 = 232\ \mu g/L$. This value for total Cr exceeds the regulatory standard. Soluble chromium concentration from operation at a slightly higher pH (~8) would meet the standard.

Anions often considered for stabilization of metals in water are hydroxide, carbonate, phosphate, and sulfide. These anions are all commonly found associated with groundwater and aquifer minerals. Figure 2.7 indicates the relative solubility of various metal salts of these anions. Several general conclusions might be drawn from this figure. Phosphate and sulfide salts are in general less soluble than hydroxide salts. Of the four, carbonate salts are the most soluble. The graph for hydroxide salts indicates one of the impacts of pH. At pH of 7 (log hydroxide concentration of −7), Fe(III) as well as Cr(III) are quite insoluble, but most of the other metals are not. Precipitation of hydroxides is better at higher pH (higher hydroxide

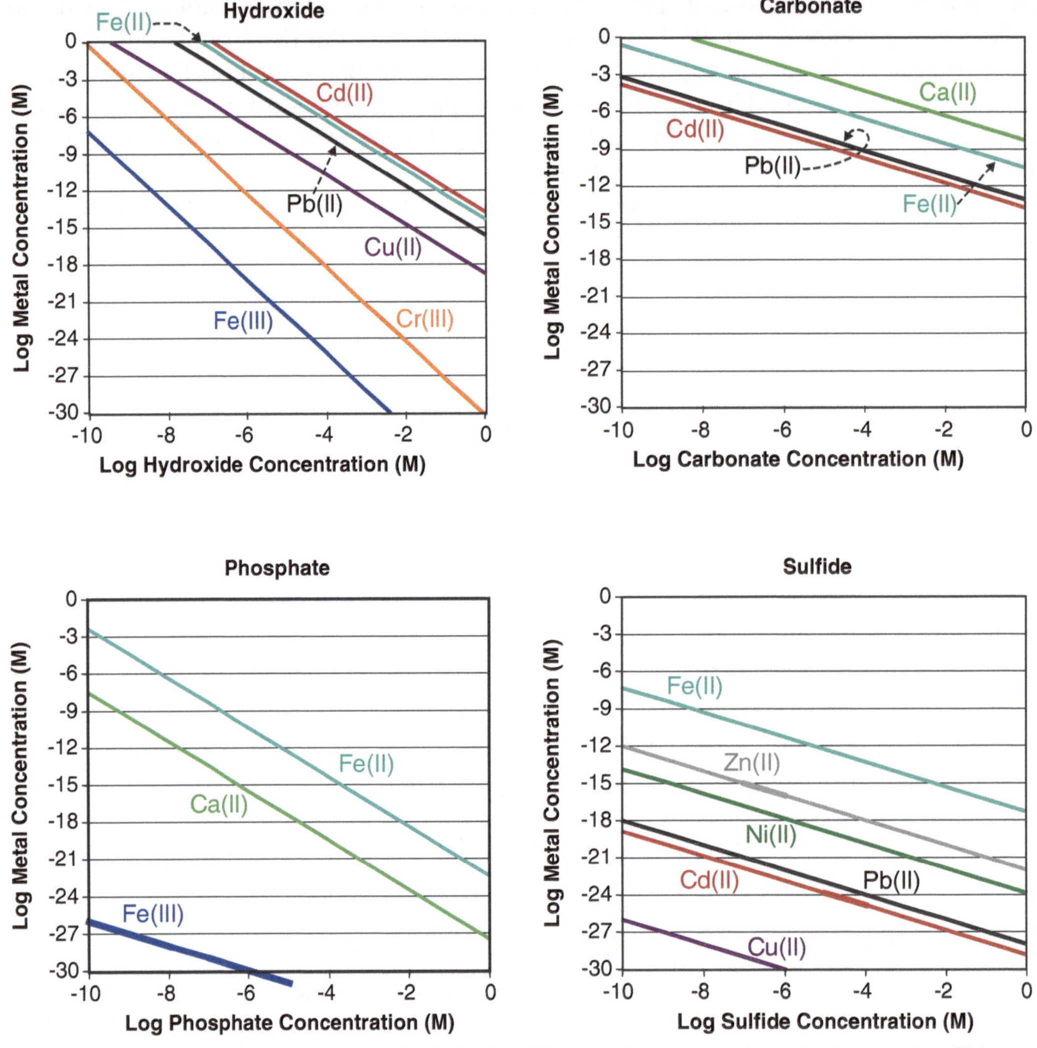

Figure 2.7. Solubility of various metal salt hydroxides, carbonates, phosphates, and sulfides based upon solubility product data. From Sawyer et al., 2003.

concentration). Fe(II) is quite soluble at pH 7, indicating why the anaerobic reduction of Fe(III) in Fe(III)-containing minerals results in the formation of soluble Fe(II).

On the other hand, the solubility of FeS, indicated in the last graph, is quite low. Thus, while anaerobic conditions may result in the reduction of Fe(III) to Fe(II), they may also result in the reduction of sulfate to sulfide so that some or all of the Fe(II) becomes stabilized in a sulfide precipitate. This illustrates the important role that microbial processes often play in the movement and fate of chemicals in groundwater. Chemicals added to an aquifer for metal stabilization may be those that react directly with the metal such as the four anions illustrated in Figure 2.7, or they may be chemicals that promote biological and redox conditions that bring about stabilization, such as reduction of soluble Cr(VI) to form the insoluble Cr(III), or the production of sulfide. Once stabilization is achieved, then chemical and biological conditions that may result in solubilization must be prevented from occurring. The many factors involved in chemical stabilization should be well understood before this is deemed an acceptable method of control.

2.5.4 pH Control

Many of the reactions listed in Table 2.5 involve hydrogen (H^+) or hydroxyl (OH^-) ions, indicating that the reactions can change pH, some reactions cause a decrease in pH and others cause an increase in pH. Reaction rates and the speciation of chemicals are greatly affected by pH, and thus it is often important for effective groundwater remediation to apply pH control. Thus, chemicals that control pH may need to be added to and be mixed with existing groundwater. The main pH buffer in groundwater is bicarbonate (HCO_3^-), so its equilibrium with carbonate (CO_3^{2-}) and CO_2 is of importance in the control of chemical speciation.

Equilibrium reactions of importance here are listed below:

$$H_2O = H^+ + OH^- \qquad \text{(Eq. 2.36)}$$

$$CO_2 + H_2O = H_2CO_3 \qquad \text{(Eq. 2.37)}$$

$$H_2CO_3 = H^+ + HCO_3^- \qquad \text{(Eq. 2.38)}$$

$$HCO_3^- = H^+ + CO_3^{2-} \qquad \text{(Eq. 2.39)}$$

Equilibrium equations of importance as derived from the above and their respective equilibrium constants at 20°C are:

$$[H+][OH^-] = K_w = 10^{-14} \qquad \text{(Eq. 2.40)}$$

$$\frac{[CO_2(g)]}{[H_2CO_3]} = K_H = 36 \text{ atm/mol} \qquad \text{(Eq. 2.41)}$$

$$\frac{[H^+][HCO_3^-]}{H_2CO_3} = K_1 = 4.8 \times 10^{-7} \qquad \text{(Eq. 2.42)}$$

$$\frac{[H^+][CO_3^{2-}]}{[HCO_3^-]} = K_2 = 4.9 \times 10^{-11} \qquad \text{(Eq. 2.43)}$$

Equation 2.41 indicates the relationship between the partial pressure of carbon dioxide and its concentration in solution. If the partial pressure of CO_2 plus that of the other gases in solution (N_2, O_2, H_2S, CH_4) exceeds 1 atm (at sea level) plus the hydrostatic pressure at a given point in an aquifer (total pressure = atm pressure + D/10.3, where D equals hydrostatic pressure in m), then the gases would exceed saturation and bubble formation is likely. This could lead to clogging of the aquifer at that point.

Concerning pH, Equation 2.42 is of importance at near neutral pH. Taking the $\log_{10}$ of both sides of Equation 2.42, and remembering that pH = −log[H+], the pH can be found as follows:

$$\text{pH} = 6.3 + \log\frac{[HCO_3^-]}{[H_2CO_3]} \qquad \text{(Eq. 2.44)}$$

Here, $[HCO_3^-]$ equals the molar concentration of bicarbonate (mg HCO_3^-/L/61,000) and $[H_2CO_3]$ is the molar solution concentration of CO_2 (mg CO_2/L/44,000). The water saturation concentration in equilibrium with 1 atm of CO_2 based upon Equation 2.41 is 0.028 M. In order to maintain a pH of 7, then according to Equation 2.42, the bicarbonate concentration would need to be 4.8 times higher or 0.134 M, which corresponds to an alkalinity of 6,700 mg/L as $CaCO_3$. If a lower pH, say 6.5 were acceptable, then the bicarbonate alkalinity would only need to be 1.58 times higher or 2,200 mg/L as $CaCO_3$. Generally, the CO_2 concentration is much less than 1 atm partial pressure, and so the need for bicarbonate alkalinity to maintain the desired pH would be proportionally less.

Some biological reactions are basic, causing pH to rise, while others are acidic, causing it to fall. The impact of various electron acceptors on bicarbonate and CO_2 concentration are illustrated by the listing in Table 2.10 using H_2 as a pH neutral electron donor. It can be seen that when oxygen or perchlorate is the electron acceptor, there is no impact of electron acceptor itself on pH. When nitrate, sulfate, bicarbonate, or ferric oxide is the electron acceptor, H^+ is consumed, and so pH tends to rise. Ferric oxidation consumes more H^+ per mole of H_2 oxidized than the other three electron donors and thus is a more basic reaction. By contrast, reductive dehalogenation of a chlorinated solvent such as PCE produces H^+ and thus is an acidic reaction, causing pH to decrease. Oxidations of many organic electron donors result in H^+ production, and thus are generally acidic as well.

2.5.4.1 Example

Problem. Lactic acid is sometimes added to accomplish reductive dehalogenation of TCE to ethene. How much lactic acid would be required to reductive dehalogenate a solution containing 1 mM TCE, and what would be the resulting pH of the solution? Assume that the initial HCO_3^- concentration is 6 mM ($6 \times 50 = 300$ mg alkalinity as $CaCO_3$/L), the initial pH is 7.0, and no other oxidation-reduction reaction occurs.

Solution. The initial aqueous CO_2 concentration (which equals the carbonic acid concentration) can be estimated using Equation 2.42:

$$[H_2CO_3] = [H^+][HCO_3^-]/4.8 \times 10^{-7} = 10^{-7}(0.006)/4.8(10^{-7}) = 0.00125 \text{ M} \quad \text{(Eq. 2.45)}$$

Lactic acid is fermented to produce acetic acid and hydrogen:

$$CH_3CHOHCOOH + H_2O = CH_3COOH + CO_2 + 2H_2 \quad \text{(Eq. 2.46)}$$

The hydrogen is used for reductive dehalogenation of TCE:

$$CCl_2{=}CHCl + 3H_2 = CH_2{=}CH_2 + 3\,HCl \quad \text{(Eq. 2.47)}$$

This equation indicates that 1.5 moles lactic acid are required to produce sufficient hydrogen for TCE reduction. The acetic acid and hydrochloric acids produced react with solution HCO_3^- to produce more carbonic acid, H_2CO_3.

$$CH_3COOH + HCO_3^- = CH_3COO^- + H_2CO_3$$

$$HCl + HCO_3^- = Cl^- + H_2CO_3$$

The overall reaction can thus be written (neglecting the donor associated with synthesis for simplicity):

$$1.5CH_3CHOHCOOH + CCl_2{=}CHCl + 4.5HCO_3^- + 3H_2O$$

$$= 1.5CH_3COO^- + 3Cl^- + CH_2{=}CH_2 + 7.5H_2CO_3$$

We see that for each millimole (mmol) of TCE dehalogenated, 4.5 mmoles bicarbonate are destroyed and 7.5 mmoles of carbonic acid are formed. With groundwater, there is no direct contact with the atmosphere and so the CO_2 formed as carbonic acid remains in solution as long as its partial pressure remains below atmospheric pressure, causing bubble formation. Checking with Equation 2.41, we see that the CO_2 formed remains in solution. Thus, our final pH would be:

$$pH = 6.3 + \text{Log}\left(\frac{6.0 - 4.5}{1.25 + 7.5}\right) = 5.5 \quad \text{(Eq. 2.48)}$$

This pH is too low for effective biological activity. Thus, bicarbonate must be added to the groundwater to maintain pH in a proper range, say 6.5 or above.

While groundwater chemistry is generally more complex than considered in this simple example, it nevertheless illustrates some of the factors involved in pH maintenance and control. Good knowledge of factors affecting pH as well as how to deliver and mix buffering chemicals if needed is often required in groundwater remediation.

2.5.4.2 Chemicals for pH Control

The major natural buffer system in groundwater is the carbonate system as governed by Equations 2.36 to 2.43. From Equation 2.42 or 2.44 it can be determined that the $[HCO_3^-]/[H_2CO_3]$ ratio must be maintained between 1.6 and 16 in order to control pH in the range of 6.5–7.5. If pH tends to be low, then it can be increased by an increase in the bicarbonate concentration. The bicarbonate levels can be increased directly, by adding sodium, potassium, or ammonium bicarbonate, or indirectly by adding a base that combines with carbon dioxide or carbonic acid to form bicarbonate. A summary of different chemicals that are commonly used in soils and aquifers to provide bicarbonate for pH control and the chemicals reactions that may be involved is provided in Table 2.13. As shown, different masses of each chemical can provide the same level of alkalinity: for example, one mole of alkaline buffer might be supplied by 50 g of $CaCO_3$ or, equivalently, by 17 g of ammonia gas.

The soluble forms of bicarbonate that can be introduced directly into an aquifer through mixing are $NaHCO_3$ and $KHCO_3$. Generally, concentrated solutions of each would be mixed with recirculating groundwater above ground or in a well. The solubility of sodium bicarbonate is about 70 g/L while that of potassium bicarbonate is quite a bit higher or about 200 g/L, thus solutions for mixing that approach these levels can be prepared for addition. NH_4HCO_3 is another possibility, but unless ammonium is needed as a nutrient for biological control, sodium and potassium salts are the better choice as ammonium adsorbs readily to clays, hindering its movement, and if oxidized, would be converted to nitrate, a soluble and hazardous chemical. Other chemicals that might be used for buffering are sodium or potassium carbonate or hydroxide. However, direct addition to an aquifer of concentrated solutions would tend to drive the pH too high, causing toxicity to microorganisms or precipitation of salts such as $CaCO_3$. Possible ways to add such chemicals while minimizing high pH problems are in an aboveground mixing chamber (used at Schoolcraft) or directly within a recirculation well where groundwater containing high CO_2 concentration can react with the chemicals as indicated in Table 2.13 to reduce pH. In such cases, there would need to be sufficient soluble H_2CO_3 in order to react with the sodium bicarbonate or hydroxide so that pH would decrease to 8.0 or below before entering the groundwater in order to avoid possible high pH problems as well as precipitation as $CaCO_3$ of Ca^{2+} that may be present in the groundwater.

In adding a chemical to prevent low pH it is also important to check for the possibility of calcium carbonate precipitation. Addition of calcium (Ca)-containing solutions such as lime is not a good idea in general as this would amplify the calcium carbonate precipitation problem near the point of injection into the aquifer. Many aquifer systems contain calcium carbonate minerals such as calcite, and the question that may then arise is whether the mineral can serve as a pH buffer by coming into solution to neutralize high acid concentrations:

$$H^+ + CaCO_3(s) = Ca^{2+} + HCO_3^- \qquad \text{(Eq. 2.49)}$$

It can to a degree, but generally at typical mineral concentrations in groundwater and at pH of 6.0 and above, the degree to which it can act as a buffer is very limited. Indeed, it is as likely to

Table 2.13. Chemicals That Might be Used to Form Bicarbonate Alkalinity for pH Control

Chemical	Reaction in formation of bicarbonate	Mass of chemical equivalent to 1 kmol (50 kg) of bicarbonate alkalinity	Comments
$CaCO_3$	$CaCO_3 + H_2CO_3 = Ca^{2+} + 2HCO_3^-$	100 g of $CaCO_3$ forms 2 moles of HCO_3^-. Thus, $100 \div 2 = 50$ kg $CaCO_3 = 1$ kmol bicarbonate	Low solubility of $CaCO_3$ limits alkalinity to 1,400–1,500 mg/L, and more so if Ca^{2+} present in groundwater
Na_2CO_3	$Na_2CO_3 + H_2CO_3 = 2Na^+ + 2HCO_3^-$	$106 \div 2 = 53$ kg	Overshooting can occur. Na deflocculates clay materials
K_2CO_3	Like Na_2CO_3	$136 \div 2 = 68$ kg	Overshooting can occur
CaO (lime)	$CaO + 2H_2CO_3 = Ca^{2+} + 2HCO_3^-$	$56 \div 2 = 28$ kg	Can result in severe precipitation of $CaCO_3$ at pH >6.8
MgO	Like CaO	$40 \div 2 = 20$ kg	Low solubility of MgO reduces chance of pH overshoot
$NaHCO_3$	$NaHCO_3 = Na^+ + HCO_3^-$	84 kg	Good but expensive. Na deflocculates clay particles
$KHCO_3$	Like $NaHCO_3$	100 kg	Good but expensive
$(NH_4)HCO_3$	Like $NaHCO_3$	79 kg	Ammonium adsorbs to clays, and if oxidized, is converted to nitrate
NaOH	$NaOH + H_2CO_3 = Na^+ + HCO_3^- + H_2O$	40 kg	Overshooting can occur, Na deflocculates clay soils
KOH	Like NaOH	56 kg	Overshooting can occur
NH_3	$NH_3 + H_2CO_3 = NH_4^+ + HCO_3^-$	17 kg	Ammonia can be toxic. Ammonium adsorbs to clays, and if oxidized, is converted to nitrate. NH_3 is naturally released when protein degrades

be removed from solution by precipitation as it is to enter solution as a buffer. One should not count on aquifer minerals or use of $CaCO_3$ itself to be good buffering solutions for pH control.

Exceptionally high pH can also be observed in groundwater, but is most likely to occur from contamination of aquifers by basic substances from industry, rather than from reactions occurring in the ground. High groundwater pH might be controlled by the addition of carbon dioxide (Equations 2.42, 2.44), or by the addition of inorganic acids such as hydrochloric acid:

$$HCl + OH^- = Cl^- + H_2O \qquad \text{(Eq. 2.50)}$$

$$HCl + CO_3^{2-} = Cl^- + HCO_3^- \qquad \text{(Eq. 2.51)}$$

In laboratory cultures, phosphate salts are often used for pH control, providing an excellent buffer near pH 7.0. However, such salts are not useful for the field because of high cost, the increased potential for causing precipitation of calcium phosphate, causing clogging problems, and the great likelihood of partitioning onto aquifer clays, preventing movement through an aquifer.

2.6 COSOLVENT AND SURFACTANT FLUSHING

Groundwater is often contaminated with a mixture of chemicals, and an early question was what effect one chemical would have on the solubility and sorption characteristics of another. Among the findings was that the presence in groundwater of highly soluble water miscible solvents such as ethanol resulted in an increased solubility and decreased sorption for another but hydrophobic chemical (Nkedikizza et al., 1985). With the growing concern about the longevity of DNAPL sources, this finding suggested one possibility in the search for new technologies with potential for DNAPL removal, and led to what is termed cosolvent flushing (Augustijn et al., 1994). Cosolvents such as methanol, ethanol, and acetone are highly soluble in water, and chlorinated solvents are much more soluble in such cosolvent mixtures than in water itself. Thus, when a cosolvent mixture containing perhaps 20% or more of the cosolvent is passed through an aquifer or injected into the groundwater near a source area, the DNAPL dissolves much more readily and can be rapidly cleansed by this process if the cosolvent solution can find its way to come into contact with DNAPL. The cosolvent/DNAPL mixture is then pumped to the surface for reuse or disposal.

2.6.1 Cosolvent Flushing

Perhaps the first application of cosolvent flushing was a field demonstration on a mixed petroleum/chlorinated solvent source area at Hill AFB, Utah in a hydraulically isolated test cell (Rao et al., 1997). Here, a cosolvent mixture consisting of 70% ethanol, 12% n-pentanol, and 18% water was pumped through the test cell over a period of 10 days, and was followed by flushing with water for 20 days. Greater than 85% mass removal of several target contaminants was observed. A pilot-scale field test of the process was later conducted for PCE removal from a site contaminated by a dry cleaner (Jawitz et al., 2000). Here, an 85% ethanol 5% water solution was pumped into the aquifer over a 3-day period, with an estimated removal of 65% of the PCE. One concern has been with the impact of the ethanol left behind with this approach, as well as with the cost of treating or disposing of the contaminated solvent removed from the aquifer. However, studies conducted 3 years after the cosolvent flushing was completed, found that the residual ethanol left behind had served as an effective electron donor for reductive dehalogenation with a significant conversion of residual PCE primarily to *cis*-DCE, but VC and ethene formation were also taking place (Mravik et al., 2003). Thus, cosolvent flushing combined with use of residual cosolvent for bioremediation emerged as a combined treatment approach for DNAPL removal.

2.6.2 Surfactant Flushing

Surfactant flushing emerged as another possible method for increasing the solubility of DNAPLs so that they could more readily be extracted from groundwater (Abdul et al., 1990; Fountain et al., 1991; Vigon and Rubin, 1989). Surfactants are organic molecules that contain a hydrophilic end with affinity for water and a hydrophobic end with an affinity for organic materials, such as chlorinated solvents. At a sufficiently high concentration of a surfactant (the critical micelle concentration), several surfactant molecules can come together to form a micelle, with the hydrophobic ends gathered together in the center and the hydrophilic ends facing out into water. Hydrophobic compounds such as the chlorinated solvents then can migrate into the hydrophobic center and hence become "solubilized." Surfactants can also lower the interfacial tension of DNAPLs, causing them to migrate downward more readily, a problem that was early recognized (Fountain et al., 1991) and one that must be prevented from occurring (Pennell et al., 1996).

One of the first field demonstrations of surfactant flushing was in a controlled test cell at the Hill AFB, where two aquifer floods were made of the petroleum/chlorinated solvent source area (Londergan et al., 2001). The reported removal of the estimated 1,300 L of residual DNAPL in this manner was 98.5%. In a subsequent demonstration of surfactant flushing for removal of a defined release of PCE DNAPL into a confined cell at Dover AFB a smaller 68% removal was obtained (Childs et al., 2006) through ten pore volumes of flushing. Here, a surfactant formulation consisting of sodium dihexyl sulfosuccinate, isopropanol and calcium chloride was used. In a pilot field-scale demonstration of surfactant flushing of PCE DNAPL under a dry cleaning facility (Abriola et al., 2005), removal of 19 L of PCE was obtained with PCE solution concentrations decreasing by two orders of magnitude at some locations (Ramsburg et al., 2005). Here, 68 cubic meters (m^3) of an aqueous solution containing 6% by weight of Tween 80, a non-ionic food grade surfactant, were injected, with 95% recovery of the injected surfactant during extraction. An interesting observation here, as in the case with solvent flushing with ethanol, was that the residual surfactant in the aquifer stimulated the growth of PCE reducing microorganisms, leading to the formation of TCE and *cis*-DCE (Ramsburg et al., 2004). This once again demonstrated the potential for combining a chemical process for removal with a biological process for transformation of residual chlorinated solvent.

2.7 INORGANIC BIOREMEDIATION EXAMPLE: OAK RIDGE FIELD RESEARCH CENTER

A pilot-scale demonstration of uranium stabilization illustrates how both physical-chemical and biological processes can be staged and integrated to enable remediation of severely contaminated sites (Wu et al., 2006a, b). From 1951 until 1984 wastes from atomic-weapon production were stored in large unlined ponds. The ponds were drained then covered with a parking lot, but groundwater continued to percolate through the contaminated soil beneath the parking lot, resulting in three separate plumes, including one that discharged to a nearby creek. The plume depth range from 9 to 30 m bgs, in a saprolite media that had fracture densities as high as 100–200 fractures/m. These fractures accounted for less than 5–10% of matrix porosity, but carried more than 95% of the flow. The surrounding highly porous aquifer materials had a low permeability and served as a sink (and continuing source) of contamination. Groundwater contaminants included 40 mg/L of depleted uranium, 540 mg/L aluminum (Al), 930 mg/L Ca, and 11–14 mg/L nickel. Disposal of nitric and sulfuric acids lowered the groundwater pH to 3.4–3.6, and resulted in extremely high concentrations of nitrate (8–10 g/L) and sulfate (~1 g/L).

Even though the soluble uranium concentrations were high (exceeding the federal drinking water standard by over 1,000-fold), most of the uranium was associated with the solid phase, with hot spots at 200–700 mg/kg. The solid phase was thus a long-term source of U(VI) groundwater contamination. Laboratory and field tests showed that uranium sorption and desorption were strongly pH dependent with the highest adsorption observed at a pH close to 6.0.

The remediation strategy focused upon converting U(VI) into sparingly soluble U(IV). Many microorganisms, including certain SRB and iron(III)-reducing bacteria (FeRB), mediate this conversion. Reduced compounds produced by these organisms, such as sulfide and green rusts can also convert U(VI) to U(IV). The basic concept was to stimulate these reductive pathways through periodic ethanol additions. But the presence of clogging agents and inhibitors factors prevented direct implementation of this approach:

- The initial soluble uranium levels were inhibitory to microbial growth.
- Nitrate levels were inhibitory to uranium reduction and caused oxidation of U(IV) back to U(VI).

- The low pH (3–4) was unfavorable for microbial activity. The high Al acidity buffered the system at this pH and, because $Al(OH)_3(s)$ precipitates at pH 4.5–5, made it difficult to increase pH to a final level better suited for microbial activity,
- The high Ca levels were prone to precipitation at pH levels above 7 and allowed formation of soluble U(VI) calcium uranyl complexes that are difficult to reduce.

The presence of clogging agents and inhibitors motivated fabrication and operation of a multi-step conditioning system designed to remove clogging agents and to create an environment favorable for microbial activity. Stepwise conditioning is useful whenever inhibitory or clogging agents are present, though the steps and methods used in each case will differ, depending upon the contaminants present and site-specific considerations.

Prior to startup of the system, a nested circulation well system containing an inner loop and an outer loop was installed to enable hydraulic control within the targeted treatment zone. Injection of clean water into the outer loop protected the inner loop from invasion of contaminated groundwater. A bromide tracer study was conducted to characterize the well-to-well connectivity and travel times between injection and extraction wells and breakthrough curves at multilevel sampling wells located between the injection and extraction wells. The subsurface was then flushed with clean water (tap water and nitrate-free water from an aboveground treatment facility) to achieve a pH of 4.0–4.5 and to remove clogging agents and inhibitors such as Al, Ca, nitrate, and volatile organics. The extracted water was treated aboveground by vacuum stripping to remove volatile organics, two-step precipitation to remove Al and Ca, and biological treatment in a fluidized bed bioreactor to remove nitrate. The treated water was reinjected into the outer recirculation loop.

After the concentration of Al in the extracted water had fallen sufficiently – i.e., so that Al was no longer judged a clogging threat when the pH was increased – a second clean water flush at pH 6–7 was carried out. The aim of this flush was to further decrease nitrate levels and to increase pH to 6–6.5. Because nitrate had diffused deep into the matrix, these flushing operations lasted for months, as predicted by computer simulations (Luo et al., 2005), but eventually nitrate levels fell from g/L levels to low mg/L levels, and pH increased to the desired range. A pH range of 6–6.5 was selected as optimal because sorption of U(VI) was highest over this range, alleviating the potential inhibitory effect of U(VI) on microbial growth, and because this pH range was more favorable for SRB that can reduce U(VI) than for methanogens that do not, but still compete for electron donor (Table 2.12).

Weekly ethanol injections over a 1-year period sequentially stimulated *in situ* denitrification of the residual nitrate diffusing from the pores of the matrix. This was followed by sulfate- and iron(III)-reduction and U(VI) reduction. Sediment samples from the treatment zone changed color from yellow-brown to dark green or black, providing further evidence of reduction and a gradual expansion of the zone of reduction. Uranium concentrations decreased to levels below the USEPA MCL (0.03 mg/L) within those zones that were hydrologically connected to the inner loop injection well where ethanol was added. Conversion of U(VI) to U(IV) was confirmed by X-ray absorption near-edge structure spectroscopy of sediment samples. Before biostimulation, no U(IV) was observed in sediment samples. After biostimulation, up to 80% of the uranium in the aquifer was reduced to U(IV).

Before addition of ethanol, only denitrifiers were detected in the groundwater, and only at an extremely low level (3 cells/mL). After ethanol addition, most probable number estimates for denitrifiers, SRB, and FeRB in sediments (cells/g dry weight) increased to 10^7–10^8. Post-treatment tests indicated that numerous microorganisms capable of reducing U(VI) to U (IV) (including SRB *Desulfovibrio*, *Desulfoporosinus*, and *Desulfotomaculum* spp. and FeRB

Geobacter and *Anaeromyxobacter* spp.) were present. The results also suggested that ethanol addition had promoted both microbial and secondary abiotic reduction of U(VI).

Very low aqueous-phase concentrations of uranium were achieved at the Oak Ridge site despite high solid-phase concentrations. This is due to the low solubility of U(IV) and to the low rates of desorption/dissolution of U(VI) species compared to the rate of reduction. Tests to evaluate the stability of the U(IV) (Wu et al., 2007) revealed that it was stable when ethanol injections were suspended for a 50-day period but anaerobic conditions were still maintained. However, additional studies demonstrated that oxygen and nitrate can remobilize uranium, indicating that long-term bioremediation will need to incorporate strategies for removal of dissolved oxygen and nitrate or development of methods to increase the stability of immobilized U(IV) upon exposure to oxidants.

2.8 SUMMARY

Chemicals are often added in groundwater remediation for a variety of different reasons and purposes. To be effective for their intended purpose, the chemicals generally need to be added in the appropriate amounts and concentrations, and mixed in a suitable manner to have the desired effect. Knowledge of reaction stoichiometry and kinetics is needed in order to apply the appropriate amount of a chemical so that remediation can be successful. This chapter provided an overview of the various remediation processes that might require chemical additions and how to determine the appropriate amounts. Some examples are provided on how chemicals might be mixed. Subsequent chapters will address processes for mixing chemicals in a much broader context and in greater detail.

REFERENCES

Abdul AS, Gibson TL, Rai DN. 1990. Selection of surfactants for the removal of petroleum products from shallow sandy aquifers. Ground Water 28:920–926.

Abriola LM, Drummond CD, Hahn EJ, Hayes KF, Kibbey TCG, Lemke LD, Pennell KD, Petrovskis EA, Ramsburg CA, Rathfelder KM. 2005. Pilot-scale demonstration of surfactant-enhanced PCE solubilization at the Bachman Road site. 1. Site characterization and test design. Environ Sci Technol 39:1778–1790.

Augustijn DCM, Jessup RE, Rao PSC, Wood AL. 1994. Remediation of contaminated soils by solvent flushing. J Environ Eng 120:42–57.

Bailey JE, Ollis DF. 1986. Biochemical Engineering Fundamentals. McGraw-Hill Companies, Inc., New York, NY, USA. 984 p.

Benjamin MM. 2002. Water Chemistry. McGraw-Hill Companies, Inc., New York, NY, USA. 668 p.

Cantrell KJ, Kaplan DI. 1997. Zero-valent iron colloid emplacement in sand columns. J Environ Eng 123:499–505.

Childs J, Acosta E, Annable MD, Brooks MC, Enfield CG, Harwell JH, Hasegawa M, Knox RC, Rao PSC, Sabatini DA, Shiau B, Szekeres E, Wood AL. 2006. Field demonstration of surfactant-enhanced solubilization of DNAPL at Dover Air Force Base, Delaware. J Contam Hydrol 82:1–22.

Cornell RM, Schwertmann U. 1996. The Iron Oxides. VCH Publishers, Weinheim, Germany. 573 p.

Dean JA. 1999. Lange's Handbook of Chemistry, 15th ed. McGraw-Hill, New York, NY, USA, pp 1521.

Dybas M, Tatara G, Knoll W, Mayotte T, Criddle CS. 1995a. Niche adjustment for bioaugmentation with *Pseudomonas* sp. strain KC. In Hinchee RE, Frederickson J, Alleman BC, eds, Bioaugmentation for Site Remediation (Bioremediation Series 3(3)), Battelle Press, Columbus, OH, USA, pp 77–84.

Dybas M, Tatara G, Criddle CS. 1995b. Localization and characterization of the carbon tetrachloride transformation activity of *Pseudomonas* sp. Strain KC. Appl Environ Microbiol 61:758–762.

Dybas MJ, Barcelona M, Bezborodnikov S, Davies S, Forney L, Heuer H, Kawka O, Mayotte T, Sepulveda-Torres L, Smalla K, Sneathen M, Tiedje J, Voice T, Wiggert DC, Witt ME, Criddle CS. 1998. Pilot-scale evaluation of bioaugmentation for in-situ remediation of a carbon tetrachloride-contaminated aquifer. Environ Sci Technol 32:3598–3611.

Dybas MJ, Hyndman DW, Heine R, Linning K, Tiedje J, Voice T, Wallace R, Wiggert D, Zhao X, Artuz R, Criddle CS. 2002. Development, operation, and long-term performance of a full-scale biocurtain utilizing bioaugmentation. Environ Sci Technol 36:3635–3644.

Ellis DE, Lutz EJ, Odom JM, Buchanan RJ, Bartlett CL, Lee MD, Harkness MR, Deweerd KA. 2000. Bioaugmentation for accelerated in situ anaerobic bioremediation. Environ Sci Technol 34:2254–2260.

ESTCP (Environmental Security Technology Certification Program). 2005. A Review of Biofouling Controls for Enhanced In Situ Bioremediation of Groundwater. ER-0429-WhtPaper.pdf. Department of Defense ESTCP, Arlington, VA, USA. 48 p. http://serdp-estcp.org/. Accessed June 14, 2011.

Fountain JC, Klimek A, Beikirch MG, Middleton TM. 1991. Use of surfactants for in situ extraction of organic pollutants from a contaminated aquifer. J Hazard Mater 28:295–311.

Gillham RW, O'Hannesin SF. 1994. Enhanced degradation of halogenated aliphatics by zero-valent iron. Ground Water 32:958–967.

Gribić-Galić D, Vogel TM. 1987. Transformation of toluene and benzene by mixed methanogenic cultures. Appl Environ Microbiol 53:254–260.

Hopkins GD, McCarty PL. 1995. Field-evaluation of in-situ aerobic cometabolism of trichloroethylene and 3 dichloroethylene isomers using phenol and toluene as the primary substrates. Environ Sci Technol 29:1628–1637.

Hyndman DW, Dybas MJ, Forney L, Heine R, Mayotte T, Phanikumar MS, Tatara G, Tiedje J, Voice T, Wallace R, Wiggert D, Zhao X, Criddle CS. 2000. Hydraulic characterization and design of a full-scale biocurtain. Ground Water 38:462–474.

Jawitz JW, Sillan RK, Annable MD, Rao PSC, Warner K. 2000. In-situ alcohol flushing of a DNAPL source zone at a dry cleaner site. Environl Sci Technol 34:3722–3729.

Jenal-Wanner U, McCarty PL. 1997. Development and evaluation of semicontinuous slurry microcosms to simulate in-situ biodegradation of trichloroethylene in contaminated aquifers. Environ Sci Technol 31:2915–2922.

Langmuir D. 1978. Uranium solution-mineral equilibria at low temperatures with applications to sedimentary ore deposits. Geochim Cosmochim Ac 42:547–569.

Lee C-H, Lewis TA, Paszczynski A, Crawford RL. 1999. Identification of an extracellular catalyst of carbon tetrachloride dehalogenation from *Pseudomonas stutzeri* strain KC as pyridine-2,6-bis(thiocarboxylate). Biochem Biophys Res Commun 261:562–566.

Leung SW, Watts RJ, Miller GC. 1992. Degradation of perchloroethylene by Fenton reagent: Speciation and pathway. J Environ Qual 21:377–381.

Levenspiel O. 1999. Chemical Reaction Engineering, 3rd ed. John Wiley & Sons, Inc., New York, NY, USA. 668 p.

Londergan JT, Meinardus HW, Mariner PE, Jackson RE, Brown CL, Dwarakanath V, Pope GA, Ginn JS, Taffinder S. 2001. DNAPL removal from a heterogeneous alluvial aquifer by surfactant-enhanced aquifer remediation. Ground Water Monit Remediat 21:57–67.

Luo J, Cirpka OA, Wu W-M, Fienen MN, Jardine PM, Mehlhorn TL, Watson DB, Criddle CS, Kitanidis PK. 2005. Mass-transfer limitations for nitrate removal in a uranium-contaminated aquifer. Environ Sci Technol 39:8453–8459.

Major DW, McMaster ML, Cox EE, Edwards EA, Dworatzek SM, Hendrickson ER, Starr MG, Payne JA, Buonamici LW. 2002. Field demonstration of successful bioaugmentation to achieve dechlorination of tetrachloroethene to ethene. Environ Sci Technol 36:5106–5116.

McCarty PL, Goltz MN, Hopkins GD, Dolan ME, Allan JP, Kawakami BT, Carrothers TJ. 1998a. Full-scale evaluation of in-situ cometabolic degradation of trichloroethylene in groundwater through toluene injection. Environ Sci Technol 32:88–100.

McCarty PL, Hopkins GD, Munakata-Marr J, Matheson VG, Dolan ME, Dion LB, Shields M, Forney LJ, Tiedje JM. 1998b. Bioaugmentation with *Burkholderia cepacia* PRI_{301} for in-situ bioremediation of trichloroethylene contaminated groundwater. EPA/600/S-98/001. U.S. Environmental Protection Agency, National Health and Environmental Effects Research Laboratory, Gulf Breeze, FL, USA. 11 p.

Morel FMM, Hering JG. 1993. Principles and Applications of Aquatic Chemistry. John Wiley & Sons, Inc., New York, NY, USA. 588 p.

Mravik SC, Sillan RK, Wood AL, Sewell GW. 2003. Field evaluation of the solvent extraction residual biotreatment technology. Environ Sci Technol 37:5040–5049.

Munakata-Marr J, McCarty PL, Shields MS, Reagin M, Francesconi SC. 1996. Enhancement of trichloroethylene degradation in aquifer microcosms bioaugmented with wild-type and genetically altered *Burkholderia (Pseudomonas) Cepacia* G4 and Pr1. Environ Sci Technol 30:2045–2052.

Nkedikizza P, Rao PSC, Hornsby AG. 1985. Influence of organic cosolvents on sorption of hydrophobic organic chemicals by soils. Environ Sci Technol 19:975–979.

NRC (National Research Council). 1994. Alternatives for Ground Water Cleanup. National Academies Press, Washington, DC, USA. 315 p.

Nyman JL, Williams SM, Criddle CS. 2005. Bioengineering for the in-situ remediation of metals. In Grassian V, ed, Environmental Catalysis. CRC Press, Boca Raton, FL, USA, pp 493–520.

O'Hannesin SF, Gillham RW. 1998. Long-term performance of an in situ 'iron wall' for remediation of VOCs. Ground Water 36:164–170.

Pennell KD, Pope GA, Abriola LM. 1996. Influence of viscous and buoyancy forces on the mobilization of residual tetrachloroethylene during surfactant flushing. Environ Sci Technol 30:1328–1335.

Puls RW, Blowes DW, Gillham RW. 1998. Emplacement verification and long-term performance monitoring of a permeable reactive barrier at the USCG Support Center, Elizabeth City, North Carolina. IAHS Publication (International Association of Hydrological Sciences) 250:459–466. http://iahs.info/redbooks/a250/iahs_250_0459.pdf. Accessed June 16, 2011.

Ramsburg CA, Abriola LM, Pennell KD, Loffler FE, Gamache M, Amos BK, Petrovskis EA. 2004. Stimulated microbial reductive dechlorination following surfactant treatment at the Bachman Road site. Environ Sci Technol 38:5902–5914.

Ramsburg CA, Pennell KD, Abriola LM, Daniels G, Drummond CD, Gamache M, Hsu H-L, Petrovskis EA, Rathfelder KM, Ryder JL, Yavaraski TP. 2005. Pilot-scale demonstration of surfactant-enhanced PCE solubilization at the Bachman Road site. 2. System operation and evaluation. Environ Sci Technol 39:1791–1801.

Rao PSC, Annable MD, Sillan RK, Dai DP, Hatfield K, Graham WD, Wood AL, Enfield CG. 1997. Field-scale evaluation of in situ cosolvent flushing for enhanced aquifer remediation. Water Resour Res 33:2673–2686.

Rard JA, Rand MH, Anderegg G, Wanner H. 1999. Chemical Thermodynamics of Technicium. North-Holland by Elsevier Science, Amsterdam, The Netherlands.

Rheinhard M, Curtis GP, Barbash JE. 1997. Natural chemical attenuation of halogenated hydrocarbon compounds via dehalogenation. In Ward CH, Cherry JA, Scalf MR, eds, Subsurface Restoration. Ann Arbor Press Inc., Ann Arbor, MI, USA, pp 397–409.

Rittmann BE, McCarty PL. 2001. Environmental Biotechnology: Principles and Applications. McGraw-Hill, New York, NY, USA. 754 p.

Sawyer CN, McCarty PL, Parkin GF. 2003. Chemistry for Environmental Engineering and Science. McGraw-Hill, New York, NY, USA. 752 p.

Schnarr M, Truax C, Farquhar G, Hood E, Gonullu T, Stickney B. 1998. Laboratory and controlled field experiments using potassium permanganate to remediate trichloroethylene and perchloroethylene DNAPLs in porous media. J Contam Hydrol 29:205–224.

Schreier CG, Reinhard M. 1995. Catalytic hydrodehalogenation of chlorinated ethylenes using palladium and hydrogen for the treatment of contaminated water. Chemosphere 31:3475–3487.

Semprini L, Dolan ME, Mathias MAB, Hopkins GD, McCarty PL. 2007. Laboratory, field, and modeling studies of bioaugmentation of butane-utilizing microorganisms for the in situ cometabolic treatment of 1,1-dichloroethene, 1,1-dichloroethane, and 1,1,1-trichloroethane. Adv Water Resour 30:1528–1546.

Smith BA, Teel AL, Watts RJ. 2006. Mechanism for the destruction of carbon tetrachloride and chloroform DNAPLs by modified Fenton's reagent. J Contam Hydrol 85:229–246.

Stumm W, Morgan JJ. 1996. Aquatic Chemistry. John Wiley & Sons, Inc., New York, NY, USA. 1022 p.

Teel AL, Watts RJ. 2002. Degradation of carbon tetrachloride by modified Fenton's reagent. J Hazard Mater 94:179–189.

Teel AL, Warberg CR, Atkinson DA, Watts RJ. 2001. Comparison of mineral and soluble iron Fenton's catalysts for the treatment of trichloroethylene. Water Res 35:977–984.

Vigon BW, Rubin AJ. 1989. Practical consideration in the surfactant-aided mobilization of contaminants in aquifers. J Water Pollut Control Fed 61:12331240.

Vogel TM, Criddle CS, McCarty PL. 1987. Transformations of halogenated aliphatic compounds. Environ Sci Technol 21:722–736.

Wadley SLS, Gillham RW. 2003. Remediation of DNAPL source zone using granular iron: A field demonstration. Proceedings, 2003 International Symposium on Water Resources and the Urban Environment, Wuhan, China, November, pp 30–35.

Weber WJ Jr, DiGiano FA. 1996. Process Dynamics in Environmental Systems. John Wiley & Sons, Inc., New York, NY, USA, 943 p.

Wu W, Carley J, Fienen M, Mehlhorn T, Lowe K, Nyman J, Luo J, Gentile ME, Rajan R, Wagner D, Hickey RF, Gu B, Watson D, Cirpka OA, Kitanidis PK, Jardine PM, Criddle CS. 2006a. Pilot-scale bioremediation of uranium in a highly contaminated aquifer I: Conditioning of a treatment zone. Environ Sci Technol 40:3978–3985.

Wu W, Carley J, Gentry T, Ginder-Vogel MA, Fienen M, Mehlhorn T, Yan H, Carroll S, Nyman J, Luo J, Gentile ME, Fields MW, Hickey RF, Watson D, Cirpka OA. Fendorf S, Zhou J, Kitanidis PK, Jardine PM, Criddle CS. 2006b. Pilot-scale bioremediation of uranium in a highly contaminated aquifer II: Evidence of U(VI) reduction and geochemical control of U (VI) bioavailability. Environ Sci Technol 40:3986–3995.

Wu W, Carley J, Luo J, Ginder-Vogel MA, Cardenas E, Leigh MB, Hwang C, Kelly SD, Ruan C, Wu L, Van Nostrand J, Gentry T, Lowe K, Mehlhorn T, Carroll S, Lou W, Fields W, Gu B, Watson D, Kemner KM, Marsh T, Tiedje J, Zhou J, Fendorf S, Kitanidis PK, Jardine PM, Criddle CS. 2007. In-situ bioreduction of uranium (VI) to submicromolar levels and reoxidation by dissolved oxygen. Environ Sci Technol 41: 5716–5723.

Yaws CL. 1999. Chemical Properties Handbook. McGraw-Hill, New York, NY, USA. 779 p.

Zhang WX, Wang CB, Lien HL. 1998. Treatment of chlorinated organic contaminants with nanoscale bimetallic particles. Catal Today 40:387–395.

CHAPTER 3

TRANSPORT AND MIXING

Peter K. Kitanidis[1]

[1]Department of Civil and Environmental Engineering, Stanford University, Stanford, CA, USA

3.1 INTRODUCTION

This chapter is a tutorial overview of physical transport processes in the subsurface, particularly as they pertain to chemical delivery and mixing. This chapter is intended as background material to facilitate the understanding of Chapters 4, 5, 6, 7, 8, 9 and 10. Readers with backgrounds in physical hydrogeology and fluid mechanics may want to skip all or parts of this chapter.

The *in situ* rates of subsurface reactions are governed by the process of mixing. Mixing is needed to bring reactants together and also to remove products that may inhibit the progress of reactions. However, mixing in geological formations is an extremely intricate and generally slow process. It is, in fact, an interrelated series of processes operating across a wide range of spatial and temporal scales. Being aware of the scale of the processes matters because the key variables, parameters, and governing equations only apply over defined scales.

This chapter discusses the key processes at three scales: pore-scale, laboratory-scale, and field-scale. After discussing how mixing affects reactions and why scale issues are critical to understand, the subsequent sections discuss the key processes at each scale. The focus in each case is on the two main mechanisms of transport – advection and diffusion – first separately and then by considering their combined effect: hydrodynamic dispersion. The goal is to provide the necessary conceptual and mathematical understanding of the fundamental physical transport and mass transfer mechanisms, with examples of how mass transfer can control *in situ* reaction rates and how fundamental principles influence the fate and behavior of contaminants in the subsurface.

For readers who would like more information and a more thorough treatment of transport processes, there are several useful books. For example, Weber and DiGiano (1996) discusses transport processes at various scales in detail, as well as issues related to reactor design. More concise and introductory treatments can be found in environmental textbooks such as Hemond and Fechner (1999). Hydrogeologic textbooks, such as Domenico and Schwartz (1998), Fetter (1998, 2000), Freeze and Cherry (1979), Bedient et al. (1999), Bear and Verruijt (1987), Charbeneau (2006), and deMarsily (1986) have detailed sections on transport processes. Several books and review articles dealing with groundwater pollution and remediation issues have appeared and contain sections on flow and transport in groundwater, such as Charbeneau et al. (1992).

Of course, we are interested in systems that involve reactions and mass transfer phenomena. This chapter focuses on physical transport and mass transfer mechanisms and deals with the issue of how mass transfer may limit reaction rates by restricting the supply or availability of at least one of the reactants. In Section 3.2, we discuss mixing and how it affects reactions. Next, we present an informal discussion of scale issues and then consider pore-scale, laboratory-scale, and field-scale descriptions.

P.K. Kitanidis and P.L. McCarty (eds.), *Delivery and Mixing in the Subsurface: Processes and Design Principles for In Situ Remediation*, doi: 10.1007/978-1-4614-2239-6_3, © Springer Science+Business Media New York 2012

3.2 MIXING

Chemists and biologists often study reactions in the laboratory in batch reactors, with liquids or mixtures of liquids, sometimes with crushed geologic media, and designed to maximize mixing rates, in order to study chemical transformation and obtain estimates of transformation rates and perhaps the thermodynamics of reactions. Such laboratory studies tend to yield *maximum reaction rates* possible in the absence of mass transfer limitations that tend to delay reactants from getting close enough together for a reaction to take place. The situation in the field, however, is that mass transfer rates are generally orders of magnitude slower than typical maximum reaction rates, primarily because diffusion coefficients are so small. Thus, the reaction rates achieved in the field tend to be much slower than the maximum rates that are measured in the laboratory.

Let us briefly review the concept of relative speeds of reactions versus mixing through a relatively simple example. Consider the reaction of compound A with compound B in an irreversible reaction; the two compounds are introduced in proportions determined based on the stoichiometry of the reaction so that A and B are completely transformed. The reactor is a cylinder with unit cross-sectional area and length d; reactant A is introduced with molar concentration C at one end and reactant B is introduced at the other end with the same concentration. The two reactants are thus initially separated by a distance d [L] and mix only through diffusion with molecular coefficient D [L^2/T]. The reaction is bimolecular with rate r, where rC has units [1/T]. Then the time scale characteristic of diffusional transport over distance d is on the order of d^2/D, while the characteristic time of reaction is on the order of $1/(rC)$. The ratio of time scales is a dimensionless number known as a Damkohler number:

$$D_a = \frac{rCd^2}{D} \qquad \text{(Eq. 3.1)}$$

For example, for $D = 10^{-9}$ m^2/s, $d = 0.1$ m, and $rC = 10^{-4}$/s, the Damkohler number becomes 1,000. When the Damkohler number is large, diffusional mass transport is slow in relation to reaction. The actual rate at which reactants are transformed should be controlled by diffusion and, thus, be much lower than the maximum one that would have been possible in a completely mixed batch reactor. Transport of reactants and chemical transformation are essentially two sequential steps in the reaction process, and the net rate is controlled by diffusion because it is the slower of the two steps.

Continuing with the same example, the transport equations and boundary conditions are shown in Equation 3.2:

$$\begin{aligned} \frac{\partial c_1}{\partial t} &= D\frac{\partial^2 c_1}{\partial x^2} - rc_1c_2, \quad c_1(t,0) = C, \ \frac{\partial c_1}{\partial t}(t,d) = 0 \\ \frac{\partial c_2}{\partial t} &= D\frac{\partial^2 c_2}{\partial x^2} - rc_1c_2 \quad \frac{\partial c_2}{\partial t}(t,0) = 0, \ c_2(t,d) = C \end{aligned} \qquad \text{(Eq. 3.2)}$$

We will consider transformation rates after sufficient time has elapsed so that steady state is achieved. One can verify that for a sufficiently high Damkohler number, the net transformation rate within the reactor is $2D\frac{C}{d}$ (which is in moles per unit time, after we consider that the reactor cross-section has unit area) and thus is independent of the reaction rate coefficient r. At the other extreme, of a very low Damkohler number, the net transformation rate is proportional to r. To obtain values for a whole range, the equations were solved numerically and the net transformation rate was computed. The solid line in Figure 3.1 shows the transformation rate within the reactor, M_t, normalized by the maximum, $2D\frac{C}{d}$, as a function of the

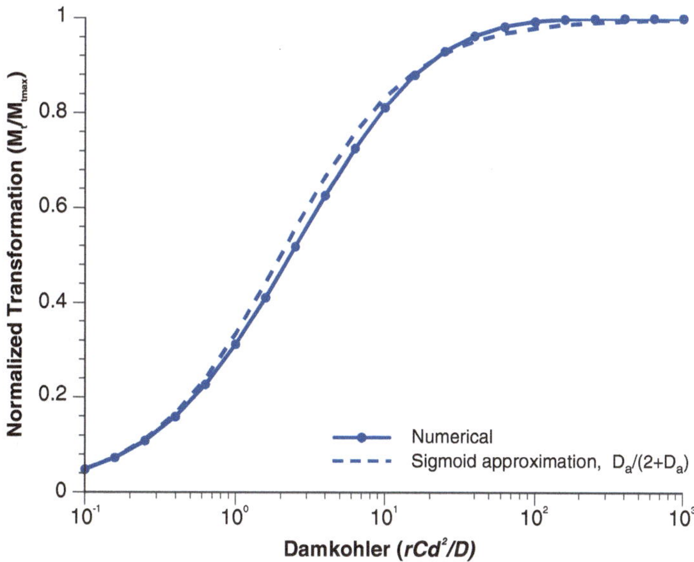

Figure 3.1. Net transformation rate as a function of the Damkohler number.

Damkohler number. This function can be approximated adequately by a simple sigmoid function (see Equation 3.3), shown as dashed line in the same figure:

$$\frac{M_t}{(2DC/d)} = \frac{D_a}{2 + D_a} \qquad \text{(Eq. 3.3)}$$

This example illustrates that the net transformation rate is controlled by the reaction-rate coefficient at low values of the Damkohler number D_a and by diffusion at large values of D_a, with a gradual transition between the two extremes.

In field applications, transport and mixing usually control reaction rates due to the slowness of transport and the larger scales that are involved. For example, the rates of monitored intrinsic remediation are often dominated by the rate of mixing of reactants in a lateral direction (transverse mixing). This occurs partly because lateral mixing is slow and key reactions occur predominantly at the fringes of the contaminant plume. Such a situation is typical for a methane plume in a natural-gradient flow field, in which electron acceptors quickly become depleted in the core of the plume (e.g., Chu et al., 2005; Cirpka and Valocchi, 2007; Huang et al., 2003; Thornton et al., 2001; Thullner et al., 2002).

Unfortunately, mixing is often hard to quantify at contaminated sites. The subsurface environment is highly heterogeneous, several processes are involved, and it can be difficult to obtain sufficiently detailed characterization data at multiple scales. As a result, it is usually hard to predict a priori reaction rates under field conditions. An interagency panel on assessing conceptual models for subsurface reactive transport (Davis et al., 2004) concluded that "A principal difficulty in conceptual model development is the identification of appropriate process models in the presence of multi-scale heterogeneities. . .Mixing processes bring into contact solutes, surfaces, and solids, that are not in chemical equilibrium. While the resulting reactions may be fast, the rate may be limited by mass transfer between adjacent zones with differing chemistries and microbial populations."

The exact way transport can limit the effective reaction rates is highly case dependent. In the following sections, we will identify and discuss the typical mechanisms or processes of mixing in the subsurface. However, it is important to realize that in most specific cases, multiple mechanisms will be impacting the overall reaction rates.

3.2.1 Mass Transfer from Separate Phases

Many contaminants are hydrophobic, and therefore tend to partition out of the aqueous phase to solid, gaseous, or non-aqueous liquid phases. For example, perchloroethene (PCE) may be sorbed on solids or stored in micro-pores, located preferentially in low-permeability formations, or in dense nonaqueous phase liquid (DNAPL) pools. For remediation to be effective, this contaminant must mix with reactants (like substrates and electron acceptors) added or present in surrounding fluid. The mixing process can be extremely slow because it is limited by diffusion with relatively mild concentration gradients. The dissolution rate of a nonaqueous phase liquid (NAPL) is slow, even if equilibrium is quickly reached at the NAPL-water interface, because it is controlled by the rate of transport in the aqueous phase. Similarly, desorption can be slow, because it is usually controlled by slow diffusional transport out of micropores and high surface area, but low conductivity domains where the contaminant has preferentially sorbed. In the case of volatilization, molecular diffusion in the aqueous phase near the water-air interface may be the rate-controlling mechanism.

3.2.2 Transverse Mixing

In many cases, the mixing that matters for reaction rates takes place in the transverse direction – the direction perpendicular to the direction of flow. For example, consider a methane contaminant plume emanating from a continuous source. Along the core near the central axis of the plume, electron acceptors like oxygen become depleted after a while. The oxidation of the contaminant happens only at the fringes of the plume where the electron donors mix with the electron acceptor. Transverse mixing through diffusion and transverse dispersion is generally quite slow, which should result in long plumes.

However, the rate of mixing is enhanced by the heterogeneity of groundwater velocities within aquifers and the fluctuating nature of subsurface flow. Heterogeneity causes streamlines to converge, thereby bringing reactants closer, and to diverge, spreading the reactants and the products. The process of mixing is highly nonuniform, but the net effect of heterogeneity is that the rate of transformation is enhanced. In addition, flow is never really steady, and even minor fluctuations can enhance mixing through oscillations of velocity in the transverse direction that cause the plume to swing sideways, enhancing the long-term rate of reactions.

3.2.3 Longitudinal Mixing and Chromatographic Mixing

When one reactant displaces another (as when one reactant is originally present in the groundwater and a solution of the other reactant is injected at a well) the reactions take place near the displacement interface where the two fluids mix. In a homogeneous medium and in the absence of sorption or other mechanisms that may retard the transport, the mixing rate slows down with time as the concentration gradient at the mixing interface gradually decreases; thus, after a while, the reaction rate diminishes. However, when the displaced reactant moves more slowly than the displacing one, higher reaction rates are maintained because the mixing is primarily due to variable degrees of adsorption, a process akin to chromatographic separation. Furthermore, mass transfer limitations, such as slow rates of sorption, cause spreading and mixing (Michalak and Kitanidis, 2000) that is often not captured by the simple advection-dispersion conceptualization but can be better represented through multi-porosity models, which we will discuss later.

3.3 SCALE DEPENDENCY

When using variables (such as concentration) or material properties (such as conductivity or dispersion coefficients), it is important to be aware that each quantity is defined, through measurement or computation, to apply at a certain scale of resolution. For example, the molecular diffusion coefficient in the aqueous phase is a measure of dispersal and mixing that describes how quickly solute molecules spread in still water; whereas the macrodispersion coefficient is also a measure of dispersal but describes the rate of spreading of a large plume in a heterogeneous geologic medium under conditions of flow. Although the diffusion coefficient and the macrodispersion coefficient have the same units and may appear in the same place in mathematical equations that describe transport, these two parameters differ by orders of magnitude and describe spreading at two dramatically different scales of resolution.

For example, let us look at the idealized and actual plumes of a conservative tracer (bromide) at a well-characterized site. Part a of Figure 3.2 shows actual iso-concentration lines of bromide on a cross-vertical section through the center of the plume at the Borden site (data from Roberts and Mackay, 1986 and analysis from Thierrin and Kitanidis, 1994). Not surprisingly, 381 days after injection, the plume is considerably more spread out than when

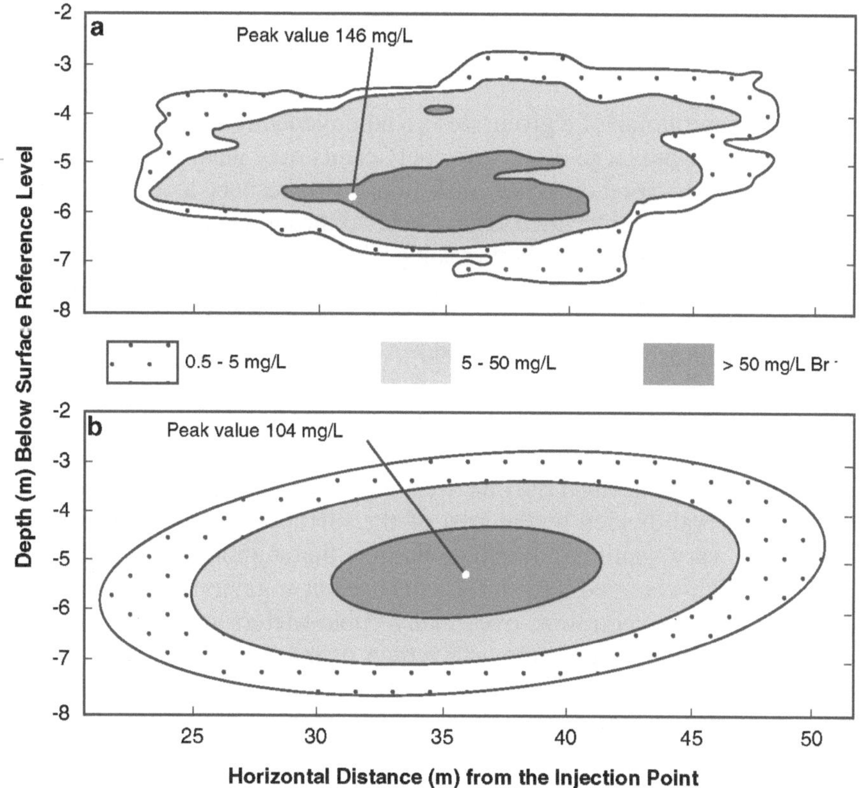

Figure 3.2. Borden field experiment day 381, vertical cross-section along the center of the plume in the flow direction: (a) actual and (b) idealized conservative tracer concentration (Thierrin and Kitanidis, 1994).

it started. The overall rate of tracer spreading can be quantified through macrodispersion coefficients; e.g., in the x direction, the macrodispersion coefficient is:

$$\frac{1}{2}\frac{\Delta\sigma_{xx}^2}{\Delta t} \qquad \text{(Eq. 3.4)}$$

where $\sigma_{xx}{}^2$ represents the mean square distance of tracer mass from the centroid of the plume, see Freyberg (1986). The longitudinal (in the direction of flow) macrodispersion coefficient computed (Freyberg, 1986) from data are many orders of magnitude larger than the molecular diffusion coefficient of bromide in water. They describe high rates of spreading, caused by variable fluid velocity that propels some mass forward while leaving some mass behind. However, the peak concentration is found to drop slower than expected from idealized models (assuming uniform velocity and that dispersion mimics diffusion with a fixed dispersion coefficient). Figure 3.2 part b shows an idealized plume with the same overall mass and the same spread as the actual plume but also with a highly regular Gaussian shape, which is expected after sufficient time elapses so that mixing evens out the variability.

Similar results were obtained at a site in Cape Cod, Massachusetts (Garabedian et al., 1991; LeBlanc et al., 1991; Thierrin and Kitanidis, 1994). In both cases, spreading is much more than one would have expected from diffusion alone, but the dilution in the actual plume is less than in the idealized Gaussian plume, as evidenced by the difference in the peak concentrations. The reason for the differences between the idealized and actual plumes is that the smoothing of concentration fluctuations, which is associated with the distribution of mass over a bigger volume, is limited by diffusive mixing at small scales. As a consequence, dilution lags behind spreading.

A simplified representation of a given site as a homogeneous one (i.e., uniform velocity and dispersion coefficients) with large dispersion coefficients may yield results that are consistent with data with respect to spreading but misleading with respect to dilution and mixing of reactants. Dispersion coefficients fitted from the breakthrough curves of two-well tracer tests reflect macrodispersive effects caused by variable advection, as the tracer is transported much faster through conductive layers than through impermeable ones. As a result, use of such large coefficients in evaluations of mixing involved in reactions may seriously overestimate the rate of reaction as discussed in Semprini et al. (1990).

Actually, all quantities encountered in the study of groundwater refer to a support volume or area, which is important to bear in mind even when the exact size of the support is not clearly defined. For example, solute concentration is equal to mass in a support volume divided by this volume; transport rates are defined over an area, e.g., flux rate is mass transported through a surface in a time interval divided by the area of the surface and the duration. The values of these variables may vary significantly depending on the support volume. For instance, the concentration in a depth-averaged transport model used in engineering applications is intended to represent the mean concentration over many cubic meters of water; the concentration measured in a small sample represents an average over cubic centimeters. The two values may differ significantly because mixing is slow and concentration can be highly nonuniform.

The scale dependency that applies to variables such as concentration also applies to parameters such as hydraulic conductivity, dispersion coefficients, and reaction rates. For example, hydraulic conductivity is defined over a volume, as we will review in the section on Darcy's law. The hydraulic conductivity depends on the geologic media, primarily the geometry and degree of interconnection of the pores and the viscosity and density of the fluid. However, for heterogeneous media, it also depends on the scale over which it is defined. For example, conductivity measured in core samples is often nearly isotropic – the same in every direction.

However, at field scales, the flow rate for a given head gradient is much faster in the direction parallel to geologic strata than in the direction normal to them. That is, the large-scale or "effective" conductivity is almost always anisotropic, being much larger in directions parallel to stratification than normal to it.

Parameters that define rates of spreading or mixing, such as diffusion and dispersion coefficients, are even more scale dependent than conductivity. The larger the scale over which dispersion (i.e., rate of spreading) is defined, the larger the dispersion coefficient. For example, Gelhar (1993) has plotted the observed macrodispersivity of plumes versus the size of the plumes. This plot (Figure 5.2 in Gelhar), indicates a definite tendency of spreading rate to increase with the size of the plume. This scale dependency can be explained by the fact that the larger the plume, the larger the scale of the velocity fluctuations that contribute to the spreading of the plume. This example also illustrates that one should exercise judgment in interpreting macroscopic parameters; although macrodispersion seems to refer to a diffusive and mixing process, it is mainly a consequence of averaging smaller-scale fluctuations in flow velocity. In addition to increasing spreading, this variability tends to increase mixing, but only after a fairly long period of time has elapsed.

Similarly, macroscopic rates of reactions are controlled primarily, and sometimes exclusively, by mass transport processes, particularly diffusive mass transfer, rather than by classical chemical kinetics of the kind that are quantified in completely mixed laboratory reactors. Thus, a macroscopic reaction rate may be more telling about the effectiveness of mixing than about the intrinsic or maximum possible chemical-reaction rate. In practice, we must estimate actual rates of transformation from field observations and not just rely on rates determined from laboratory studies involving highly mixed reactors. This may necessitate performing reactive tracer tests, conducting pilot studies, and analyzing the performance and fine-tuning the operation during full-scale implementation.

In the next three sections, we will review processes, starting with the finest or most detailed scale.

3.4 PORE SCALE

At the pore-scale of resolution, we distinguish between the pore and the solid phases, see Figure 3.3. We assume that the pore space is filled with water (the "aqueous phase"), and we consider fluid flow, solute advection, and solute diffusion in the aqueous phase within the pore space.

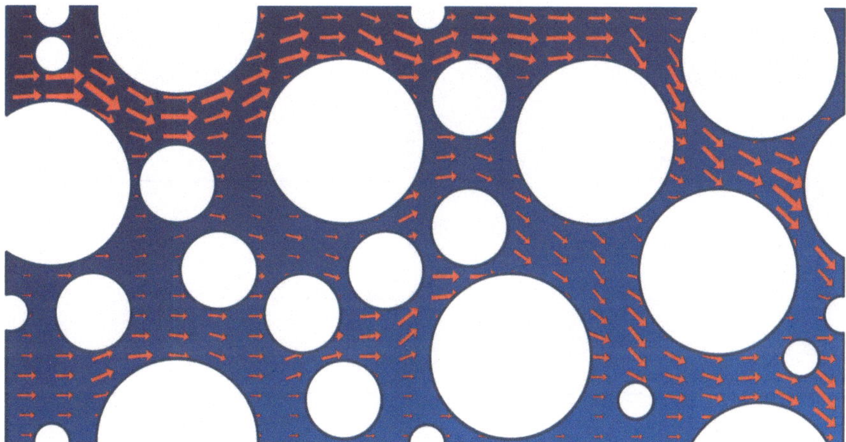

Figure 3.3. Porous network in an idealized porous medium.

3.4.1 Flow

Unlike the flow in rivers, the flow of liquids in porous media is dominated by the effects of viscosity. The effects of viscosity can be manifested by comparing how the more viscous honey flows on an inclined surface compared to the less viscous water. The viscous nature of flow in pores is most important as it significantly reduces the potential for mixing of water and chemicals within it. Thus, it is important to develop a fundamental understanding of factors affecting flow.

Viscosity acts as internal friction within a fluid and is responsible for converting mechanical energy into heat; the rate of conversion of mechanical energy is proportional to the *dynamic* (sometimes called *absolute*) *viscosity coefficient,* μ, with dimensions $[ML^{-1}T^{-1}]$. Common units of μ in applications are the poise (1P = 0.1kilogram per meter per second [kg/m/s]) and centipoise (cP = 0.001kg/m/s). The viscosity of water at 20 degrees Celsius (°C) is about 1cP and decreases with temperature. The ratio of dynamic viscosity to density is known as the *kinematic viscosity coefficient,* $v = \frac{\mu}{\rho}$, where μ is dynamic viscosity, and ρ is density. The dimensions of v are $[L^2T^{-1}]$ and a common unit is the centistoke, cSt, where 1cSt = 10^{-6} m/s. The kinematic viscosity of water at 20°C is roughly 1cSt.

Water, like many other fluids, conforms to *Newton's law of friction* that postulates a linear relation between shear stress and the rate of deformation in the fluid. Consider the case of rectilinear flow, with straight streamlines parallel to the x-axis, shown in Figure 3.4. For velocity components u, v, and w in the x, y, and z directions respectively, rectilinear flow means $u = u(y)$, $v = 0$, $w = 0$. The shear is related to the gradient of the velocity according to:

$$\tau=\mu\frac{\partial u}{\partial y} \quad \text{(Eq. 3.5)}$$

where τ is shear stress. The energy dissipation per unit volume (in units of Joule/s/m^3) is given by the expression:

$$P = \mu\left(\frac{du}{dy}\right)^2 \quad \text{(Eq. 3.6)}$$

In porous media, typical average velocities are on the order of 0.01–1 m/day. The velocity within the pore space is highly heterogeneous because the velocity on solid boundaries, the surfaces of solid grains, must be zero due to adhesion effects (the requirement of zero

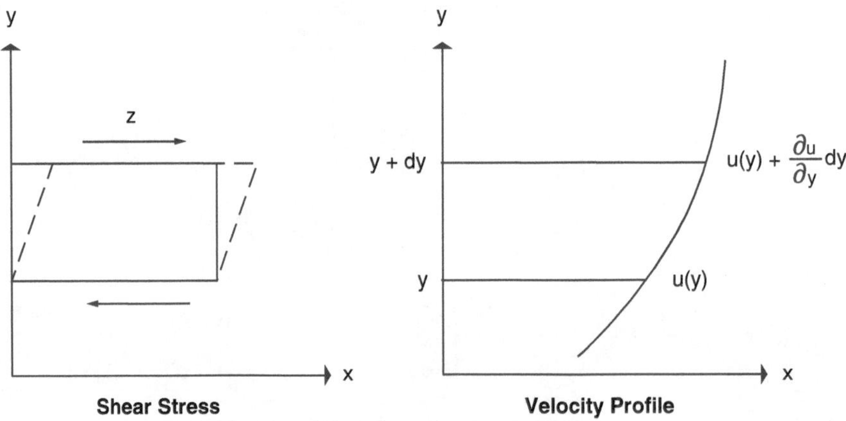

Figure 3.4. Rate of transformation due to shear stress.

velocity on a stationary solid boundary is known as the no-slip condition) while it takes its highest value near the center of the channel. The velocity must change from zero to its highest value within a very short distance, such as 0.1 millimeter (mm), which means that $\frac{du}{dy}$ is large even though u is small. Thus, viscous energy dissipation is a dominant feature of flow in porous media.

The energy needed to maintain the flow comes from loss of mechanical energy, kinetic and potential, that turns into heat through viscous energy dissipation. Kinetic energy in groundwater is negligible due to the low velocities. Potential energy has two components: one is related to elevation and the other is related to pressure. The fluid gives up potential energy as it moves from a higher elevation to a lower one within the earth's gravitational field. This gravitational potential energy normalized per unit volume of fluid is ρgz, where ρ is density, g is acceleration of gravity (the nominal or average value, known as standard gravity, is 9.81 m/s^2) and z is the elevation above a datum. The choice of the datum is unimportant because only changes in elevation matter in applications. Potential energy is also given up when fluid moves from higher to lower normal pressure, p. This potential energy due to pressure, normalized by volume, is p. In practice, the two types of potential energy need to be added to determine the *total potential energy*, $p + \rho gz$. The flow tends to be from points of higher to points of lower total potential energy.

Often density differences can be neglected and then it is more practical to normalize further by the specific weight, ρg, of the fluid. What is obtained is potential energy per unit weight, which has units of length and is called *hydraulic head*, ϕ:

$$\phi = \frac{p}{\rho g} + z \qquad \text{(Eq. 3.7)}$$

Flow tends to be from higher to lower hydraulic head.

The dynamics of flow within the pores is governed by the Stokes equation, which is obtained from the incompressible Navier–Stokes equation by dropping the nonlinear momentum-flux terms, which are negligible with flow in porous media compared to the term associated with viscosity. The absence of the nonlinear terms results in groundwater's non-turbulent flow (also known as creeping flow and, in some contexts, as laminar flow). The absence of turbulence reduces the potential for mixing.

The ratio of momentum to viscous terms can be quantified by Reynolds number:

$$R_e = \frac{UL}{\nu} \qquad \text{(Eq. 3.8)}$$

where U is mean velocity, L is a length representative of the scale over which velocity varies (which is an average pore or grain size), and ν is kinematic viscosity. In essence, the Reynolds number represents the balance of momentum, or inertia, and the drag imposed by the viscosity of the fluid. At Reynolds numbers greater than 1, momentum is of greater importance, leading to turbulent flow and creating additional energy losses.

For the most part, subsurface flow is dominated by viscosity effects, and flow is therefore non-turbulent. For example, for $U = 10^{-5}$ m/s, $L = 10^{-4}$ m, and $\nu = 10^{-6}$ m^2/s, $R_e = 10^{-3}$. This would be a typical value for Reynolds number in commonly encountered porous geological media, except perhaps near extraction or injection wells, where R_e may be of the order of 1. The Reynolds number can also be 1 or higher in coarse-gravel beds, in fractures in rock, or in karstic channels created through dissolution of limestone.

Although water is certainly not as viscous as other fluids, such as glycerol or molasses, it must be emphasized that what matters in terms of flow behavior is the dimensionless Reynolds number. So, at Reynolds numbers less than 1, water in porous media behaves in a very viscous way, reminiscent of molasses dripping from a spoon. Even though velocity in a porous medium is highly variable due to the complexity and tortuosity of flow paths in the pore space, the flow is not turbulent. The absence of turbulence in such a flow suppresses mixing. Thus mixing in the aqueous phase in porous media is orders of magnitude slower than mixing in rivers, lakes, or oceans.

The flow of gases in the unsaturated zone can be studied following similar principles. However, while water is practically incompressible, the compressibility of air must be taken into account.

3.4.2 Advection

Advection is, in terms of rates, the most important transport mechanism in the subsurface for carrying water and the chemicals it may contain. Advection is the transport of a solute (e.g., a contaminant or an oxidant) along with the flow of the solvent (water). Consider the horizontal flow of water through a vertical plane normal to the flow. The transport of solute mass Q_a, expressed as mass per time, is:

$$Q_a = c\, u\, A \qquad \text{(Eq. 3.9)}$$

where c [M/L^3] is solute concentration, A [L^2] is the area of the surface, and u[L/T] is velocity normal to the surface.

The maximum rate of advection is limited by the solubility because it governs the contaminants' maximum concentration. Let us put some numbers into this equation by considering the advection of dissolved oxygen. The solubility of oxygen in freshwater at 15°C and atmospheric pressure is about 10 mg/L. Take $u = 0.1$ m/day (a typical natural gradient average flow rate) and consider the flux through an area equal to 1 m^2. Then,

$$Q_a = 10\ \text{g/m}^3 \times 10^{-1}\ \text{m/day} \times 1\text{m}^2 \sim 1\text{g/day}$$

This maximum rate of advection of oxygen might be too low for many remediation applications. In engineered remediation systems, advection can be enhanced by increasing the velocity and using more soluble additives. Nitrate salts, which can be effective oxidizers, are quite soluble compared to oxygen, and most sulfate salts are also. However, there may be mass-transport limitations caused by the products formed from the reactants. It is possible to have a high concentration of nitrate, which acts as an oxidant, but the product – nitrogen gas (N_2) – is about as insoluble as oxygen. The excess nitrogen can form bubble accumulation, which is generally undesirable because it blocks the flow of water through the pore space. Sulfate can be reduced to form toxic sulfide.

It is common in mathematical modeling to use the specific discharge rate q_a, which is discharge normalized per area, and to express it at a point in the x, y and z directions:

$$q_{ax} = c\, u,\ q_{ay} = c\, v,\ q_{az} = c\, w \qquad \text{(Eq. 3.10)}$$

where u, v, w are the velocity components in the x, y, z directions.

3.4.3 Molecular Diffusion

Molecular diffusion is a consequence of random movements of molecules that manifests itself in the net flux of mass from areas of higher concentration (relative surplus of molecules) to areas of lower concentration (relative deficit). The diffusive flux, Q_d, through a surface of area, A, is expressed in Fick's first law:

$$Q_d = -D_d \frac{dc}{dn} A \qquad \text{(Eq. 3.11)}$$

where Q_d [M/T] is mass per unit time; D_d [L^2/T] is the molecular diffusion coefficient; c [M/L^3] is solute concentration; A [L^2] is surface area; and n [L] is a coordinate along an axis normal to the surface. The rate of change of concentration in the direction normal to the surface (gradient) is expressed through $\frac{dc}{dn}$. The negative sign in Equation 3.11 indicates that the discharge takes place in the direction of decreasing concentration. Thus, if the concentration gradient is negative, the mass flux is positive.

The diffusion coefficient depends on molecular size and temperature. One may consult standard handbooks, such as Linde et al. (2006), or may visit websites such as the USEPA On-line Tools for Site Assessment Calculation (http://www.epa.gov/athens/learn2model/part-two/onsite/estdiffusion.html) to find estimates of diffusion coefficients of various substances dissolved in water. At first approximation, for many commonly encountered solutes with lower molecular weights of 100 or so, one can use $D_d = 10^{-9}$ m^2/s as a representative value of aqueous-phase molecular diffusion. Higher molecular weight molecules have lower diffusion coefficients.

Combining Fick's law with the law of conservation of mass, we find that concentration in one dimensional space follows a well-studied equation (see, for example, Crank, 1975), which is sometimes referred to as "Fick's second law" or "heat-conduction equation":

$$\frac{\partial c}{\partial t} = D_d \frac{\partial^2 c}{\partial x^2} \quad \text{(Eq. 3.12)}$$

To illustrate an important feature, consider that initially all the mass M of a given solute is distributed uniformly over a cross section at $x = 0$, in quiescent water and away from boundaries and other disruptions. Then, at time t, the concentration follows a bell-shaped (known as Gaussian) distribution:

$$c(t,x) = \frac{M}{\sqrt{4\pi D t}} \exp\left(-\frac{x^2}{4Dt}\right) \quad \text{(Eq. 3.13)}$$

By plotting this expression as a function of location x and following how the mass spreads over time, it is enlightening to evaluate how slow the spreading is, as a consequence of the smallness in the molecular diffusion. To quantify the spreading, we can use the mean square distance of mass from the center of symmetry of the distribution, which follows the simple formula, which can be derived from Equation 3.13,

$$\sigma^2 = 2D_d t \quad \text{(Eq. 3.14)}$$

The quantity σ has units of length and characterizes the spread of the mass distribution. Let us see how much time it takes for σ to attain certain values. To become 1 mm, it takes about 8 min; for 1 cm, it takes about 14 h; for 1 m, it takes about 16 years! It is clear that molecular diffusion is an ineffective spreading or mixing mechanism, unless the distances involved are quite small. Even in small laboratory-scale batch reactors, complete mixing cannot be achieved simply through diffusion alone, but requires rigorous mechanical mixing.

The specific diffusive discharge can be obtained by normalizing the mass discharge of a solute by the area through which it passes and can be written in terms of three components corresponding to the coordinate system x, y, z:

$$q_{dx} = -D_d \frac{\partial c}{\partial x}, \; q_{dy} = -D_d \frac{\partial c}{\partial y}, \; q_{dz} = -D_d \frac{\partial c}{\partial z} \quad \text{(Eq. 3.15)}$$

By combining with mass conservation, we obtain the shortened vector representation:

$$\frac{\partial c}{\partial t} = D_d \nabla^2 c \qquad \text{(Eq. 3.16)}$$

In applications, accounting for both advective and diffusive transport produces the *advection-diffusion equation:*

$$\frac{\partial c}{\partial t} + \nabla(\mathbf{u}c) - D_d \nabla^2 c = 0 \qquad \text{(Eq. 3.17)}$$

in vector-notation form or, in the equivalent scalar-notation form

$$\frac{\partial c}{\partial t} + \frac{\partial (uc)}{\partial x} + \frac{\partial (vc)}{\partial y} + \frac{\partial (wc)}{\partial z} - D_d \left(\frac{\partial^2 c}{\partial x^2} + \frac{\partial^2 c}{\partial y^2} + \frac{\partial^2 c}{\partial z^2} \right) = 0 \qquad \text{(Eq. 3.18)}$$

Because molecular diffusion in liquids is so slow, it is often the step that limits the rate of many important processes. A list of processes, systems, and reactions where molecular-diffusion mass transfer may play a central role is given in Table 1.1 of Weber and DiGiano (1996). One example of an important situation where molecular diffusion is the rate-limited step is the diffusion of oxygen (or other dissolved gases) at air-water interfaces. Even though the equilibration between the air and gas phases may be practically instantaneous, the net rate of mass transfer through the interface is limited by the slow molecular diffusion of oxygen in water.

Another example of molecular diffusion control of process rates is adsorption and desorption on solid phases. In this case, the process of attachment or detachment of ions at exchange sites may be fast, but the net rate of adsorption or desorption may be quite slow because ions must be transported in the aqueous phase through diffusion, often through very small apertures. A third example, as we have already seen, is bimolecular reactions (such as the degradation of contaminants by chemical oxidants) that are fast once the two reactants are mixed, as is typical through stirring in a batch reactor.

3.5 LABORATORY-SCALE PROCESSES

Consideration of flow and transport processes at the pore-scale is instructive and perhaps appealing because it involves first principles and processes that are well understood. However, description of processes at such a fine scale of resolution is not practical in applications because (1) there is not enough information to describe in detail the pore structure, (2) it is impossible to perform computations involving so much complexity, and (3) practitioners are interested in bulk averages that better represent the overall behavior of a natural or engineered system.

Consider support volumes with dimensions of at least a few centimeters so that they contain a very large number of grains and pores. Such may be the volume of the porous medium in a permeameter, which is a device used to evaluate the overall ability of a porous medium to convey a liquid, or the volume of a laboratory column, used to evaluate transport and reaction properties of solutes. It could also be the volume of an individual element used in the mathematical modeling technique known as finite elements; in this method, the actual domain is subdivided into a mosaic consisting of tiles known as "elements." Instead of representing the microstructure within these volumes, the material is treated as homogeneous with properties (such as the ability to contain water or to allow flow) that describe bulk properties.

More generally, while it is understood that a granular porous material is composite (consisting of solid grains and pore spaces filled with water) at the microscopic scale, it is found expedient to represent it as a simple material with continuously varying features.

This macroscopic representation makes it possible to adopt standard continuum modeling techniques, such as partial differential equations to describe the evolution of flow and transport variables.

The idealized porous medium would be something like clean sand with grains made of a relatively inert material like quartz – precipitation and dissolution reactions can be neglected and the medium is unaffected by water flow or solute concentrations – and relatively homogeneous, such as the sand found in sand dunes or aeolic formations. Because of the tremendous complexity in the flow and variability in concentration values, we focus on volume or surface averages. These averages are over regions that are much larger than the characteristic dimension of a grain and, thus, each average is not affected by the details of the specific region but is representative of the medium.

If we take a sample of aquifer with total volume V, it will include a volume V_p of interconnected pores. The ratio

$$\eta = \frac{V_p}{V} \qquad \text{(Eq. 3.19)}$$

is called the porosity. A typical value of porosity is 0.35, but in every application case-specific values need to be determined through laboratory tests or inferred from field data (API, 1956; Bass, 1987; Dullien, 1979; Hearst et al., 2000). The porosity is useful in evaluating how much water is stored within a volume of porous material saturated with water, but it is not indicative of how easily the flow takes place within this volume.

Before we examine the question of conductivity, how easily a fluid can flow through a porous medium, let us define some useful quantities. Consider flow in a cylinder packed with a porous medium (Figure 3.5). The flow through the cross-section with total area A, in m^2, is

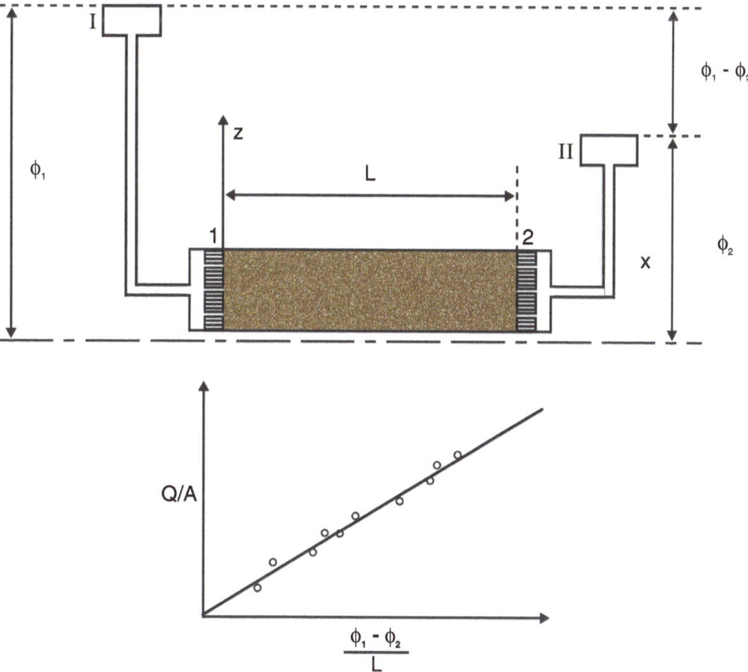

Figure 3.5. Constant-head permeameter test.

described by the (total) discharge Q, in m^3/s. Next, divide the discharge by the area to determine the specific discharge:

$$q = \frac{Q}{A} \tag{Eq. 3.20}$$

The specific discharge is an areal average of discharge. Although the specific discharge has units of velocity, m/s, and is often referred as Darcy velocity, it is not really a velocity (in the sense of swiftness of movement). The reason is that q has been obtained by dividing by the total cross-sectional area, A, and not the area over which flow takes place, which is ηA. To calculate what is known as the mean linear velocity, use:

$$u_a = \frac{Q}{\eta A} = \frac{q}{\eta} \tag{Eq. 3.21}$$

The specific discharge can be defined over a volume or over a cross-section. Similarly, the hydraulic head can be averaged over a volume or a surface.

3.5.1 Darcy's Law

Figure 3.5 shows a typical setup for evaluating the conductivity of a porous medium. The porous medium is contained within a cylinder with porous end caps that provide little resistance to flow. Each end is connected to a constant-head tank with different surface elevations. Practically all energy losses take place within the porous medium while the interconnected vessels that do not contain any porous medium satisfy (for all practical purposes) the hydrostatic relation, i.e., the hydraulic head ϕ is constant.

The porous medium specimen, with cross-sectional area A and length L is subjected to head difference, $\Delta\phi = \phi_1 - \phi_2$. Repeated experiments reveal that the discharge is proportional to A and the head difference, $\Delta\phi$, and is inversely proportional to the length L:

$$Q = kA\frac{\Delta\phi}{L} \tag{Eq. 3.22}$$

or

$$q = k\frac{\Delta\phi}{L} \tag{Eq. 3.23}$$

Darcy's law effectively states that the specific discharge, $q = \frac{Q}{A}$, is proportional to the hydraulic head gradient, $\frac{\Delta\phi}{L}$; the coefficient of proportionality, k, is called the hydraulic conductivity and has dimensions of velocity, such as meters per second (m/s). The value of the hydraulic conductivity depends on the characteristics of both the porous medium and the fluid that flows through it. Darcy's law can be justified on the basis of experimental evidence, but it is also an immediate consequence of Stokes flow and can be obtained from it through a process known as upscaling (or homogenization or coarse-graining), e.g., Whitaker (1986). Derivation through upscaling is useful because it verifies that Darcy's law is consistent with pore-scale hydrodynamics and the first principles of mass conservation, Newton's law of friction, and Newton's second law of motion, under conditions of low Reynolds number and gradually varying flow. These conditions are satisfied to a reasonable degree for almost all porous-media flows encountered in practice. The derivation also reveals that the conductivity can be expressed as follows:

$$k = \frac{\kappa g}{\nu} \tag{Eq. 3.24}$$

The parameter κ, which has units of length square, is called the intrinsic permeability and depends on the size distribution and packing (arrangement) of the grains. The hydraulic conductivity is proportional to the intrinsic permeability of the porous medium and is inversely proportional to the kinematic viscosity of the fluid. Thus, the conductivity is affected somewhat by temperature, since the kinematic viscosity of water decreases at higher temperature (see Linde (1990) and other handbooks or online tools and compilations of physical data). The admixture of certain chemicals may affect the viscosity. For example, addition of certain water-soluble polymers can increase the kinematic viscosity of water and thus reduce the hydraulic conductivity.

For isotropic media, conductivity is the same in all directions and the flow is along the path where the hydraulic head decreases the fastest (the negative gradient of the head). The hydraulic conductivity can be represented through a single number. However, most porous media are anisotropic due to the orientation of grains or micro layers of heterogeneities (e.g., clay laminae in sand). The conductivity is highest in the direction parallel to the stratification and lowest in the direction transverse to the stratification. The conductivity here cannot be represented through a single number (a scalar) but rather through a set of numbers that comply with certain conditions (a tensor). In anisotropic media, the flow tends to be in a direction between the direction of steepest decrease of hydraulic head and the direction of highest conductivity.

Textbooks (e.g., Bear, 1972; Freeze and Cherry, 1979) contain tables of representative hydraulic conductivity values for subsurface media. Because these values vary over many orders of magnitude, in any specific application, one must determine conductivity values based on representative data. Conductivity can be determined in the laboratory using constant- or falling-head permeameter tests, using a principle similar to the one shown on Figure 3.5. However, due to the challenges associated with obtaining undisturbed core samples and the fact that the samples yield values that may not be representative of the conductivity at a larger scale, conductivities are often estimated from well tests or other indirect measurements.

Note that Darcy's equation describes flow at a coarser scale than Stokes' equation. While Stokes equation is expressed in terms of the finely resolved and highly variable flow velocity within pores, $\mathbf{u}$, Darcy's equation is expressed in terms of the coarsely resolved and smoothly varying specific discharge. In practical applications, Darcy's law is the foundational law of flow in porous media.

3.5.2 Diffusion

We now turn our attention to solute transport at the laboratory scale. Consider a block of a porous medium, saturated with water but without flow. Then, diffusive mass transport of a non-sorbing solute is due to a difference in concentration at two opposing ends. The net diffusive flux is (see Bear, 1972; Freeze and Cherry, 1979; Maerki, et al., 2004):

$$Q = -\frac{D_d \eta}{\tau_p} \frac{\Delta c}{\Delta L} A \qquad \text{(Eq. 3.25)}$$

where D_d is the aqueous-phase molecular diffusion coefficient, η is the porosity, τ_p is a dimensionless number larger than 1, called the tortuosity coefficient, Δc is the concentration difference between the two opposing surfaces of area A that are at distance L. The expression $D_d\, \eta/\tau_p$ is the effective diffusion coefficient that accounts for the transport taking place only in the aqueous phase, which occupies a volume equal to the porosity times the total volume, as well as for the effect of tortuosity of paths within the porous medium (travel paths are, in fact, longer as molecules are obstructed by and must move around mineral grains). For relatively impervious formations, such as clays or nonporous rocks with a hydraulic conductivity of $k < 10^{-6}$ m/s, molecular diffusion is the foremost transport mechanism. In these cases, it is important to correctly estimate the effective diffusion coefficient. In more permeable formations, advection dominates transport, and diffusion is less important than dispersion (which will be discussed next), so the simplifying assumption that $\tau_p = 1$ is often made.

3.5.3 Advection-Dispersion Equation

At the coarser scale of resolution we called the laboratory scale, transport of nonreactive solutes can be broken down into advection, which means transport with the mean flow velocity, and dispersive (as contrasted with diffusive) transport, which enhances spreading and mixing. Under certain conditions, the spreading and mixing take place in a way analogous to diffusion by following a Fickian type law. So, transport is described through the advection-dispersion equation:

$$\frac{\partial c}{\partial t} + \nabla \cdot (\mathbf{U}c - \mathbf{D}\nabla c) = 0 \qquad \text{(Eq. 3.26)}$$

Note that this equation appears quite similar to the advection–diffusion equation (Equation 3.17). However, the following differences must be kept in mind:

- The concentration in this equation is defined over a bigger support volume and thus varies more slowly than the concentration in the pore-scale advection–diffusion equation.
- The velocity $\mathbf{U}$ is averaged over a volume that contains many pores, so it should not be confused with the highly fluctuating velocity $\mathbf{u}$ within individual pores.
- The rate of dispersion is quantified by a tensor of the dispersion coefficients, $\mathbf{D}$, which generally is dominated by variability in advective velocity $\mathbf{u}$, but is also affected by diffusion.

The influences of advection and diffusion on dispersion at this scale can be conceptualized by considering two molecules or ions that start in the same streamline, or on neighboring streamlines. Due to advective forces, the average distance between two particles in neighboring streamlines will increase over time, because the average velocity is different along the two streamlines. However, the effect of diffusion is apparent when we realize that even two particles that start on the same streamline will end up on different streamlines because diffusion shifts them over small distances; even these small effects can result in two particles that start from the same location to experience completely different travel paths.

The dispersion tensor expresses that dispersion is highly anisotropic (direction-dependent), being affected by the direction and correlation structure of velocity fluctuations $\mathbf{u}$–$\mathbf{U}$. Rates of dispersion are much higher in the direction of mean flow (known as the longitudinal direction) than in the other direction that are orthogonal to the direction of flow (known as transverse directions). That is why, over time, plumes of conservative solutes in natural-gradient flow fields become elongated with time (see Mackay et al., 1986 for example). The tensor is second-order symmetric, positive definite (it has positive eigenvalues and the directions of maximum and minimum dispersion rates form a Cartesian coordinate system) and increases with the mean velocity. More details can be found in Bear (1972) and Scheidegger (1974). A useful parameterization is that by Scheidegger (1961), in terms of the mean velocity $U = \|\mathbf{U}\|$ and dispersivity lengths, which are properties of the medium. In the principal system where the first axis is aligned with the flow and the other two axes are the axes of maximum transverse and minimum transverse dispersion:

$$
\begin{aligned}
D_L &= \alpha_L U + \frac{1}{\tau_p} D_d \\
D_{T1} &= \alpha_{T1} U + \frac{1}{\tau_p} D_d \\
D_{T2} &= \alpha_{T2} U + \frac{1}{\tau_p} D_d
\end{aligned}
\qquad \text{(Eq. 3.27)}
$$

where α_L is the longitudinal dispersivity, α_{T1} and α_{T2} are transverse dispersivities, D_d is the aqueous-phase diffusion coefficient, and τ_p accounts for tortuosity. The lengths α_L, α_{T1}, α_{T2} and the dimensionless number τ_p are parameters of the porous medium; D_d of the solute/solvent pair; and U depends on the flow field.

The larger the scale over which homogenization is performed, the larger the value of the longitudinal dispersion because it represents the effects of more velocity variability that is averaged out. However, the transverse dispersivity does not increase much by increases in the scale of the problem (Dagan, 1989; Gelhar, 1993). That is part of the reason there is disparity in the estimates of the ratio $\frac{\alpha_L}{\alpha_T}$ reported in the literature. Representative values can be found in Anderson (1979).

3.5.4 Dual-Porosity Models

The advection-dispersion equation (Equation 3.26) can be obtained through a process of averaging (known as upscaling, homogenization, or coarse-graining) from the pore-scale advection–diffusion equation (Equation 3.17) and the geometry of the micro-structure. One then obtains a macroscopic description that involves variables, such as concentrations, that vary much more gradually that the variables involved in the microscopic description. However, to obtain Equation 3.26, certain conditions must be met. One of them is that the concentration values at the inflow boundary change slowly so that diffusion at the pore scale can be effective in distributing the mass among low- and high-velocity regions within the support volume over which homogenization is implied. Then, the advective velocity of the solute mass is identical to the mean flow velocity and the rate of spreading may be expressed through the dispersion coefficients.

However, there are many cases in practice that do not satisfy these requirements and, as a consequence, Equation 3.26 is not an accurate representation of transport. A prominent example is fractured media, where most of the flow takes place in a network of fractures, the mobile zone, while most of the water is stored in a low-conductivity mineral matrix, the immobile zone. When solute mass first gets into a block of such a fractured medium through the inflow boundary, almost all of the mass is concentrated within the fractures where velocity is relatively high. The average velocity experienced by the mass is higher than the mean velocity of water in the block, and this is reflected in the fast breakthrough of mass at the exit of the block. Given enough time, under constant concentration at the inflow boundary, the concentration levels in the mobile and the immobile zone equilibrate and the mean advective velocity is the same as the water velocity averaged over the whole block. But then, when the concentration at the inflow boundary is diminished, the mass diffuses back into the mobile zone slowly, resulting in mass appearing at the outflow boundary long after the concentration has become zero at the inflow boundary. This slow diffusion is responsible for the long-term "tailing" of breakthrough curves that is commonly observed in tracer tests (see Haggerty et al., 2000).

The time needed for equilibration between the mobile and immobile zones in the case of fractured media can be on the order of months or years. If conditions change more rapidly, the simple advection-dispersion model may fail to capture essential features of the system behavior. The same holds true if chemical transformations prevent equilibration between mobile and immobile zones; for example, transformations of a solute may take place in mobile zones only while the solute source may be in immobile zones. These issues are not limited to strictly fractured media. For partially consolidated media and even for media that at first appearance are characterized as unconsolidated granular porous media, a relatively long time (hours or days) may be needed for equilibration through diffusion of concentration in preferential flow paths and in the relatively stagnant dead-end zones, caused by small-scale heterogeneities.

For such cases, a transport model that involves dual (or multiple) porosity zones, originally proposed by Warren and Root (1963), is useful, practical, and comes with some justification based on first principles (Arbogast et al., 1990; Saez, et al., 1989). This is a dual-continuum model, where one conceptualizes two types of materials filling the space simultaneously: the mobile zone and the immobile zone, with porosities η_m and η_i, respectively. All flow and dispersive transport is assumed to take place in the mobile zone, with advective velocity the specific discharge divided by the mobile-zone porosity. The immobile zone serves only for mass storage. The diffusion-dominated mass transfer between the mobile and immobile zone is represented through a phenomenological relation, often a linear law in which the mass-transfer rate from the mobile to the immobile zone is $\alpha(c_m - c_i)$, where α is a mass transfer coefficient, and c_m, c_i are the concentrations in the mobile and immobile zones, respectively. This model requires two parameters in addition to those of the basic advection-dispersion (the mass transfer coefficient and the break-up of the total porosity into two components).

3.5.5 Sorption

In many applications, the mass exchange between the solid phase and aqueous phase is a crucial process that must be adequately described. Sorption is a term that describes various processes that involve the association of mass with the solid mineral phase of geologic formations. This includes adsorption (onto the media), which indicates that mass adheres to solid surface; absorption (into), which denotes that the bulk of the solid particles is involved, e.g., through diffusion of mass into micropores or organic materials on mineral surfaces; ion exchange, as when positively charged cations replace another cation attracted to a negatively charged mineral surface; chemisorption, when solute mass is incorporated into the mineral through chemical reaction; precipitation and dissolution; and so forth. In this review, we consider adsorption and absorption as they affect transport. As similar models are used to describe them, we will refer to them collectively as sorption.

The most basic models are equilibrium ones that consider that a one-to-one relation (or *isotherm*) is established between the aqueous and solid concentration of a chemical. The linear isotherm results when the ratio between the solid concentration and the aqueous concentration is a constant, called the partition coefficient, commonly denoted by k_d. Other, nonlinear, models include the Freundlich and Langmuir isotherms.

The most prominent effect of equilibrium sorption on transport is retardation (for example, see Bedient et al., 1999; Domenico and Schwartz, 1998). In simple terms, if mass M_a is dissolved in the mobile aqueous phase and mass M_s is associated with the immobile solid phase, then the total mass is retarded (compared to fluid) by a factor equal to $R = \dfrac{M_s + M_a}{M_a}$. For example, at equilibrium, if there would be nine times more mass in the solid than in the aqueous phase, the retardation coefficient is 10; as a consequence, mass balance equations show that the solute plume migrates at a speed ten times slower than the groundwater. The advection-dispersion equation, then, requires only a slight modification, the storage term $\frac{\partial c}{\partial t}$ needs to be replaced by $R\frac{\partial c}{\partial t}$.

The equilibrium sorption model is easy to work with and is often used to make predictions. For example, in a pump-and-treat system, this model suggests that R times more water volume would have to be extracted for treatment compared to the case of no sorption. Unfortunately, in most cases, this prediction turns out to be wildly overoptimistic. The rate of desorption is generally anything but instantaneous, and it might take a long time for equilibrium to be reached between the concentrations in the solid and aqueous phases. It has been observed during pump-and-treat systems that the concentration of a contaminant in the aqueous phase may be reduced to below the maximum contaminant level (MCL) while most of the contaminant concentration is still in the solids. Once the pumping stops, the concentration in the aqueous phase gradually rebounds as equilibrium is slowly approached through transfer of mass from the solid phase.

The rate of desorption may be the rate-limiting mechanism for many remediation technologies that involve removing and treating contaminants at the surface or *in situ* treatment. The kinetics of sorption can be described through phenomenological models in ways analogous to those for treating immobile aqueous phases, and a practical empirical model is a reversible linear mass transfer in which the rate of transfer from the aqueous to the solid is $\beta(k_d c_a - c_s)$, where β is a mass transfer coefficient, c_a, c_s are concentrations, and k_d is the partition coefficient (for equilibrium described by a linear isotherm).

The rate-limited sorption behavior has several differences from the equilibrium one (e.g., see Michalak and Kitanidis, 2000). First, the solute retardation is not constant. Consider that a treatment technology may remove mass from the aqueous phase; then, most of the remaining mass is in the solid phase causing the effective retardation coefficient to become very large. That is why pump-and-treat is ineffective for slowly desorbing contaminants. The removal rate is controlled not by the rate of pumping but by the kinetics of desorption, which is the rate-limiting mechanism. A second important difference from the equilibrium case is that kinetically controlled sorption processes enhance spreading and mixing. Thus, sorption may affect to some extent the rate of bimolecular reactions that are controlled by the rate of mixing of the reactants.

In addition to sorption, which is exchange between the aqueous and a solid mineral phase, transport may be affected by the presence of separate gaseous or liquid phases. In this case, the portioning is between the mobile aqueous phase and an immobile gaseous phase (such as nitrogen gas bubbles) or liquid phase (such as a separate pool of trichloroethene [TCE] or vegetable oil).

3.6 FIELD-SCALE PROCESSES

In many practical applications, flow and transport processes must be represented and understood at scales much larger than the laboratory scale of a permeameter or a packed column. Such applications include studies of intrinsic and engineered remediation at specific sites. Analytical models often represent large domains of geologic media as homogeneous, with equations that usually mimic equations applicable to the laboratory scale (like the advection-dispersion equation or dual-porosity models) and with parameters that somehow capture the overall behavior of the system. Numerical models, such as finite volume or finite element models, discretize the domain into supposedly homogeneous blocks that are also quite large and actually consist of heterogeneous media.

Considerable progress has been made recently in improving our understanding regarding the applicability of conventional models to express large-scale behavior with constant or piece-wise constant parameters. A variety of mathematical approaches have been used (for example, Cushman, 1997; Dagan, 1989; Gelhar, 1993; Rubin, 2003) and field experiments have been performed to evaluate the validity of such models (for example, Garabedian et al., 1991; Mackay et al., 1986). Some important practical lessons can be gleaned from these works.

First, there is a significant difference between flow and transport problems: generally, flow problems are more tractable. In other words, the requirement for a large block to effectively satisfy Darcy's law is that flow conditions change slowly compared to the time required for head fluctuations to dissipate over the volume of the block (see Dagan, 1982a, b; Kitanidis, 1990). Because head fluctuations propagate relatively quickly (for example, the diffusion coefficient of such fluctuations in the aqueous phase is given by the kinematic viscosity, which is about 10^{-6} m^2/s), this requirement is met in many of the cases encountered in practice.

In solute transport problems, however, the speed with which fluctuations diffuse is controlled by molecular diffusion (the molecular diffusion coefficient in the aqueous phase being about 10^{-9} m^2/s, three orders of magnitude less than kinematic viscosity). It is worth noting that although advection is an important transport mechanism, by itself it does not distribute mass over bigger volumes or smooth out fluctuations in concentration. The slowness in diffusion, which is responsible for so many other phenomena, is also a major contributor to difficulties in upscaling.

For example, macrodispersion theories (Dagan, 1989; Gelhar, 1993) have demonstrated that in natural-gradient flow in a formation with statistical regularity in the rise and fall in values of hydraulic conductivity, as for example in the Borden aquifer (Freyberg, 1986), transport can be macroscopically described through an advection-dispersion equation. However, such conditions are often not satisfied in other applications; even when they are, the time scales required for model results to become good approximations of real conditions are quite long as they are controlled by diffusion (Kapoor and Gelhar, 1994; Kapoor and Kitanidis, 1998). As a consequence, the macro-advection-dispersion equation with constant velocity and dispersion coefficients may be a poor model for simulating transport in many remediation problems.

Another important issue is that the type of measurements that are most often used in applications are representative of large-scale behavior and may provide little definite information about heterogeneity in the system. For example, the effective porosity and dispersion coefficients obtained from a two-well tracer test are highly dependent on the way the test was conducted and the data were fitted, rather than be intrinsic physical properties of a supposedly homogeneous formation. For instance, a fast breakthrough in a two-well tracer test may be interpreted as indicative of low porosity when in reality it may be caused by stratification that results in a high degree of non-uniformity of advection velocities. In another example, a high dispersion coefficient may be fitted to a model that assumes one-dimensional transport with constant velocity when the actual transport is in a three-dimensional domain with highly nonuniform velocity. Simplified macroscopic models should be used with caution.

Numerical models can offer more realistic representations of actual processes, including accounting for heterogeneity in properties or non-equilibrium in sorption, but they may have many parameters that are hard to determine based on data. In practice, one may need to employ both simple and elaborate models, collect and interpret data carefully, and remain mindful of complexity. More information about models can be found in Chapter 4.

3.7 CONCLUDING REMARKS

A sound understanding of flow and physical transport mechanisms is important not only in order to interpret correctly conservative nonreactive tracer tests, but also because flow and transport often control, through mixing processes, the reaction rates observed in the field. In particular, the overall reaction rates are often controlled by diffusive transport of one type or another. The success of remediation schemes may hinge on correctly appreciating and overcoming these limitations.

REFERENCES

Anderson MP. 1979. Using models to simulate the movement of contaminants through groundwater flow systems. CRC Crit Rev Environ Control, pp 97–156.

API (American Petroleum Institute). 1956. API Recommended Practice for Determining Permeability of Porous Media. RP-27. API, Dallas, TX, USA. 30 p.

Arbogast T, Douglas J Jr, Hornung U. 1990. Derivation of the double porosity model of single phase flow via homogenization theory. SIAM J Math Analysis 21:823–836.

Bass DM Jr. 1987. Properties of reservoir rocks. In Bradley HB, ed, Petroleum Engineering Handbook. Society of Petroleum Engineers of AIME, Richardson, TX, USA.

Bear J. 1972. Dynamics of Fluids in Porous Media. American Elsevier, New York, NY, USA. 764 p.

Bear J, Verruijt A. 1987. Modeling Groundwater Flow and Pollution with Computer Programs for Sample Cases. Reidel, Dordrecht, The Netherlands. 414 p.

Bedient PB, Rifai HS, Newell CJ. 1999. Ground Water Contamination: Transport and Remediation. Prentice Hall PTR, Upper Saddle River, NJ, USA. 604 p.

Charbeneau RJ. 2006. Groundwater Hydraulics and Pollutant Transport. Waveland Press Inc., Long Grove, IL, USA. 593 p.

Charbeneau RJ, Bedient PB, Loehr RC. 1992. Groundwater Remediation, CRC Press, Boca Raton, FL, USA. 188 p.

Chu M, Kitanidis PK, McCarty PL. 2005. Modeling microbial reactions at the plume fringe subject to transverse mixing in porous media: When can the rates of microbial reaction be assumed to be Instantaneous? Water Resour Res 41, W06002, 10.1029/2004WR003495.

Cirpka OA, Valocchi AJ. 2007. Two-dimensional concentration distribution for mixing-controlled bioreactive transport in steady state. Adv Water Resour 30:1668–1679.

Crank J. 1975. The Mathematics of Diffusion, 2nd ed. Oxford University Press, Oxford, United Kingdom. 347 p.

Cushman JH. 1997. The Physics of Fluids in Hierarchical Porous Media: Angstroms to Miles. Kluwer Academic Publishers, Dordrecht, The Netherlands. 467 p.

Dagan G. 1982a. Analysis of flow through heterogeneous random aquifers: 2. Unsteady flow in confined formations. Water Resour Res 18:1571–1585.

Dagan G. 1982b. Stochastic modeling of groundwater flow by unconditional and conditional probabilities: 2. The solute transport. Water Resour Res 18:835–848.

Dagan G. 1989. Flow and Transport in Porous Formations. Springer-Verlag, New York, NY, USA. 465 p.

Davis JA, Yabusaki SB, Steefel CI, Zachara JM, Curtis GP, Redden GD, Criscenti LJ, Honeyman BD. 2004. Assessing conceptual models for subsurface reactive transport of inorganic contaminants. EOS Trans Am Geophys Union 85:449–455.

deMarsily G. 1986. Quantitative Hydrogeology: Groundwater Hydrology for Engineers. Academic Press, San Diego, CA, USA. 440 p.

Domenico PA, Schwartz FW. 1998. Physical and Chemical Hydrogeology, 2nd ed. John Wiley, New York, NY, USA. 506 p.

Dullien FAL. 1979. Porous Media: Fluid Transport and Pore Structure. Academic Press, San Diego, CA, USA. 396 p.

Fetter CW. 1998. Contaminant Hydrogeology, 2nd ed. Prentice Hall Inc, Upper Saddle River, NJ, USA. 500 p.

Fetter CW. 2000. Applied Hydrogeology, 4th ed. Prentice Hall Inc, Upper Saddle River, NJ, USA. 598 p.

Freeze RA, Cherry JA. 1979. Groundwater. Prentice-Hall Inc, Englewood Cliffs, NJ, USA. 604 p.

Freyberg DL. 1986. A natural gradient experiment on solute transport in a sand aquifer: 2. Spatial moments and the advection and dispersion of nonreactive tracers. Water Resour Res 22:2031–2046.

Garabedian SP, LeBlanc DR, Gelhar LW, Celia MA. 1991. Large-scale natural gradient tracer test in sand and gravel, Cape Cod, Massachusetts: 2. Analysis of spatial moments for a nonreactive tracer. Water Resour Res 27:911–924.

Gelhar LW. 1993. Stochastic Subsurface Hydrology. Prentice Hall, Englewood Cliffs, NJ, USA. 390 p.

Haggerty R, McKenna SA, Meigs LC. 2000. On the late-time behavior of tracer test breakthrough curves. Water Resour Res 36:3467–3479.

Hearst JR, Nelson PH, Paillet FL. 2000. Well Logging for Physical Properties, 2nd ed. Wiley, New York, NY, USA. 483 p.

Hemond HF, Fechner EJ. 1999. Chemical Fate and Transport in the Environment, 2nd ed. Academic Press, San Diego, CA, USA. 433 p.

Huang WE, Oswald SE, Lerner DN, Smith CC, Zheng C. 2003. Dissolved oxygen imaging in a porous medium to investigate biodegradation in a plume with limited electron acceptor supply. Environ Sci Technol 37:1905–1911.

Kapoor V, Gelhar LW. 1994. Transport in three-dimensionally heterogeneous aquifers: 2. Predictions and observations of concentration fluctuations. Water Resour Res 30:1789–1801.

Kapoor V, Kitanidis PK. 1998. Concentration fluctuations and dilution in aquifers. Water Resour Res 34:1181–1193.

Kitanidis PK. 1990. Effective hydraulic conductivity for gradually varying flow. Water Resour Res 26:1197–1208.

LeBlanc DR, Garabedian SP, Hess KM, Gelhar LW, Quadri RD, Stollenwerk KG, Wood WW. 1991. Large-scale natural gradient tracer test in sand and gravel, Cape Cod, Massachusetts: 1. Experimental design and observed tracer movement. Water Resour Res 27:895–910.

Linde D. 1990. CRC Handbook of Chemistry and Physics, 71 ed. CRC Press, Boca Raton, FL, USA.

Linde N, Finsterle S, Hubbard S. 2006. Inversion of tracer test data using tomographic constraints. Water Resour Res 42, W04410, doi:10.1029/2004WR003806.

Mackay DM, Freyberg DL, Roberts PV, Cherry JA. 1986. A natural gradient experiment on solute transport in a sand aquifer: 1. Approach and overview of plume movement. Water Resour Res 22:2017–2029.

Maerki M, Wehrli B, Dinkel C, Mueller B. 2004. The influence of tortuosity on molecular diffusion in freshwater sediments of high porosity. Geochimica et Cosmochimica Acta 68:1519–1528.

Michalak AM, Kitanidis PK. 2000. Macroscopic behavior and random-walk particle tracking of kinetically sorbing solutes. Water Resour Res 36:2133–2146.

Roberts PV, Mackay DM. 1986. A Natural Gradient Experiment on Solute Transport in a Sand Aquifer. Technical Report No 292, Department of Civil Engineering, Stanford University, Stanford, CA, USA.

Rubin Y. 2003. Applied Stochastic Hydrogeology. Oxford University Press, Oxford, United Kingdom. 391 p.

Saez AE, Otero CJ, Rusinec I. 1989. The effective homogeneous behavior of heterogeneous porous media. Transport in Porous Media 4:213–238.

Scheidegger AE. 1961. General theory of dispersion in porous media. J Geophys Res 66: 3273–3278.

Scheidegger AE. 1974. The Physics of Flow Through Porous Media, 3rd ed. University of Toronto Press, Toronto, Canada. 353 p.

Semprini L, Roberts PV, Hopkins GD, McCarty PL. 1990. Field evaluation of in-situ biodegradation of chlorinated ethenes: Part 2, Results of biostimulation and biotransformation experiments. Ground Water 28:715–727.

Thierrin J, Kitanidis PK. 1994. Solute dilution at the Borden and Cape Cod groundwater tracer tests. Water Resour Res 30:2883–2890, doi:10.1029/94WR01983.

Thornton SF, Quigley S, Spence MJ, Banwart SA, Bottrell S, Lerner DN. 2001. Processes controlling the distribution and natural attenuation of dissolved phenolic compounds in a deep sandstone aquifer. J Contam Hydrol 53:233–267.

Thullner M, Mauclaire L, Schroth MH, Kinzelbach W, Zeyer J. 2002. Interaction between water flow and spatial distribution of microbial growth in a two-dimensional flow field in saturated porous media. J Contam Hydrol 58:169–189.

Warren JE, Root PJ. 1963. The behavior of naturally fractured reservoirs. Soc Petroleum Eng J 3:245–255.

Weber WJ Jr, DiGiano FA. 1996. Process Dynamics in Environmental Systems. Wiley, New York, NY, USA. 848 p.

Whitaker S. 1986. Flow in porous media I: A theoretical derivation of Darcy's law. Transp Porous Media 1:3–25.

CHAPTER 4

HYDROGEOCHEMICAL MODELS

Albert J. Valocchi[1]

[1]Department of Civil & Environmental Engineering, University of Illinois at Urbana-Champaign, Urbana, IL, USA

4.1 INTRODUCTION

Mathematical models are tools to integrate the processes affecting transport and fate of contaminants in the subsurface. The transport processes of advection and dispersion have been described in the previous chapter, while biogeochemical reaction processes were presented in Chapter 2. Here we adopt the standard continuum or Darcy-scale representation of a porous medium (Bear, 1979) and use the mass balance principle to couple all relevant processes within the quantitative framework of the advection-dispersion-reaction equation. In theory, any reaction or mass-transfer process can be incorporated into the mass balance equations, as long as the process can be described with a suitable mathematical relationship. Mathematical models that couple both hydrologic transport and biogeochemical reaction processes are called "hydrogeochemical models" or "reactive transport models." These models are invaluable tools to aid groundwater management and remediation design decisions. They can be used to improve understanding of the coupling between mixing and reaction processes, and they serve as a framework for interpreting, integrating and synthesizing laboratory and field information. In practice, hydrogeochemical models can be applied to aid site-specific assessments of alternative remediation designs (e.g., different reagent delivery strategies) and make predictions of future system behavior. For site-specific applications, the hydrogeochemical model is a mathematical representation of the site conceptual model, which describes all the key features of the geology, hydrogeology, groundwater flow system, system boundaries, contaminant sources and distribution. See Anderson and Woessner (1992) for further discussion. For reacting chemicals, the conceptual model also must include a description of all the important site-specific geochemical and microbiological reactions (Davis et al., 2004).

In this chapter, we provide an overview and assessment of some of the many available simulation codes for hydrogeochemical modeling. To motivate the need to couple transport, mixing, and reaction processes, we begin by describing some biogeochemical reactions that are important in specific remediation technologies. Then in the following section (Section 4.3), we concisely present the governing equations that are solved in hydrogeochemical modeling codes. Because of the need to couple many interacting reaction processes, these codes can be computationally demanding and so we include a brief discussion of numerical solution strategies. There are numerous commercially available or public domain software codes that can be used for hydrogeochemical modeling; many of these codes are flexible and allow the user to add new reactions. Section 4.4 summarizes a few of the software codes that are available without cost. This section also includes a brief discussion of analytical "screening type" models, which are restricted to simplified hydrogeology (e.g., uniform flow and a homogeneous aquifer) and simplified reactions (e.g., first-order transformation).

P.K. Kitanidis and P.L. McCarty (eds.), *Delivery and Mixing in the Subsurface: Processes and Design Principles for In Situ Remediation*, doi: 10.1007/978-1-4614-2239-6_4, © Springer Science+Business Media New York 2012

Before a hydrogeochemical model can be run to generate simulation results, all parameter values must be specified. Therefore, in Section 4.5 we discuss the challenging issues of parameter estimation, model calibration and model validation. This chapter closes with a presentation of several case studies where hydrogeochemical models have been applied to specific sites.

4.2 MIXING AND REACTION PROCESSES

4.2.1 Overview

As discussed in earlier chapters, mixing of chemicals can play an important role in many remediation operations. To accomplish destruction or transformation of contaminants requires that the contaminants be brought into contact with one or more chemical reactants. In the case of permeable reactive barrier walls (see Chapter 7), a commonly used reactant is a solid phase (e.g., zero-valent iron) that is physically emplaced downgradient of the dissolved contaminant plume. In this case, the contact is ensured as long as the overall groundwater hydraulics continues to force the plume to flow through the barrier, although there are documented cases where preferential flow can occur within the barrier thereby reducing overall contact time and effectiveness (e.g., Benner et al., 2001; Jeen et al., 2007). In most other situations, the contact between contaminant and reactants occurs by mixing of groundwater fluids with different compositions. For some cases of intrinsic biodegradation, the contaminants within the plume mix along the plume boundary with other reactants that are naturally occurring in the ambient groundwater (Tuxen et al., 2006). Many active engineered remediation strategies require direct input of reactants into the contaminated groundwater zone via injection wells, infiltration trenches, physical emplacement of solid phases that slowly release reactants, etc. The overall effectiveness of remediation depends upon both the efficiency of mixing and the rate of relevant geochemical and microbiological reactions that transform the contaminants. Therefore, quantitative models must have the capability to simulate complex three-dimensional, transient flow in heterogeneous porous media as well as a variety of equilibrium and kinetic reactions.

This section gives an overview of some of the important mixing and reaction processes that arise in remediation and hence need to be included in hydrogeochemical models. Many of these have already been introduced in Chapters 2 and 3. In Section 4.2.2 a few remediation methods are described in more detail in order to better illustrate the need to model coupled transport-geochemical-microbiological processes.

Chemical reactants are often introduced using multi-level recirculating well networks, and pulsed injection can be used to enhance mixing and prevent bio-clogging (Gandhi et al., 2002a; Hyndman et al., 2000). Therefore, hydrogeochemical models should be capable of simulating three-dimensional transient flow conditions to handle the full spectrum of field problems, although simplification may be appropriate in specific circumstances. It is well known that aquifer hydraulic conductivity is highly variable over small spatial scales, even in granular unconsolidated aquifers. Moreover, this small-scale variability can have a major impact upon the effectiveness of mixing, since preferential flow paths can develop leading to bypass of any contaminant that is in the lower permeability zones. For intrinsic biodegradation (see Chapter 9) where mixing is primarily due to transverse dispersion along the plume fringes, heterogeneity may lead to enhanced mixing and reaction, and hence a more favorable remediation (Bauer et al., 2009; Cirpka et al., 1999). Transient seasonal changes in the groundwater flow magnitude and direction can also lead to enhanced mixing and reaction (Cirpka, 2005; Prommer et al., 2002), another scenario for which transient flow simulation capability is required.

Another critical process in subsurface transport is inter-phase mass transfer. Water-gas mass transfer is key to remediation by soil-vapor extraction and air sparging (see Chapter 8 for further details). Liquid–solid reactions like sorption and ion-exchange are very common due to the large solid-water interfacial area of porous media. Organic and inorganic contaminants that have been in the subsurface for long periods of time may slowly diffuse into grains and aggregates and sorb to surfaces that are not directly accessible to the flowing pore fluid. During cleanup, contaminant concentrations in the pore fluid can drop relatively quickly, but contaminant concentrations in grains and aggregates may slowly decline due to very slow diffusion and strong sorption processes (Ball and Roberts, 1991). This poses serious problems when remediation requires these slowly desorbing contaminants to mix with injected reactants. For example, Luo et al. (2007) report that desorption of uranium played a significant role in controlling the overall effectiveness of an *in situ* bioreduction experiment at the Oak Ridge Field Research Center facility. On the other hand, under some scenarios sorption processes may theoretically lead to enhanced remediation due to the so-called "chromatographic mixing" effect (Janssen et al., 2006; Oya and Valocchi, 1997). This occurs when the retardation factor of the contaminant differs from that of the injected reactants; this leads to different effective velocities for the different chemical species which can be exploited to mix the reactants and contaminant much more effectively than would be possible by dispersion alone. In particular, when the contaminant has a larger retardation factor than the injected chemical (e.g., in the case of injection of electron acceptors like oxygen or nitrate to stimulate aerobic degradation), the injected front travels at a larger velocity than the retarded contaminant front, leading to a spatial zone where the two fronts overlap with high concentrations of both the contaminant and injected chemical, resulting in enhanced reaction.

Precipitation/dissolution reactions are another important category of mass-transfer reaction. Reactions between input chemical reactants and contaminants or natural constituents may possibly lead to precipitation/dissolution. For example, *in situ* chemical oxidation (ISCO) commonly uses injection of potassium permanganate ($KMnO_4$) to oxidize chlorinated ethenes such as perchloroethene (PCE; also termed perchloroethylene or tetrachloroethylene), trichloroethene (TCE) and dichloroethene (DCE) (see following subsection). This can lead to the precipitation of manganese oxides, which can potentially reduce aquifer permeability (Li and Schwartz, 2004b; Schroth et al., 2001). Iron and manganese solid phases serve as important terminal electron acceptors for biodegradation of petroleum hydrocarbons (Lovley et al., 1989; Wiedemeier et al., 1999). These same solid phases can also react with electron donors that are injected to stimulate reductive dehalogenation reactions (Evans and Koenigsberg, 2001; Pavlostathis et al., 2003). Strategies for remediation of metal-contaminated groundwater strive to change the oxidation state of the metal to a form that precipitates as a relatively insoluble solid. This can be done using *in situ* redox barriers (see following subsection) and reactive permeable barriers (see Chapter 7). Secondary mineral precipitation reactions within permeable barriers used for treatment of organic contaminants can also be important (Yabusaki, 2001).

Mass transfer between entrapped nonaqueous phase liquid (NAPL) and flowing water is important if source zone remediation is being considered (NRC, 2005). This is an area that has been studied extensively and many different models have been proposed to simulate NAPL dissolution (Christ and Abriola, 2007; Parker and Park, 2004; Saenton and Illangasekare, 2007). However, these models all require assumptions about the amount (i.e., volume fraction) and form (i.e., ganglia and pools) of the NAPL source, information that is not generally known in practice. Further discussion on models of NAPL dissolution will be presented in Chapter 10.

4.2.2 Example Remediation Technologies

This section briefly describes some of the important mixing and reaction processes for a few remediation technologies. We only consider a few select technologies, since a comprehensive discussion is beyond the scope of this chapter and many of the relevant mixing processes are similar for different technologies. The first technology is enhanced *in situ* biodegradation in which certain chemicals are deliberately introduced in order to create favorable conditions for biological transformation of target contaminants. For example, through addition of electron donors and nutrients under anaerobic conditions it is possible to stimulate bacteria to transform recalcitrant chlorinated solvents via the process of reductive dechlorination (McCarty, 1997; Wiedemeier et al., 1999). It is also possible to inject microorganisms that are specialized for degradation of target contaminants; this is termed bioaugmentation (Dybas et al., 1998). Although electron donors like hydrogen and lactate can be input via aqueous solutions, numerous studies have explored the use of lower cost carbohydrates like molasses (Lee et al., 2004), or inexpensive low-maintenance polymeric organic material (e.g., wood chips, sawdust, chitin) that can be emplaced in boreholes or trenches perpendicular to the contaminant flow path (Brennan et al., 2006; Kao et al., 2003). Fermentation of these complex carbohydrates and organic materials produces electron donors like acetate and hydrogen that in turn stimulate the transformation of chlorinated hydrocarbons. Based upon extensive research into the key microbial processes (Fennell and Gossett, 1998), the following reactions have been used in the reactive transport model by Hammond et al. (2005), which is similar to that by Christ and Abriola (2007):

- Fermentation of butyrate:

$$Butyrate + 2H_2O \xrightarrow{Butyrate\ Fermenters} 2Acetate + H^+ + 2H_2 \tag{Eq. 4.1}$$

- Reductive dechlorination by two different microbial populations:

$$PCE + H_2 \xrightarrow{Dechlorinator1} TCE + H^+ + Cl^- \tag{Eq. 4.2}$$

$$TCE + H_2 \xrightarrow{Dechlorinator1} DCE + H^+ + Cl^- \tag{Eq. 4.3}$$

$$DCE + H_2 \xrightarrow{Dechlorinator2} VC + H^+ + Cl^- \tag{Eq. 4.4}$$

$$VC + H_2 \xrightarrow{Dechlorinator2} ETH + H^+ + Cl^- \tag{Eq. 4.5}$$

- Methanogenesis which competes with reductive dechlorination for the hydrogen:

$$Acetate + H_2O \xrightarrow{Acetotrophic\ Methanogens} CH_4 + HCO_3^- \tag{Eq. 4.6}$$

$$HCO_3^- + 4H_2 + H^+ \xrightarrow{Hydrogenotrophic\ Methanogens} CH_4 + 3H_2O \tag{Eq. 4.7}$$

Kinetic equations based upon the Monod expression are developed for all of the reactions above, as well as population mass balance equations for each of the biomass types. The kinetic degradation equations are modified to account for hydrogen thresholds for the different bacterial populations. Since the study by Christ and Abriola (2007) focuses upon NAPL source zones, their model also includes bacterial inhibition due to high concentrations of PCE. The model by Hammond et al. (2005) includes buffering reactions with solid phase calcite, dolomite and magnesite.

The overall success of enhanced *in situ* bioremediation ultimately depends upon mixing of the electron donor and acceptor, since they must both be present at the same location for the reaction to proceed. Therefore, electron donor input strategies should be planned to achieve high mixing efficiency with contaminated plumes. An interesting new method for electron donor delivery is to use emulsions with food-grade vegetable oil; the emulsion can be injected into an aquifer to develop a relatively large zone where immobilized oil serves as a slow-release carbon and electron donor source (Jung et al., 2006). Coulibaly et al. (2006) modeled the transport and fate of these oil emulsions using colloid filtration theory.

In situ cometabolic treatment of aquifers contaminated with chlorinated aliphatic hydrocarbons has also been successfully demonstrated at the field scale (Roberts et al., 1990; Semprini et al., 2007). This is usually done under aerobic conditions and requires delivery of co-substrates such as toluene, methane or butane. Because aerobic biodegradation reactions are usually rapid with high biomass yields, it is often necessary to add a source of oxygen, and to use chemical pulsing strategies to prevent excessive biomass growth near the injection points as well as to enhance mixing of the electron acceptor (i.e., oxygen) and the co-substrate (e.g., toluene) with contaminated groundwater. Because of the need to control mixing and prevent bioclogging, many of the successful operations reported in the literature utilize careful hydraulic control systems with recirculating wells (McCarty et al., 1998). Comprehensive reactive transport models for cometabolic degradation include transport equations for contaminant, co-substrates, and oxygen, as well as a population balance for the biomass and various forms of inhibition (Goltz et al., 2001; Semprini et al., 2007). *In situ* cometabolic treatment of TCE contaminated groundwater is one of the case studies presented in Section 4.6 (Gandhi et al., 2002a).

In addition to treatment of organic pollutants, there has been interest in the use of biological processes to address metals and radionuclide contamination problems (Hazen and Tabak, 2005). Inorganic species like iron(III) (Fe(III)), chromium(VI) (Cr(VI)), and uranium (VI) (U(VI)) can serve as an electron acceptor if there is an appropriate carbon source and electron donor. Microbially mediated reduction transforms chromium and uranium from the soluble valence state (VI) to relatively insoluble valence states (Cr(III) and U(IV)). Therefore the harmful metal species precipitates and is immobilized, as long as re-oxidation does not occur. *In situ* bioreduction of chromium and uranium is being studied extensively by the U.S. Department of Energy (USDOE), and there have been several pilot studies conducted at the Oak Ridge National Laboratory, Tennessee, where ethanol is injected as the electron donor (Istok et al., 2004; Luo et al., 2007). At the USDOE Hanford, Washington site, a pilot study was conducted where lactate was added to stimulate iron reducers to produce dissolved Fe(II) which then reduces Cr(VI) to insoluble Cr(III) (Faybishenko et al., 2008). Quantitative modeling requires coupling multiple organic and inorganic processes since metals and radionuclides can undergo speciation, redox, sorption and precipitation/dissolution reactions. For example, Scheibe et al. (2006) developed a model to investigate a scenario where acetate is injected into a heterogeneous aquifer for biostimulation of iron-reducing bacteria which also reduce uranium. The model includes 32 reactions for carbonate chemistry, uranium and iron speciation, sorption of Fe(II) and U(VI) onto iron oxide surfaces, microbial reduction of Fe(III) to Fe(II), microbial reduction of U(VI) to U(IV), and biomass population balance. This work also explored the key role that physical and chemical heterogeneity plays in mixing the injected acetate and the U-contaminated groundwater.

As noted above, biostimulation of iron reducing bacteria has been tested at the Hanford site to address the problematic hexavalent chromium plumes that are discharging into the Columbia River. In this case, a permeable reactive barrier was created by injecting nutrients and electron

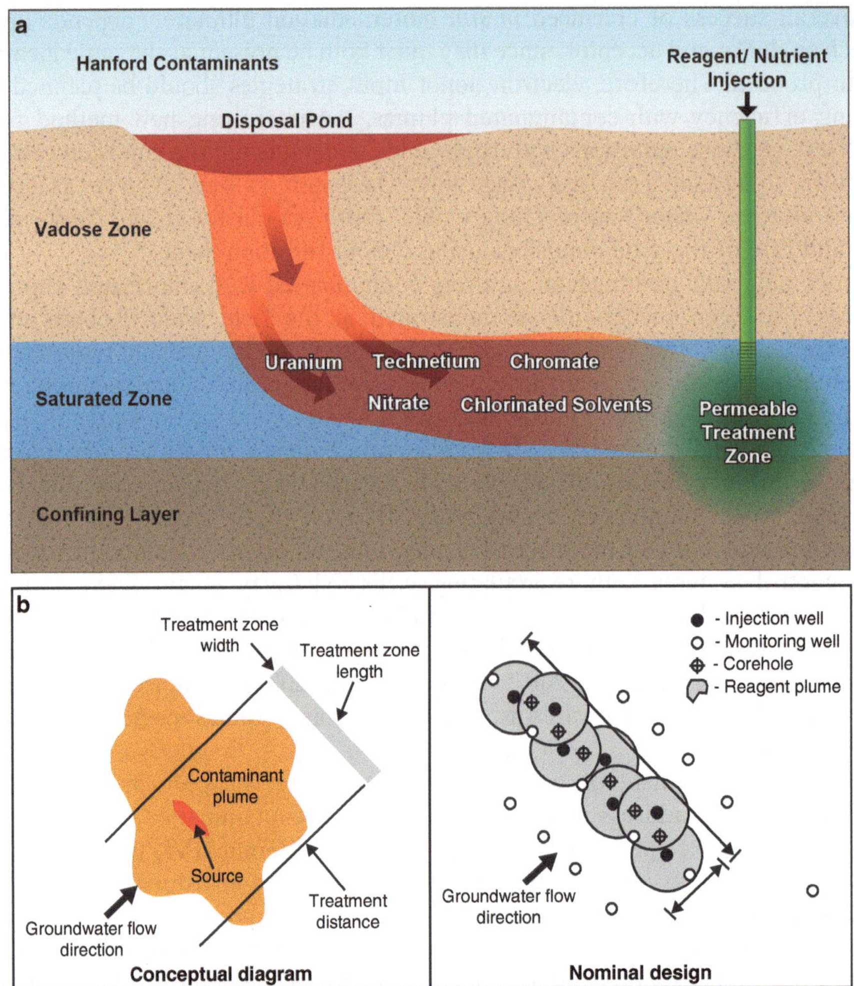

Figure 4.1. Schematic diagram of *in situ* permeable redox barrier at the USDOE Hanford site, Washington: (a) adapted from Innovative Technology Summary Report (2000); (b) from Innovative Technology Summary Report (2000).

donors into the saturated zone. This is an example of so-called *in situ* redox manipulation, a novel remediation concept developed by researchers at the USDOE Pacific Northwest Laboratory, Washington (Fruchter et al., 2000). A schematic is shown above in Figure 4.1. At the Hanford site, this concept has also been tested for Cr(VI) remediation using chemical redox manipulation (Chilakapati et al., 2000; Istok et al., 1999). A strong reductant, in this case sodium dithionite, is injected into the aquifer to convert iron (III) oxides to Fe(II), which re-adsorbs onto the sediment thus creating a Fe(II) barrier. As chromium contaminated groundwater migrates through the barrier, the Fe(II) reduces Cr(VI) to Cr(III) which precipitates as a chromium hydroxide. This precipitate has very low solubility and thus re-oxidation of chromium back to the mobile hexavalent form is unlikely under typical aquifer conditions (Chilakapati et al., 2000; Patterson et al., 1997). Chilakapati et al. (2000) developed a simplified model for design of the redox barrier. The first stage is creation of the Fe(II) zone; the key reaction is reduction of Fe(III) oxides by dithionite, but there is also an unavoidable dithionite disproportionation reaction that also consumes the reagent. The second stage is re-oxidation of the Fe(II) barrier by reactions with Cr(VI) and any dissolved oxygen present in the plume.

The design model incorporates economic factors (i.e., installation, operations, maintenance and materials costs) to optimize the number and spacing of injection wells (e.g., see Figure 4.1b), the dithionite reagent mass injection rate, and the time interval for periodic regeneration of the barrier. Limited operating experience from a full-scale barrier at the 100D area at Hanford indicates some early breakthrough due to heterogeneity of both hydraulic conductivity and Fe(III) mineral abundance (Oostrom et al., 2007; Szecsody et al., 2005). This same barrier technology has been used for Cr(VI) remediation at the Frontier Hard Chrome Superfund site in Vancouver, Washington (Vermeul et al., 2004), and TCE remediation at the Fort Lewis Logistics Center near Tacoma, Washington (Szecsody et al., 2000).

The final example technology we consider in this section is ISCO. Many dissolved organic contaminants can be transformed to harmless end products via reaction with strong chemical oxidants such as peroxide, ozone, or permanganate (Yan and Schwartz, 1999). In practice it is common to inject an aqueous solution of potassium permanganate ($KMnO_4$), since it is relatively inexpensive and easy to handle. ISCO is often applied to treat dense nonaqueous phase liquid (DNAPL) source zones (NRC, 2005). Zhang and Schwartz (2000) have presented a reactive transport model that includes the following processes: oxidation of PCE, TCE and DCE by permanganate; reactions between permanganate and oxidizable species in both the aquifer solid phase and background pore water; and NAPL dissolution. The model results were compared with column and flow cell experiments, and the model was also used to simulate the field-scale treatment of TCE contamination at the USDOE Portsmouth Gaseous Diffusion Plant, Ohio (Cline et al., 1997; NRC, 2005).

The reaction of $KMnO_4$ with TCE can be written as (NRC, 2005):

$$2KMnO_4 + C_2HCl_3 \longrightarrow 2CO_2 + 2MnO_{2(s)} + 2KCl + HCl \qquad \text{(Eq. 4.8)}$$

The reaction produces solid manganese dioxide (MnO_2), which precipitates in the soil, and carbon dioxide (CO_2), which can potentially outgas from solution. Both of these processes can plug the porous medium thereby causing the injected solution to bypass contaminated zones (Li and Schwartz, 2004a; Schroth et al., 2001). Natural heterogeneity also can adversely affect the efficiency of technologies like ISCO that rely on injection of chemical reactants (Ibaraki and Schwartz, 2001). At the Portsmouth field site a combination of aquifer plugging and heterogeneity caused some localized areas to show high TCE reduction, while other areas had essentially none (NRC, 2005). To partially alleviate these problems, Lee and Schwartz (2007) have recently proposed the use of a polymeric resin for controlled slow release of $KMnO_4$; this material can be placed in wells or in reactive barriers upgradient of DNAPL source zones. A recent study by Henderson et al. (2009) demonstrating the application of a hydrogeochemical model to a field pilot test of TCE oxidation by injection of permanganate is one of the case studies presented in Section 4.6.

This section has summarized a few of the many treatment technologies for organic and inorganic contaminants which require mixing of the contaminated water with other reactants that may either be naturally present in the ambient groundwater or deliberately injected into the subsurface. The mixing processes are dependent upon complex flow patterns that can be a consequence of small-scale spatial variability of permeability. The ensuing degradation reactions can generate numerous other secondary geochemical reactions that may affect the progress of remediation. Mathematical models are a useful tool to integrate the physical, geochemical, and biological processes into a quantitative framework for understanding, designing, and management of remediation. The basic principles and governing equations for these models are discussed in the next section.

4.3 HYDROGEOCHEMICAL MODEL GOVERNING EQUATIONS

Many contaminated sites contain complex mixtures of organic and inorganic species that are strongly affected by both transport and reaction processes. Some of these reaction processes were described in the previous section, and include both interactions among the dissolved species as well as reactions with immobile constituents present as solid minerals, adsorbed species, attached bacteria, or trapped NAPLs. Therefore, hydrogeochemical models must be capable of coupling transport of multiple interacting dissolved chemical species with geochemical and biological processes. Due to the need to simulate a large number of chemical species and reaction processes, hydrogeochemical models can be computationally demanding.

The governing transport and reaction equations are based upon mass balance. There are many excellent books and review papers that give a thorough derivation of these governing equations under a variety of conditions, including multiphase flow and fractured media (Barry et al., 2002; Bethke, 2008; Lichtner, 1996; Simunek and Valocchi, 2002; van der Lee and De Windt, 2001). In order to keep the presentation here concise and simple, we assume single-phase saturated groundwater flow in porous media, with constant fluid density and viscosity. Note that many field contamination problems require assessment of water flow in the vadose zone or flow of separate nonaqueous phase liquids, and some of the codes reviewed below in Section 4.4 include this capability. For single-phase saturated groundwater flow, the governing equation is:

$$S_s \frac{\partial h}{\partial t} = \nabla \cdot (\mathbf{K} \nabla h) + Q \qquad \text{(Eq. 4.9)}$$

where S_s is the specific storage coefficient that represents the elastic storage properties of the aquifer, h is the piezometric head, $\mathbf{K}$ is the hydraulic conductivity tensor, and Q is the water source rate (volume of water added/volume of aquifer-time). For a properly posed problem where all parameters are known, Equation 4.9 can be solved for the piezometric head as a function of space and time. Then Darcy's Law is used to compute the specific discharge or Darcy flux:

$$\mathbf{q} = \phi \mathbf{v} = -\mathbf{K} \nabla h \qquad \text{(Eq. 4.10)}$$

where $\mathbf{q}$ is the Darcy flux vector, $\mathbf{v}$ is the pore water velocity or average linear velocity (mean speed of water movement) vector, and ϕ is the porosity. The pore water velocity is a key parameter since it not only determines the advective transport of dissolved species, but it also controls the mechanical dispersion process, as will be shown below.

As noted above, we consider both dissolved and immobile species. Let us assume that there are N mobile (dissolved) species with concentrations denoted by the vector $\mathbf{c} = (c_1, c_2, \ldots c_N)^t$, and $\bar{N}$ immobile species with concentrations denoted by the vector $\bar{\mathbf{c}} = (\bar{c}_1, \bar{c}_2, \ldots, \bar{c}_{\bar{N}})^t$. The superscript t denotes the vector transpose, i.e., $\mathbf{c}$ and $\bar{\mathbf{c}}$ are each column vectors. We will adopt the convention that the dissolved species have units of moles per volume of water, while the immobile species have units of moles per mass of solids. For simplicity, here we assume that the immobile phase is associated with the aquifer solids and hence the immobile species refer to adsorbed or precipitated chemicals, or attached microorganisms. The governing equations would need to be modified slightly if the immobile phase were composed of trapped nonaqueous phase liquids (Barry et al., 2002; Christ and Abriola, 2007). These mobile and immobile species participate in N_R reactions with rates denoted by $R_r(\mathbf{c}, \bar{\mathbf{c}})$. The mass balance equations for the N dissolved and $\bar{N}$ immobile chemical species are:

$$\frac{\partial \phi c_i}{\partial t} + L(c_i) = \sum_{r=1}^{N_R} v_{ir} R_r \qquad i = 1, 2, ..., N \tag{Eq. 4.11}$$

$$\frac{d\rho \bar{c}_i}{dt} = \sum_{r=1}^{N_R} \bar{v}_{ir} R_r \qquad i = 1, 2, ..., \bar{N} \tag{Eq. 4.12}$$

where v_{ir} and $\bar{v}_{ir}$ are stoichiometric coefficients denoting the moles of mobile and immobile species i, respectively, produced in reaction r, *L( )* is the advection-dispersion operator (i.e., a shorthand notation representing terms arising from considering advective and dispersive transport, see below), and ρ is the soil bulk density (mass of solids/volume of porous media). The right hand side of the mass balance equations represents the reaction source/sink term, giving the moles of species i produced or consumed per unit volume of porous media per unit time. The advection-dispersion operator appearing in the dissolved-phase transport Equation 4.11 can be written as:

$$L(c_i) = \nabla \cdot (\mathbf{q} c_i - \mathbf{D} \nabla c_i) \tag{Eq. 4.13}$$

where **q** is the specific discharge vector computed by Darcy's Law (Equation 4.10), and the classical model for hydrodynamic dispersion gives the components of the dispersion tensor **D** as the sum of molecular diffusion (which depends upon the molecular diffusion coefficient of species i) plus mechanical dispersion. The mechanical dispersion components can be written as a function of the specific discharge vector **q**, and the longitudinal and transverse dispersivities of the porous medium. The form of **D** is not given here since it can be found in several texts (Bear, 1979) as well as in Chapter 3. Note that we also assume in Equation 4.11 that the dispersion coefficients do not depend on the particular chemical species; this is a commonly adopted simplification that is acceptable in most circumstances.

As described previously in Section 4.2, there are a variety of geochemical and biological reactions that can occur during remediation operations. These include mass-transfer processes like mineral precipitation/dissolution and NAPL dissolution that are slow relative to the time scale of transport. On the other hand, aqueous speciation reactions occurring among dissolved-phase species are typically very fast. Therefore, hydrogeochemical models applied to mixing and remediation problems must be capable of handling both equilibrium and kinetic reactions. Since microbes in the subsurface are usually assumed to be attached onto solid surfaces, biodegradation reactions are also typically modeled as kinetic due to the time scales of bacterial metabolism and mass transfer of reactants from the bulk fluid to solid surfaces.

To more clearly illustrate application of the governing equations (Equations 4.11, 4.12), we use a simple example of aerobic degradation of groundwater contaminated with benzene (C_6H_6). This case is summarized below in Table 4.1. Benzene is a known carcinogen and is one of the more soluble components of gasoline, and hence it is present at many sites where petroleum products have leaked from storage containers. Although benzene can be degraded under anaerobic conditions using terminal electron acceptors such as nitrate and sulfate, or even through methanogenesis without the need for an external electron acceptor, for simplicity we only consider the thermodynamically most favorable case with dissolved oxygen as the electron acceptor. The reaction for complete degradation of benzene is given by Equation 4.14 in Table 4.1, taken from Wiedemeier et al. (1999). In this reaction, biomass is represented by the commonly used chemical formula $C_5H_7O_2N$ (Rittmann and McCarty, 2001). The bacteria use benzene as the electron donor and carbon source, and use oxygen as the electron acceptor to synthesize new biomass. As noted previously, we also assume that the biomass is attached as biofilms to surfaces of the aquifer solids and are thus immobile (Rittmann, 1993).

Table 4.1. Summary of Reaction Terms in the Mass Balance Equations for Aerobic Degradation of Benzene (Equations 4.11, 4.12)

Reactions ($N_r = 2$): Aerobic degradation of benzene (R_1): $C_6H_6 + 2.5\,O_2 + HCO_3^- + NH_4^+ \xrightarrow{biomass} C_5H_7O_2N + 2H_2CO_3$ (Eq. 4.14) Biomass decay (R_2): $C_5H_7O_2N + 5\,O_2 + 3H_2O \longrightarrow 5HCO_3^- + NH_4^+ + 4H^+$ (Eq. 4.15)	
Mobile species (N = 2): Benzene, C_6H_6 Oxygen, O_2	**Immobile species** ($\bar{N}$=1): Biomass, $C_5H_7O_2N$
Reaction rate laws: $R_1(c_{benzene}, c_{oxygen}, \bar{c}_{biomass})$ Dual Monod kinetics (Eq. 4.14) $R_2(\bar{c}_{biomass}, c_{oxygen})$ First-order biomass decay with O_2 consumption (Eq. 4.15)	
Stoichiometric coefficients – mobile: $\nu_{11} = -1$ $\nu_{21} = -2.5$ $\nu_{12} = 0$ $\nu_{22} = -5$	**Immobile:** $\bar{\nu}_{11}$ = yield coefficient $\bar{\nu}_{12} = -1$

Bacteria undergo natural decay, which also consumes oxygen, which can be written as Equation 4.15 in Table 4.1, taken from Barry et al. (2002).

We assume here that our main focus is modeling the degradation of benzene, so we only consider the two reactions, Equations 4.14 and 4.15; hence N_R =2 in Equations 4.11 and 4.12. In the following section we will consider some secondary reactions involving other chemical species appearing in Equations 4.14 and 4.15. With this assumption, there are two mobile and one immobile species (N = 2 in Equation 4.11, and $\bar{N}$=1 in Equation 4.12), with $\mathbf{c} = (c_{benzene}, c_{oxygen})^t$ and $\bar{\mathbf{c}} = \bar{c}_{biomass}$. The degradation reaction (Equation 4.14) is often modeled using the dual Monod kinetic rate law which depends nonlinearly on the concentration of benzene and oxygen, and linearly upon the biomass concentration (see Chapter 2, Equations 2.10 and 2.11, as well as the more general discussion under Section 2.4.3). Therefore, the reaction rate for Equation 4.14 can be expressed symbolically as $R_1(c_{benzene}, c_{oxygen}, \bar{c}_{biomass})$ and has units of moles of benzene degraded per unit time per unit volume of aquifer. The biomass decay reaction (Equation 4.15) is often modeled as first-order with respect to the biomass concentration multiplied by a nonlinear function of the available oxygen; it can be expressed symbolically as $R_2(\bar{c}_{biomass}, c_{oxygen})$and has units of moles of biomass decayed per unit time per unit volume of aquifer. The stoichiometric coefficients on the right-hand side of Equations 4.11 and 4.12 for the mobile and immobile species mass balance equation are given in Table 4.1; ν_{11} =−1 because benzene is consumed in Equation 4.14, the values of ν_{21} =−2.5 and ν_{22} =−5 are found from Equation 4.14 and 4.15, respectively. For the immobile species mass balance equation, $\bar{\nu}_{11}$ is the yield coefficient which equals the moles of biomass synthesized per mole of benzene degraded, and $\bar{\nu}_{12}$=−1.

4.3.1 Solution of Governing Equations

The governing equations given by Equations 4.11 and 4.12 are a system of partial and ordinary differential equations that are coupled through the nonlinear reaction terms on the right hand side. In general, this system must be solved approximately using numerical techniques. If certain simplifying assumptions are made, however, it is possible to develop analytical solutions. Generally, these assumptions include uniform flow in a homogeneous aquifer and linear kinetic reactions. These analytical solutions have been published by several authors, including Quezada et al. (2004), Sun et al. (1999), Jones et al. (2006), and Christ et al. (1999). These analytical solutions form the basis for several popular screening models (e.g., BIOSCREEN, BIOCHLOR), which will be summarized in Section 4.4.1. In this chapter, greater emphasis will be placed on numerical models, which can handle general nonlinear equilibrium and kinetic reactions among an arbitrary number of chemical species.

Numerical solution of the coupled system (Equations 4.11, 4.12) is extremely challenging due to the potentially large number of chemical species and the highly nonlinear nature of the reaction rate expressions. Accordingly, there is a sizeable body of literature devoted to efficient solution of this problem. This literature can be divided into two categories – methods to reduce the number and complexity of the equations, and methods for coupling the transport and reaction calculations during the numerical solution stage. The first category entails reformulation of the system of equations (Equation 4.11) using new dependent variables such that: (1) some of the resulting partial differential equations are linear and uncoupled, and (2) the number of coupled nonlinear equations is significantly smaller than for the original system. A common example of this approach is the use of the concept of chemical components, which form the basic building blocks of the aqueous system. As defined by Westall et al. (1976), components are chosen so that every chemical species can be represented as a combination of the components, and no component can be represented as a combination of the other components. We usually write the chemical reactions as follows where an aqueous complex is written in terms of components:

$$\sum_{i=1}^{N_c} \tilde{v}_{ji}\hat{c}_i = \hat{x}_j \qquad j = 1, 2, ..., N_x \qquad \text{(Eq. 4.16)}$$

where $\hat{c}_i$ represents the chemical formula for component i, $\hat{x}_j$ represents the chemical formula for aqueous complex j, N_x is the number of aqueous complexes, and N_c is the number of components which can be shown to equal the total number of species N minus the number of linearly independent reactions (Lichtner, 1996). The stoichiometric coefficient in Equation 4.16, $\tilde{v}_{ji}$, is the number of moles of component i in complex j. We furthermore consider the case where all aqueous-phase speciation reactions are at equilibrium, which is usually reasonable since these are fast reactions. Since the reactions (Equation 4.16) are at equilibrium, the mass action law can be used to write the concentration of each complex, x_j, as a function of the concentration of the components, c_i (Lichtner, 1996; Tebes-Stevens et al., 1998):

$$x_j = K_j \prod_{i=1}^{N_c} c_i^{v_{ji}} \qquad \text{(Eq. 4.17)}$$

where K_j is the equilibrium constant for the reaction given by Equation 4.16. We can thus partition the vector of dissolved species as $\mathbf{c} = (c_1, \ldots, c_{N_c}; x_1, ..., x_{N-N_c})^t$, where the aqueous complex concentration x_j can be computed from knowledge of the aqueous component concentrations c_i; for this reason, the components are sometimes referred to as the "primary dependent variables" while the complexes are called the "secondary dependent variables." Therefore, we only need to solve N_c instead of N advection-dispersion-reaction equations. Rather than solving for c, some models instead formulate the governing transport equations in terms of the total dissolved concentration of the component, defined as:

$$C_i = c_i + \sum_{j=1}^{N-N_c} \tilde{v}_{ji}x_j \qquad \text{(Eq. 4.18)}$$

We illustrate Equations 4.16, 4.17, and 4.18 by extending the example of aerobic benzene degradation considered above in Table 4.1. It can be seen that Equations 4.14 and 4.15 involve other species (e.g., carbonate species) that can participate in other reactions. In fact, these secondary reactions can generate "footprints" or "patterns" in the geochemistry that are often used as part of the "lines of evidence" to document the extent of natural attenuation processes in the field (Maurer and Rittmann, 2004b; NRC, 2000). Table 4.2 lists the components, complexes and secondary reactions relevant to aerobic benzene degradation; we also include calcite ($CaCO_{3\,(s)}$) as a mineral phase since it is widely present in aquifer material and reacts with the carbon dioxide that is produced by aerobic degradation. Values of the equilibrium constants appearing in Table 4.2 are those reported in Maurer and Rittmann (2004a). Note that a much more extensive and complicated set of secondary reactions is required when biodegradation involves other terminal electron acceptors such as nitrate, iron, and sulfate (Maurer and Rittmann, 2004b; Prommer et al., 1999).

Table 4.2. Components, Complexes and Secondary Reactions for the Aerobic Benzene Degradation Example

Components	Complexes	Minerals
$Ca^{+2}, H_2CO_3, H_2O, H^+$	$CO_3{}^{-2}, HCO_3{}^{-}, CaCO_3{}^0, CaHCO_3{}^+, OH^-$	$CaCO_{3(s)}$
Aqueous equilibrium complexation reactions (Equation 4.16)		**Log K_j (Equation 4.17)**
$H_2CO_3 - 2H^+ = CO_3{}^{-2}$		−16.68
$H_2CO_3 - H^+ = HCO_3{}^{-2}$		−6.35
$Ca^{+2} + H_2CO_3 - 2H^+ = CaCO_3{}^0$		−13.46
$Ca^{+2} + H_2CO_3 - H^+ = CaHCO_3{}^+$		−3.67
$H_2O - H^+ = OH^-$		−14.0
Kinetic precipitation/dissolution reaction		
$CaCO_{3(s)} = Ca^{+2} + CO_3{}^{-2}$		
Total dissolved concentration of components (Equation 4.18)		
$C_{Ca^{+2}} = c_{Ca^{+2}} + c_{CaCO_3^0} + c_{CaHCO_3^+}$		
$C_{H_2CO_3} = c_{H_2CO_3} + c_{HCO_3^-} + c_{CO_3^{-2}}$		
$C_{H^+} = c_{H^+} + 2c_{H_2CO_3} + c_{HCO_3^-} - c_{OH^-}$		

The total dissolved concentration defined in Equation 4.18 represents the total mass of a particular component in the dissolved phase per unit volume of water. The aqueous complexation reactions (Equation 4.16) distribute the components c_i among complexes x_j but do not affect the total mass of the component in the dissolved phase. Therefore, for any component that does not participate in decay reactions or reactions with immobile species (e.g., precipitation/dissolution or adsorption/desorption reactions), the total dissolved concentration C_i is a conserved quantity and is thus governed by the linear advection-dispersion reaction. Otherwise, the mass balance equation for C_i will have some nonlinear reaction source/sink terms on the right hand side, as already demonstrated above in Table 4.1 for species that undergo biodegradation. For the simple example illustrated in Tables 4.1 and 4.2, none of the components are conserved: Ca^{+2} participates in the calcite precipitation/dissolution reaction, and H_2CO_3, and H^+ participate in the biodegradation reactions (Equations 4.14, 4.15). Given C_i, all aqueous component and complex concentrations (i.e., c_i and x_j) can be computed by solving a system of nonlinear algebraic equations, as implemented in several commonly used software packages for chemical speciation modeling in batch systems (Allison et al., 1991; Parkhurst et al., 1980). The reaction source/sink terms in the species mass balance equations (Equations 4.11, 4.12) are normally functions of these concentrations. Salvage and Yeh (1998) show how to extend the above formulation to the case where some of the aqueous phase reactions are kinetic.

Several other investigators have presented more general methods for systematically transforming the governing equations (Equations 4.11, 4.12) into a reduced system of equations (Kräutle and Knabner, 2007; Molins et al., 2004; Steefel and MacQuarrie, 1996; Zhang et al., 2007). It can be shown that the reduced system is mathematically equivalent to the original system, so there is no error introduced by these reduction methods. For certain test problems, Kräutle and Knabner (2005) have shown that these methods can result in computer execution time savings of up to a factor of 10.

The second large body of work devoted to efficient solution of hydrogeochemical model equations considers numerical strategies for coupling the transport and reaction calculations. Issues related to coupling arise regardless of the numerical technique (i.e., finite difference, finite element) used to discretize the advection-dispersion operator Equation 4.13. The most accurate and robust strategy solves the entire coupled system (Equations 4.11, 4.12) (or the reduced system discussed above) and is called "Global Implicit" (GI). The main drawback of GI is its computational demand, since for multicomponent geochemical transport problems even the reduced system can require solution of a large number of coupled nonlinear partial differential equations. Numerical techniques based on the iterative Newton–Raphson method are normally used for coupled nonlinear equations. Several studies have examined novel numerical methods to solve the coupled system in a more efficient way (Hammond et al., 2005; Robinson et al., 2000). To reduce computational effort, it is possible to de-couple the individual mobile and immobile species, solving each mass balance equation separately (see Equations 4.11, 4.12) followed by iteration; this is called the "Sequential Iterative Approach" or SIA (Tebes-Stevens et al., 1998; Yeh and Tripathi, 1991). The simplest approach is to use "operator splitting" (OS), also denoted the "Sequential Non-iterative Approach," for which the advective-dispersive transport is solved separately from the reaction. In essence, each discrete grid block is treated as a "batch" chemical reactor. These OS methods are conceptually and computationally straightforward, and they allow the use of existing software packages for solving the coupled batch reaction equations. However, OS methods are subject to additional "splitting errors" that can be large for moderately fast reactions. For problems where the dissolved species are strongly retarded due to sorption or precipitation/dissolution reactions, GI methods often allow large time steps and are more efficient than OS methods (Hammond et al., 2005). Further discussion and evaluation of GI and OS methods is given by several authors, including Barry et al. (2002), Valocchi and Malmstead (1992), and Steefel and MacQuarrie (1996).

4.4 SURVEY OF AVAILABLE HYDROCHEMICAL MODELS

There have been many different models developed in recent years that solve the fundamental reactive transport equations outlined in the previous section. Although in theory it is possible to create a single all-purpose simulator, in practice a variety of models have been developed to focus upon specific applications. Some of these specialized models will be presented later in Chapters 7 and 8. We do not attempt to give a comprehensive survey here, but rather summarize and compare some of the more commonly used models for remediation applications. For more comprehensive reviews, see Barry et al. (2002), Brun and Engesgaard (2002), and van der Lee and De Windt (2001).

4.4.1 Analytical Models

As stated previously, if certain assumptions are made (i.e., homogeneous aquifer, uniform and steady flow, linear reaction kinetics), then it is possible to solve the reactive transport equations analytically. For example, the work by Quezada et al. (2004) presents a solution for transport of multiple species subject to equilibrium adsorption with inter-species transformation governed by an arbitrary first-order reaction network. The simplified solution for chain-decay type reactions is the basis for the popular BIOCHLOR screening model, which is widely applied to assess natural attenuation at chlorinated solvent release sites (Wiedemeier et al., 1999). BIOCHLOR solves the sequential reductive dechlorination of PCE through ethene (ETH),

$$PCE \xrightarrow{k_1} TCE \xrightarrow{k_2} DCE \xrightarrow{k_3} VC \xrightarrow{k_4} ETH \qquad \text{(Eq. 4.19)}$$

The model can accommodate a constant or decaying source of PCE, as well as one, two, or three-dimensional dispersion. For multidimensional domains, the popular approximate Domenico solution is utilized for advective–dispersive transport; the limitations of this solution are clearly elucidated by Srinivasan et al. (2007). The BIOCHLOR model can be downloaded from the U.S. Environmental Protection Agency (USEPA) Center for Subsurface Modeling Support web site: www.epa.gov/ada/csmos/models/biochlor.html. An example application is described by Clement et al. (2002).

Another popular screening model used for natural attenuation assessment of hydrocarbons at petroleum spill sites is BIOSCREEN, also available from the USEPA Center for Subsurface Modeling Support: www.epa.gov/ada/csmos/models/bioscrn.html. This model is for either a two or three-dimensional groundwater system and again uses the Domenico solution. Either first-order or instantaneous reaction is modeled between the petroleum contaminant and multiple electron acceptors, which are utilized sequentially according to the thermodynamics of the terminal electron accepting processes (Wiedemeier et al., 1999).

Mixing limitations are not explicitly accounted for in the BIOCHLOR or BIOSCREEN models. Although the former allows for spatial zones with different first order decay constants, the sequential first-order decay chain reaction (Equation 4.19) assumes that there is complete mixing between the electron accepting chlorinated ethenes and the electron donor. BIOSCREEN has an option for instantaneous reaction between the electron donor (i.e., petroleum hydrocarbons represented as benzene, toluene, ethylbenzene and total xylenes (BTEX)) and different electron acceptors (e.g., oxygen, nitrate, iron, sulfate). Again, complete mixing is assumed between the BTEX contaminants emanating from the source and the various electron acceptors present in the background groundwater. However, it is possible to develop analytical solutions for mixing-limited reactions between two reactants, as long as the reaction rate is assumed to be much faster than the mixing rate. Under this assumption, both reactants cannot be simultaneously present at the same location in the porous medium. A schematic is shown in (Figure 4.2) for a case where biodegradation occurs along the plume fringes where transverse dispersion mixes electron acceptor (oxygen) present in the ambient groundwater with the dissolved contaminant that emanates from a source zone. Several authors have shown that this case can be converted mathematically to a linear conservative transport equation, which can be solved analytically for uniform groundwater flow (Ham et al., 2004; Liedl et al., 2005). An analytical solution that accounts for first-order decay within the plume core in addition to mixing-controlled degradation along the plume fringe has recently been presented by Gutierrez-Neri et al. (2009). In addition to their use as screening tools, these analytical models also provide useful physical insight into the parameters and processes controlling the overall dynamics of contaminant plume migration.

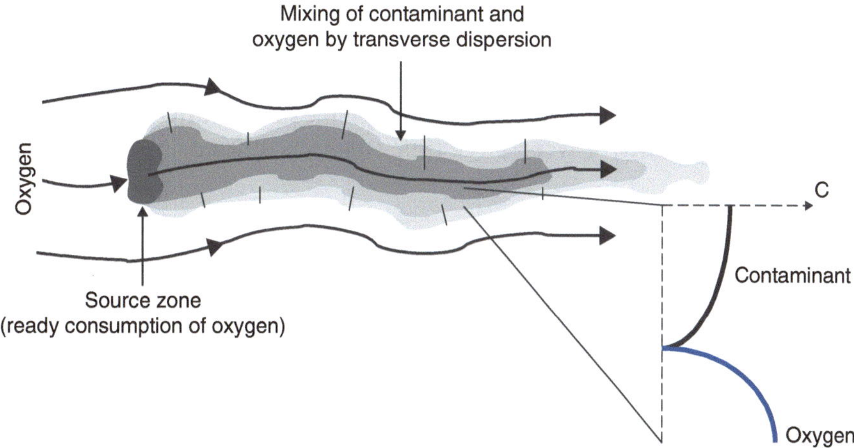

Figure 4.2. Schematic of mixing-controlled instantaneous reaction between contaminant from source and oxygen in background groundwater. Blow-up of transverse concentration profile along plume fringe shows that the reaction is assumed to occur at a single point. Adapted from Cirpka and Valocchi, 2007.

4.4.2 Numerical Models

As stated above, there are a multitude of numerical reactive transport models that have been developed and described in the literature. Many of these are not widely used in practice, either because they are specialized research codes or are proprietary. Here we select some of the more flexible and widely used codes, which are summarized and compared below in Table 4.3. We only include codes that are available on line and can be downloaded from the world wide web.

Table 4.3. Overview of Selected Numerical Hydrogeochemical Models

	MT3D	RT3D	SEAM3D	PHT3D	HBGC123D	UTCHEM	FEHM	STOMP	TOUGH-REACT
Unsaturated/multiphase flow	N	N	N	N	U	U/M	U/M	U/M	U/M
Separate NAPL phase with mass transfer[a]	N	Y	Y	N	N	Y	N	Y	N
Spatial discretization[b]	FD/FV	FD/FV	FD/FV	FD/FV	FE	FD	FE	FV	FV
Coupling scheme[c]	OS	OS	OS	OS	OS	OS	GI/SI	OS	SI/OS
Geochemical database	N	N	N	Y	N	N	N	N	Y
User defined reactions	Y	Y	N	Y	Y	N	Y	Y	N
Graphical user interface	Y	Y	Y	Y	N	Y	N	N	N

[a]Indicates if code has special option for NAPL dissolution
[b]*FD* finite difference, *FE* finite element, *FV* finite volume
[c]*OS* operator splitting, *SI* sequential iterative approach, *GI* global implicit

We now summarize briefly each of the codes appearing in the above table, and provide appropriate citations and links to further information. We begin with a series of codes in the well-known MODFLOW family; these codes solve the contaminant mass balance equations (Equations 4.11, 4.12), and are linked with MODFLOW which solves the groundwater flow (Equation 4.9) and Darcy equation (Equation 4.10) to provide the specific discharge q which is required input for the reactive transport equations. These models have the advantage that they are widely used and tested, and are supported by several popular graphical user interfaces by virtue of their linkage with MODFLOW. However, because these models combine separate modules for transport and reaction, they all use the simple non-iterative OS technique which has the potential for numerical error when large time steps are utilized.

MT3DMS

Primary contact:	Dr. C. Zheng, University of Alabama
Documentation:	http://hydro.geo.ua.edu/mt3d/
Basic references:	Zheng and Wang (1999); Prommer et al. (2003)
Remarks:	This code developed from the non-reactive single species model MT3D which included several advanced numerical techniques to accurately solve advection-dominated transport problems. The code has a flexible, modular structure to allow the user to develop add-on reaction packages. In fact, the following three codes are built upon MT3D's nonreactive transport modules.

RT3D

Primary contact:	Dr. P. Clement, Auburn University
Documentation:	http://bioprocess.pnl.gov/rt3d.htm
Basic references:	Clement et al. (1998, 2000)
Remarks:	RT3D is a set of reaction packages for MT3DMS. The pre-programmed reactions packages include: (1) six-species, first-order, rate-limited, BTEX degradation using sequential electron acceptors (e.g., oxygen (O_2), nitrate (NO_3^-), Fe^{2+}, sulfate (SO_4^{2-}), CO_2); (2) rate-limited sorption; (3) sequential first-order decay (up to four species, e.g., PCE/TCE/DCE/VC); and (4) aerobic/anaerobic chlorinated ethene dechlorination. The user has the flexibility to specify additional add-on reaction packages.

SEAM3D

Primary contact:	Dr. M. Widdowson, Virginia Tech
Documentation:	http://el.erdc.usace.army.mil/elpubs/pdf/trel00-18.pdf
Basic references:	Waddill and Widdowson (2000); Brauner and Widdowson (2001)
Remarks:	Like RT3D, SEAM3D is a set of reaction packages for MT3DMS. These consist of packages for biodegradation, NAPL dissolution, co-metabolic biodegradation, and reductive dechlorination.

PHT3D

Primary contact:	Dr. H. Prommer, CSIRO Land and Water Centre, Australia
Documentation:	www.pht3d.org/
Basic references:	Prommer et al. (2000, 2003)
Remarks:	The code combines MT3DMS with the geochemical reaction model PHREEQC-2 (Parkhurst and Appelo, 1999). The latter code includes an extensive database file of aqueous chemical species and solid minerals which gives the user a flexible method to specify equilibrium or kinetic reactions without developing separate add-on packages. The PHREEQC-2 database can be extended to include other immobile kinetically reacting species which provides a method to include NAPL dissolution and biodegradation coupled with bacterial growth. A whole host of water-rock reactions can be readily simulated, including equilibrium aqueous-phase speciation, equilibrium or kinetic precipitation/dissolution, and equilibrium ion exchange.

The next two codes are also available within a graphical user interface system, namely, GMS which was developed under the auspices of the U.S. Army Corps of Engineers. However, these codes can handle non-isothermal cases and are not restricted to saturated flow conditions.

HBGC123D

Primary contact:	Dr. J. P. (Jack) Gwo, US Nuclear Regulatory Commission, Office of Nuclear Material Safety and Safeguards
Documentation:	http://hbgc.emsgi.org/
Basic references:	Gwo et al. (2001); Yeh et al. (1998)
Remarks:	The code is designed for flexible simulation of transport with coupled equilibrium and kinetic reactions, including aqueous complexation, sorption, ion exchange, precipitation/dissolution and biodegradation. All reactions can be defined by the user, but there is no pre-defined thermodynamic database like in PHT3D described above. A hybrid Eulerian–Lagrangian method is used to yield high accuracy for advection-dominated problems with sharp fronts. For transient flow scenarios, it is necessary to interface with another flow code to solve for the time-varying velocity and moisture content; the related code FEMWATER (Yeh and Ward, 1980) is recommended. This code is built upon the original HYDROGEOCHEM model by Yeh and Tripathi (1990), which has very recently been updated to version 5 which includes most of the features in HBGC123D (Yeh et al., 2004).

UTCHEM

Primary contact:	Dr. G. Pope, University of Texas at Austin
Documentation:	http://www.cpge.utexas.edu/utchem/
Basic references:	Brown et al. (1994); Jin et al. (1997)
Remarks:	This model was originally developed to address problems in petroleum engineering, and therefore has sophisticated capability for simulating multi-phase organic-water-gas flow with phase partitioning. The code has been used for surfactant enhanced remediation applications. Other reactions include precipitation/dissolution, cation exchange, sorption, and biodegradation. All speciation, ion exchange, and precipitation/dissolution reactions are assumed to be at equilibrium.

The final three codes were developed at USDOE laboratories and can also handle non-isothermal and multi-phase flow conditions. These codes were developed to address subsurface contamination problems in the USDOE complex, as well as other energy-related applications.

FEHM

Primary contact:	Dr. George A. Zyvoloski, Los Alamos National Laboratory
Documentation:	http://fehm.lanl.gov/
Basic references:	Viswanathan et al., 1998; Zyvoloski et al., 2003
Remarks:	This model was originally developed to study flow and mass transport in the saturated and unsaturated zones at the proposed high-level radioactive waste repository at Yucca Mountain. Thus, the code has the capability to include coupled hydrologic-thermal processes like boiling and condensation, and can represent complex geological heterogeneities with unstructured meshes. Pre-defined reaction modules are used for up to ten solutes; reactions include equilibrium aqueous speciation or sorption; kinetic sorption, precipitation/dissolution,

(continued)

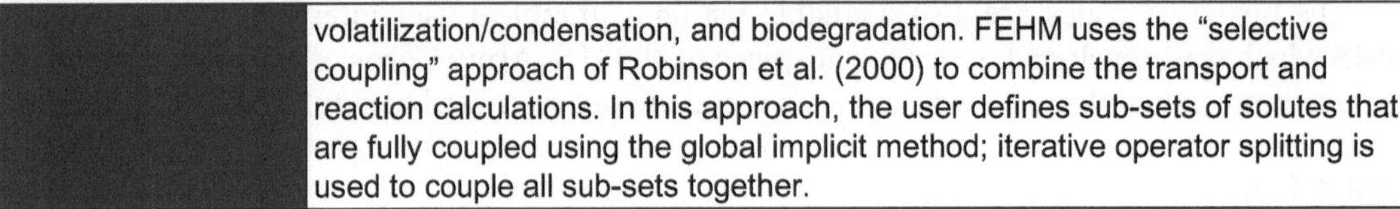

	volatilization/condensation, and biodegradation. FEHM uses the "selective coupling" approach of Robinson et al. (2000) to combine the transport and reaction calculations. In this approach, the user defines sub-sets of solutes that are fully coupled using the global implicit method; iterative operator splitting is used to couple all sub-sets together.

STOMP

Primary contact:	Dr. Mark White, Pacific Northwest National Laboratory
Documentation:	http://stomp.pnl.gov/
Basic references:	White and Oostrom, 1998, 2006
Remarks:	This model was originally developed to study contamination by organic liquids at the Hanford site. It has been extended to model coupled hydrologic-thermal-reactive processes, and has been applied to problems involving dense brines, methane hydrates and injection of supercritical carbon dioxide. A recent version of the code incorporates multi-species reactive transport by coupling with an equilibrium/kinetic geochemical module based on the work by (Fang et al., 2003).

TOUGHREACT

Primary contact:	Drs. Tianfu Xu and Karsten Pruess, Lawrence Berkeley National Laboratory
Documentation:	http://esd.lbl.gov/TOUGHREACT/
Basic references:	Xu et al. (2004, 2006)
Remarks:	This code can simulate chemically reactive multiphase and non-isothermal flows; the code and applications to geological sequestration of carbon dioxide are described further in (Xu et al., 2005, 2006; Xu and Pruess, 2001). TOUGHREACT has an option to use either the SIA or OS methods to couple the non-isothermal, multi-phase flow and transport model TOUGH2 with a mixed equilibrium/kinetic geochemical simulator; it does not include biodegradation processes. A polygon-based finite volume spatial discretization method is used in the TOUGH family of codes; this allows much greater flexibility for irregular grids, unlike codes such as MODFLOW which are based on the finite difference method and are restricted to rectangular meshes.

As already noted, the above list in Table 4.3 is incomplete and there are many additional codes that are described in the literature. The U.S. Geological Survey (USGS) has several reactive transport models available for download, including PHAST and RATEQ. PHAST3D is similar to PHT3D described above in that it also uses operator splitting to couple an existing flow and nonreactive transport solver with the batch geochemical code PHREEQC; whereas PHT3D used MT3DMS, PHAST uses the finite element model HST3D (Parkhurst et al., 2004). The model RATEQ uses operator splitting to couple MT3D with a batch equilibrium/kinetic geochemical model that has the capability to use the thermodynamic database of PHREEQC but in addition has capability for biodegradation and surface complexation sorption reactions; the model has been extensively applied to study fate and transport at the Naturita, Colorado, uranium mill tailings site (Curtis, 2005; Curtis et al., 2006).

Finally, there are many research codes that have been described in the literature. One example is the code MIN3P (Mayer et al., 2002) which is capable of simulating mixed equilibrium/kinetic reactions in saturated/unsaturated flow systems; it has been extensively applied to assess the performance of *in situ* reactive barriers (Jeen et al., 2007; Mayer et al., 2001), and has recently been extended to *in situ* chemical oxidation (Henderson et al., 2009). This latter study is included as one of the case studies in Section 4.6.

4.5 CALIBRATION AND VALIDATION

The literature has several excellent discussions of the groundwater modeling process, including the key role of the conceptual model and the importance of calibration and validation (Anderson and Woessner, 1992; Hassan, 2004b; Hill and Tiedeman, 2007). Here we will emphasize special difficulties that arise for hydrogeochemical models that couple biogeochemical reactions with transport. Any given model will have numerous parameters that need to be specified before the simulation model can be run. Some of these parameters can be fixed based upon prior knowledge or experience. If the model is being used to explore hypothetical scenarios or gain insight into coupled processes, then all parameters can be specified and varied as required. However, if the model is applied to study a particular location, then there will always be site-specific parameters that need to be determined. Calibration is the process of determining the values of unknown parameters so that the model fits observed data. Calibration is also referred to as the "inverse problem." Validation is a much more nebulous term. Hassan (2004a, b) describes validation as a long-term iterative process to assess that the model is adequate for its intended purpose. For our discussion, validation is the process of testing the predictive ability of the calibrated model.

From the presentation of the model governing equation in Section 4.3, it is clear that there are three general types of parameters: (1) groundwater flow (i.e., hydraulic conductivity, specific yield, recharge); (2) dispersivity (i.e., longitudinal dispersivity α_L and transverse dispersivity α_T, and possibly both horizontal and vertical transverse dispersivities); and (3) reaction (e.g., all kinetic rate parameters, biomass growth and decay coefficients, sorption/mass transfer parameters). Although all these parameters can be potentially space dependent, in practice only hydraulic conductivity is assumed to be spatially heterogeneous. In addition to specifying values for all parameters, it is also necessary to specify initial and boundary conditions. This latter category includes quantitative representation of the initial contaminant plume and contaminant source zone, which in practice are highly uncertain. In fact several modeling studies have demonstrated that uncertainty about the contaminant source can have a significant impact upon simulation results, comparable to the effects of uncertainty about parameter values (Clement et al., 2000; Lu et al., 1999; Schafer and Therrien, 1995).

Calibration of comprehensive hydrogeochemical models requires a large quantity of high quality data. The lack of adequate data may make numerical models less useful and instead favor simpler analytical screening models, like those summarized in Section 4.4.1. When adequate data are available, most calibration exercises reported in the literature proceed in stages – first, the hydraulic parameters like conductivity and recharge are calibrated to observations of piezometric head or discharge; then, dispersivity parameters are calibrated to observations of tracer transport; and finally, reaction parameters may be calibrated to observations of various chemical species. In many cases, reaction parameters are fixed at literature values or are determined from laboratory experiments using site-specific samples. Although in some cases it is reasonable that certain parameters (e.g., biomass yield, thermodynamic equilibrium constants) can be fixed at literature values, there is often no other choice since there are not enough data to calibrate all the reaction parameters. Although there are some cases where laboratory-determined reaction parameters yield adequate predictions of field-scale transport (Schirmer et al., 2000), many reaction processes are scale dependent and hence parameters determined from laboratory-scale experiments may not apply at the field scale. This is particularly true for mass transfer processes that are affected by physical heterogeneity at multiple scales. In some of the case studies described in the following section, several reaction parameters are fixed to literature or laboratory values while others are calibrated to field observations. For field conditions, the rate-limiting mechanism is often governed by mass-transfer and mixing rather than the intrinsic rate of the reaction, and hence the overall simulation results are not sensitive to the particular values chosen for the reaction rate. This has been investigated for the case of a steady-state plume with mixing-controlled reactions along the plume fringe (see Figure 4.2) by Chu et al. (2005), Cirpka and Valocchi (2007) and Knutson et al. (2007).

Although calibration is often still done manually by trial-and-error adjustment of parameters, there are now effective software tools for automatic calibration, for example, PEST (Doherty, 2002; Doherty, 2003) and UCODE (Poeter and Hill, 1999). These tools treat calibration as a non-linear regression problem. Most applications of automatic calibration are restricted to groundwater flow parameters, though Gandhi et al. (2002b) also consider spatially uniform dispersivity and some reaction parameters. This study is summarized in greater detail below in Section 4.6. Inversion software is an invaluable tool for calibration since, in addition to automatically searching the parameter space, it also provides useful diagnostic information like sensitivity coefficients. If certain statistical assumptions are made (i.e., that the relationship between the model output and parameters is linear, and that the model error is Gaussian), then these tools also yield the covariance and confidence intervals for the estimated parameters, which can be propagated to yield uncertainty estimates of the model predictions (Hill and Tiedeman, 2007; Tonkin et al., 2007). Sensitivity and covariance measures may indicate that certain parameters cannot be identified uniquely given the available data; hence values for these parameters must be assigned by the user (e.g., from the literature or from laboratory experiments). Even if automatic calibration is not used, sensitivity analyses should always be performed. All of the case studies summarized in the next section (and many more in the literature that are not reported here) include sensitivity analyses. Such analysis enables systematic evaluation of effects of parameter uncertainty on model performance, provides insight into the relative importance of individual reactions within a complex network of biogeochemical and mass-transfer processes, and can help guide additional investigations to more accurately estimate the most critical parameters. Although sensitivity analysis can be done by simple parameter perturbation and re-simulation, more formal methods are available (Tebes-Stevens and Valocchi, 2000; Wang et al., 2003).

The groundwater inverse problem is known to be "ill posed" which means that the estimated parameter values are non-unique and are highly sensitive to small changes in the observed data. There has been extensive research conducted on this issue; the most common way to remedy this problem is to reduce, effectively, the number of estimated parameters (Carrera et al., 2005; McLaughlin and Townley, 1996). In the context of the groundwater flow problem, this implies that it is only possible to estimate large-scale smooth trends of the hydraulic conductivity field despite the fact that it is well known that K can vary dramatically over short distances. Unfortunately, this small-scale variability can have a major impact upon active remediation operations where reagents are input via wells (Ibaraki and Schwartz, 2001), as well as upon intrinsic remediation where reactants mix along the plume fringes under natural flow conditions (see Figure 4.2) (Bauer et al., 2009). In these cases, predictions (e.g., the impact of changing the reagent injection rate or adding new injection wells) based upon the calibrated model can be highly uncertain because the calibrated model does not adequately capture the small-scale variability of the K field. (See the last paragraph in this section for further discussion.) Moreover, due to the non-uniqueness of inverse problems, multiple large-scale K fields may yield equally good fits to the observed head data. Several authors have advocated a "multiple models" philosophy where probabilistic predictions are made using an ensemble of calibrated models (Moore and Doherty, 2006; Neuman, 2003; Poeter and Anderson, 2005).

As defined here, model validation would entail comparing the observed response to a significant perturbation of the actual remediation system with predictions made by the calibrated model. This is rarely done in practice, even for groundwater flow and nonreactive transport models. In fact, there is on-going debate regarding the predictive capability of models for even these relatively simple conditions (de Marsily et al., 1992; Konikow and Bredehoeft, 1992). A few examples of more limited model testing and validation have been reported in the literature. One approach is to use calibration for the hydraulic conductivity and dispersivities, and then use laboratory or literature values for all reaction parameters. If the model simulations are close to observed data for chemically reactive species, then there would be some potential that the model would have predictive power assuming the calibrated K and dispersivities apply for other flow scenarios. However, as noted previously and discussed further in the following paragraph, laboratory-determined reaction rate parameters may not apply directly to field-scale problems in which the rate-limiting factor is mixing. Thus, some investigators start by using all laboratory or literature values, but then adjust (either manually or via automatic calibration) the values of a few reaction parameters to fit the field data obtained during remediation (Gandhi et al., 2002b; Phanikumar et al., 2005). The more stringent validation exercise would be to compare field data measured after the commencement of remediation with simulations conducted with a model calibrated using pre-remediation data; unfortunately this is rarely reported in the literature. An example of more limited validation is reported by Luo et al. (2006) who studied the transport of bromide and ethanol in a controlled flow system established by a set of injection and extraction wells. The recirculating well system was established for biostimulation experiments on the reduction and immobilization of uranium. A short-term tracer test was conducted and the travel-time model approach (see Chapter 5) was used. The non-reactive bromide response at the extraction well was used to fit the travel-time distribution, which lumps the impact of all flow heterogeneity and dispersion processes. The ethanol response from the tracer test was used to fit a first-order decay rate. Then this same model was used successfully to predict the ethanol breakthrough for a longer-term biostimulation experiment under the same flow conditions as the short-term tracer test. However, predictions based on this simple travel-time model would probably not be accurate for different flow or mixing conditions, since the fitted travel-time distribution cannot be used if the flow rates change. Also, due to the recirculating flow system, all the injected ethanol is completely mixed with the other reacting chemicals within the injection well; that is, since mixing of the reactants is not occurring *in situ*, the travel-time approach is appropriate because it is not necessary to accurately model spatially dependent dispersion processes.

Recent research findings call into question the common practice of using dispersion coefficients based upon observations of nonreactive transport in the governing transport equation for reactive species. It is generally accepted that field-scale dispersion coefficients are scale-dependent and are much larger than laboratory-scale coefficients (Gelhar et al., 1992). According to the governing equations (Equations 4.11, 4.13), the transport processes for a nonreactive solute are advection, governed by the groundwater specific discharge vector $\mathbf{q}$, and dispersion, which measures the spreading of solutes caused by fluctuations in the velocity at spatial scales smaller than the scale at which **q** is resolved. The model calibration techniques described above estimate large-scale variability of permeability (which, from Darcy's Law (Equation 4.10), will yield the large-scale, smoothly varying **q** field), but they cannot capture the small-scale spatial heterogeneity that is ubiquitous at real field sites. The small-scale fluctuations in groundwater velocity will be reflected by large field-scale dispersion coefficients estimated from tracer observations. Although these dispersion coefficients accurately represent the spread of the nonreactive tracer, in general they will over-estimate the mixing of reactive solutes (Kitanidis, 1994). Figure 4.3 illustrates these concepts schematically for the case of horizontal flow in a heterogeneous aquifer where vertically-averaged solute concentrations are measured in a fully-screened well. This figure represents a remediation scenario in which reactant A is input into an aquifer contaminated with species B; when the two species mix due to molecular diffusion a reaction occurs resulting in contaminant mass reduction. The actual reaction zone is narrow and irregular, being controlled by molecular diffusion across the irregular mixing interface due to the small-scale velocity fluctuations. However, in most practical cases only the vertically-averaged velocity can be estimated, and thus a large dispersion coefficient is needed to match the spread of the measured vertically-averaged concentration profiles or breakthrough curves. As shown in Figure 4.3, use of these large dispersion coefficients will over-estimate the actual mixing and hence over-estimate the overall extent of reaction. This may partially explain why laboratory-measured reaction rates must sometimes be reduced for field-scale applications. Some investigators have proposed modifying the reaction rate expressions with terms that depend upon stochastic representation of the small-scale heterogeneity (Kapoor et al., 1997), while others have proposed modifying the dispersion coefficients used for reactive transport (Cirpka, 2002; Cirpka and Kitanidis, 1999). There is much current and ongoing research on this and related issues of upscaling reactive transport in heterogeneous porous media (Binning and Celia, 2008; Cirpka et al., 2008; Fernàndez-Garcia et al., 2008; Luo et al., 2008).

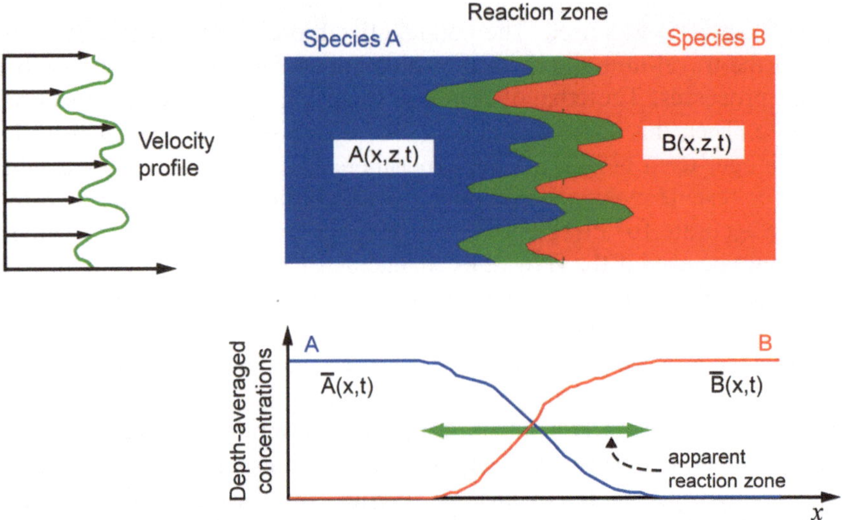

Figure 4.3. Schematic illustration of horizontal flow in a heterogeneous aquifer, with mixing of two reactive chemical species. Use of dispersion coefficients based on vertically-averaged concentrations will over-estimate mixing and subsequent reaction.

4.6 CASE STUDIES OF MODEL APPLICATIONS

This section provides a summary of a few of the many case studies where reactive transport models have been applied to assess remediation at actual field sites. The case studies selected focus on model calibration and validation.

4.6.1 Natural Attenuation of Organic Pollutants

Several authors have applied hydrogeochemical models to simulate natural attenuation processes at sites contaminated by petroleum hydrocarbons and chlorinated solvents. In particular, Clement and co-workers (Clement et al., 2000; Lu et al., 1999) have applied the RT3D model to several field sites. This section describes the application to a chlorinated solvent contamination site at Dover Air Force Base (AFB), Delaware (Clement et al., 2000). This application nicely illustrates the complexity of real field sites and how they can be approximated by simplified models. Sampling data was available to delineate plumes of PCE, TCE, DCE, vinyl chloride (VC), as well as oxygen and methane. Based upon this and other geochemical data, it was assumed that TCE, DCE and VC could degrade via an aerobic pathway as well as anaerobic reductive dehalogenation. First-order degradation and constant microbial biomass were assumed. The latter assumption is reasonable for natural attenuation scenarios where plumes are approximately at steady state and microbial growth and decay should be balanced. Based on site geochemistry, four different spatial reaction zones were identified–two anaerobic zones near the DNAPL source, a transition zone, and a downgradient aerobic zone; different first-order decay rates apply within each zone.

A two-dimensional, confined, steady-state MODFLOW model was constructed for the site; the grid size was 30.5 by 30.5 meters (m) (100 by 100 feet [ft]). Trial-and-error adjustment was used to fit a spatially varying pattern of transmissivity to observed head data. The MT3D/RT3D code was used to simulate nonreactive and reactive transport. Relatively large dispersivity values were selected to represent typical field-scale conditions; longitudinal and transverse dispersivity values were 12.2 m (40 ft) and 2.4 m (8 ft), respectively. At the Dover AFB site, like most typical contaminated sites, there were no data available for nonreactive transport, so the dispersivity values were estimated based upon the compilation of Gelhar et al. (1992) and then adjusted by comparing simulations to field observations of the contaminant plumes. Although theoretical justification for such large dispersivity values is controversial (Gelhar et al., 1992; Neuman, 1990), this is common practice in field applications lacking detailed observations of tracer transport. Another critically important unknown input is the source zone location and strength. The approach taken here is to represent each node in the general source zone area as a mass injection well. The mass inflow rate at each source zone node and the first-order decay rates are determined by calibration to the observed DNAPL plume. Trial-and-error adjustment was used, although the degradation rates were varied within reasonable ranges based on literature values. Recognizing the non-uniqueness of the calibration, Clement et al. (2000) also conducted a sensitivity analysis and found that the plume length was most sensitive to the transmissivity field, while the organic contaminant mass was most sensitive to the decay constants. As noted by Clement et al. (2000), the modeling exercise serves to improve general understanding about the relevant natural attenuation processes at the Dover AFB site, but it is not possible to use the calibrated model to conduct any predictive assessments of future scenarios where groundwater hydraulics or transport conditions may change.

In the Dover AFB study described above, mixing processes were not significant to the overall reaction because it was assumed that there was excess dissolved organic matter to serve as electron donor in the anaerobic zones and excess oxygen in the aerobic zones. A recent study by Prommer et al. (2006) explicitly accounts for mixing processes in the natural attenuation of micro-pollutants in municipal landfill leachate plumes. Field studies at a mature landfill in Denmark demonstrated the presence of phenoxy acid herbicides (in particular, mecoprop [MCPP]) in the leachate. MCPP is recalcitrant under anaerobic conditions, but can degrade rapidly under aerobic conditions. Accordingly, detailed vertical sampling reported by Tuxen et al. (2006) revealed degradation of these herbicides in the plume fringes where there is a transition from anaerobic to aerobic conditions. Prommer et al. (2006) applied the PHT3D model to study these processes quantitatively.

A steady-state, three-dimensional groundwater flow and tracer transport model was established initially, and the hydraulic conductivity field was adjusted using a combination of manual and automatic calibration to yield a good fit to observed piezometric head data. The grid spacing was 5 m (16.4 ft) for the longitudinal and horizontal direction, and 0.2 m (0.6 ft) for the vertical direction. Dispersivity values were determined through trial-and-error calibration to detailed vertical profiles of chloride and bromide. The results for one of the multi-level sampling wells are shown below in Figure 4.4 where it can be seen that the simulation results are highly sensitive to the value of transverse dispersivity. It is also instructive to compare the small value with the much larger dispersivities used in the Dover AFB case study. In the latter, a two-dimensional coarse-grid vertically-averaged model is used while in the former a three-dimensional fine-grid model is implemented. See also Figure 4.3 for a schematic illustration of why larger dispersivities are used in vertically-averaged models. In Figure 4.4 below, the zone of elevated chloride corresponds approximately to the leachate plume at this location.

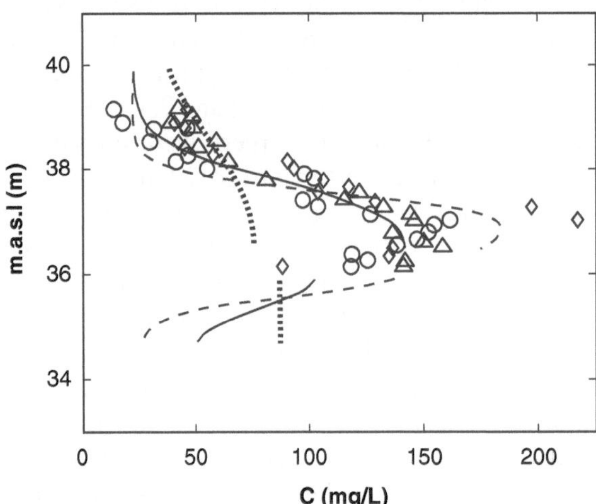

Figure 4.4. Comparison of measured and simulated vertical profiles of chloride at one of the multilevel samplers. The *dashed*, *solid*, and *dotted lines* correspond to simulations with transverse dispersivity equal to 0.3 millimeter (mm), 3 mm, and 3 centimeter (cm), respectively. Taken from Prommer et al., 2006.

In order to save computation time, all PHT3D reactive transport simulations were conducted for a two-dimensional vertical cross-section along the main down-gradient flow path that was extracted from the three-dimensional model. The vertical grid spacing was further reduced in the 2D model to 0.1 m (0.3 ft). The following reactions were considered: (1) mineralization of the leachate dissolved organic carbon (DOC) using oxygen, nitrate and manganese as electron acceptors; (2) nitrification of ammonium present in the leachate by dissolved oxygen; (3) mineralization of MCPP by dissolved oxygen using nitrate as a nitrogen source; and (4) precipitation/dissolution with calcite, goethite, and several manganese minerals. Microbial growth and decay were considered for the MCPP-degrading bacteria, but not for the DOC degraders. Several of the reaction rate parameters were fixed at literature values while others were calibrated manually by fitting the simulation results to the observed vertical profiles of DOC, nitrate, ammonium, alkalinity, sulfate, manganese, oxygen, and MCPP. Some of the modeling results are shown below in Figure 4.5. The flow is generally from right to left, the landfill location is denoted by the red box on top of the water table, and the irregular aquifer bottom is indicated. In the left panel the MCPP reaction zone is clearly indicated by the narrow band of degraders along the boundary between the MCPP plume emanating from the landfill and the oxygen that recharges from the water table. The comparison between simulated and measured vertical profiles of oxygen and MCPP on the right panel dramatically demonstrates the impact of transverse mixing on the narrow reaction zone. Reactive transport modeling helped to elucidate the role of the various interacting processes; for example, it was found that nitrification of the

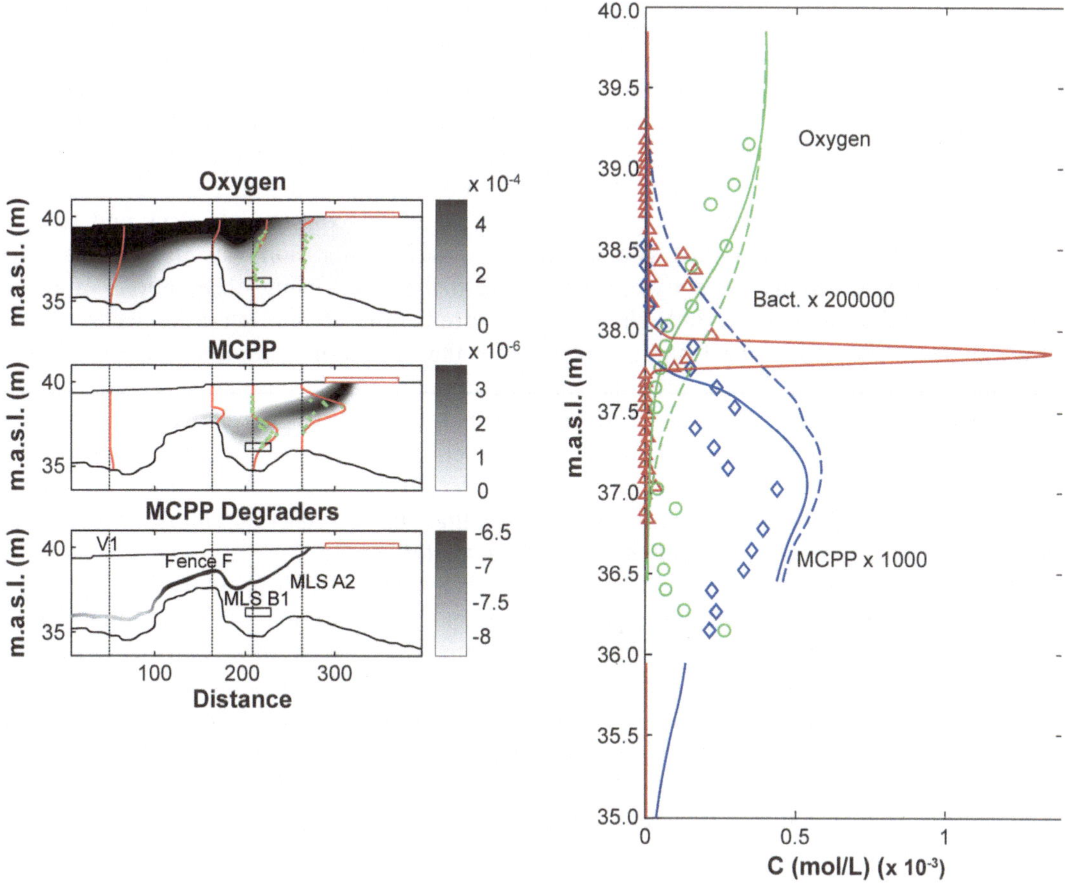

Figure 4.5. ***Left panel*** **shows the results from the 2D PHT3D simulations for oxygen, MCPP, and MCPP degraders. Concentrations are in moles/liter and the MCPP degrader scale is logarithmic.** ***Right panel*** **shows a comparison between simulated and measured vertical profiles for oxygen, MCPP, and MCPP-degraders at multilevel sampler MLSB1 (indicated near distance 200 in the** ***left panel*****). Taken from Prommer et al., 2006.**

ammonium within the leachate plume also occurred in the fringe area and that this reaction competed with MCPP degradation for the available oxygen that was supplied by transverse dispersion. Prommer et al. (2006) also conducted a formal sensitivity analysis by simulating the impact of parameter perturbations. Although it was found that several of the reaction rate parameters were important, the overall degradation of MCPP was particularly sensitive to transverse dispersivity.

4.6.2 Enhanced *In Situ* Cometabolic Degradation of TCE

As noted above in Section 4.2.2, chlorinated aliphatic compounds like TCE can be cometabolically degraded in the presence of oxygen and a primary substrate, such as toluene, methane, phenol or butane. A pilot-scale study of aerobic cometabolic degradation of TCE was conducted at Moffet Federal Air Field, California (Roberts et al., 1990) and model simulations were reported by (Semprini and McCarty, 1991). Here we summarize the modeling of the full-scale field demonstration at Edwards AFB, California reported by (Gandhi et al., 2002b). A background discussion for this demonstration is provided in Chapter 2, Section 2.4.6. The hydrogeology and contamination at the site are somewhat unique; TCE is present in both an upper and lower aquifer that is separated by a low permeability aquitard. This led to use of a recirculation well system (refer to Figure 2.6 in Chapter 2). In this case all of the mixing occurs within the well–the contaminant TCE is drawn into the injection zone by extraction from the adjacent aquifer, and oxygen and the co-substrate toluene are supplied from the surface. Oxygen is continuously added, and toluene is added via regular pulsing. Hydrogen peroxide is added both as an additional oxygen source and to prevent bio-fouling around the injection screen. Bio-active zones are established around each of the injection zones.

A three-dimensional finite-element model was developed to simulate flow and reactive transport in the aquifer-aquitard system and automatic calibration was used to estimate the hydraulic conductivity, effective porosity, and dispersivity using observations of piezometric head and bromide breakthrough curves measured at different observation wells (Gandhi et al., 2002a). The use of the finite element rather than the more common finite difference technique, allowed for a flexible irregular spatial grid required to accurately represent the flow in the vicinity of the recirculation wells. It should be noted that there is a relatively large amount of data available in this study, unlike for most practical large-scale remediation operations. Head observations were available at 38 wells and there were 32 breakthrough curves from a nonreactive tracer (bromide) experiment. It was found that a relatively simple model where each aquifer is homogeneous could fit the data adequately.

The reactive transport model incorporated hydrogen peroxide disproportionation, oxygen outgassing, and inhibition of biomass growth by peroxide into the mass balance equations for TCE, toluene, oxygen, hydrogen peroxide, and biomass. There were a total of 23 reaction parameters. It was found that simulation results were not sensitive to some of the parameters, so these were fixed at literature values. Certain other biomass growth and decay parameters were set equal to values determined from the earlier pilot study at Moffet Field. In the end, there were six parameters that were estimated by automatic calibration; again the observation data set was extensive and consisted of over 14,000 measurements of TCE, dissolved oxygen, and toluene at 41 monitoring locations. The calibrated model was able to capture all of the major trends observed in the data (see Gandhi et al., 2002b for details comparing the observed and simulated time history of concentrations at selected monitoring wells). An example of the simulation model results is shown below in Figure 4.6. The top panel shows the simulated TCE concentrations in the upper aquifer toward the end of the demonstration project, while the bottom panel shows the simulated biomass concentration. The lack of biomass near the

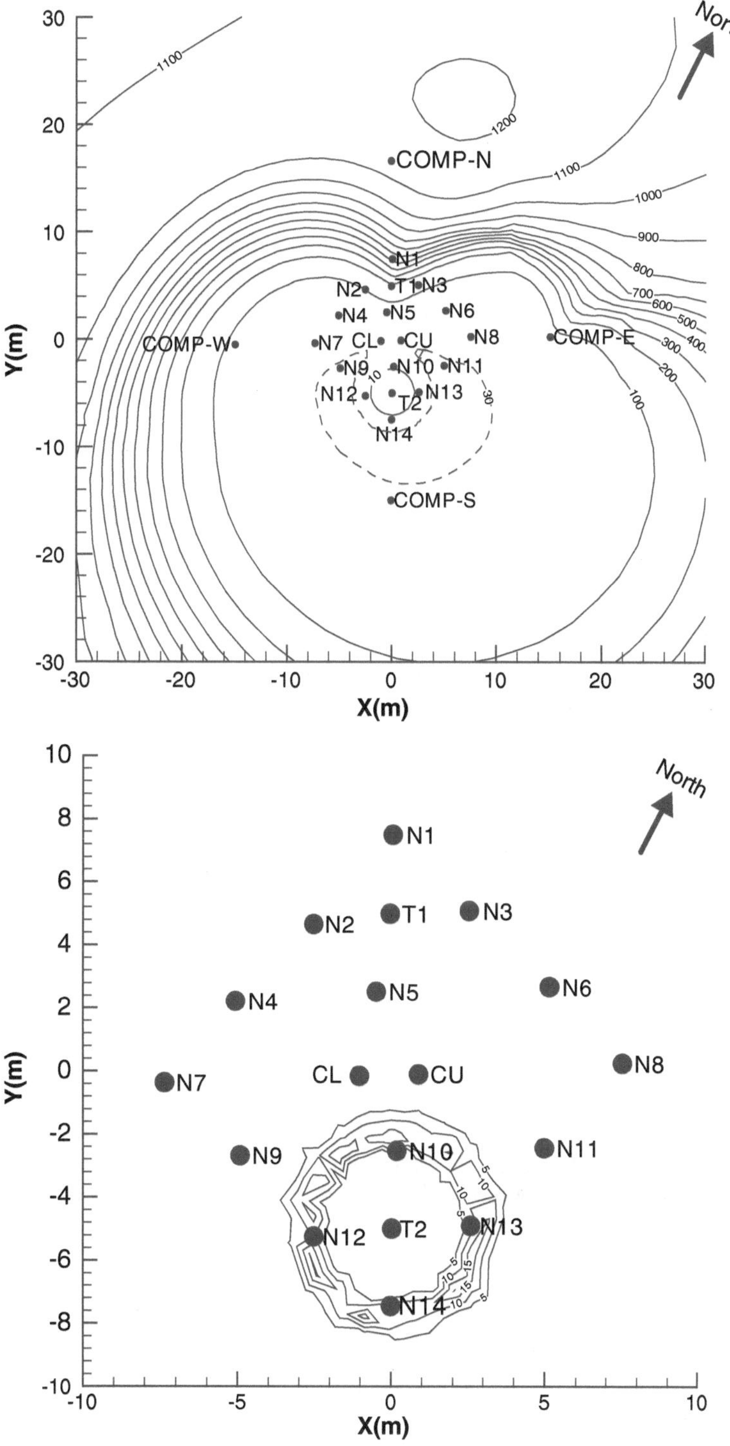

Figure 4.6. Simulated TCE concentration (*top panel*) in micrograms/L and biomass concentration (*bottom panel*) in mg/L on day 444 of the pilot study. From Gandhi et al., 2002b.

injection well (well T2) is due to the inhibitory effect of hydrogen peroxide addition. It can be seen that the bioactive zone is capable of degrading TCE, as evidenced by the reduced concentrations shown in the top panel.

4.6.3 *In Situ* Chemical Oxidation of TCE by Potassium Permanganate

This final case study presents use of the inorganic oxidation reaction Equation 4.8 to reduce TCE contamination through delivery of potassium permanganate ($KMnO_4$). We summarize the recent modeling study of Henderson et al. (2009), which considers the density contrast between the ambient groundwater and the potassium permanganate fluid as a method for reactant delivery and mixing. The general concept is shown below in Figure 4.7, which is a schematic of a pilot field test at an industrial site in Connecticut. The TCE-NAPL source zone is located near the bottom of a sand aquifer, above a low permeability aquitard; this is a common situation since TCE is denser than water. Typical doses of $KMnO_4$ have densities that are also greater than water; hence, if the oxidant solution is injected above the TCE source zone it will sink, spread laterally and react with dissolved TCE. This is represented by the two-stage injection-treatment phases shown in Figure 4.7. In this particular field test, off-site migration was avoided by isolating the area surrounding the source zone within a steel sheet pile keyed into the underlying aquitard. The $KMnO_4$ solution was injected in two 8-h periods 4 days apart. The resulting migration of permanganate and changes in the concentration of TCE and other aqueous species were monitored at 15 multi-level samplers located along a cross section through the contaminated portion of the site.

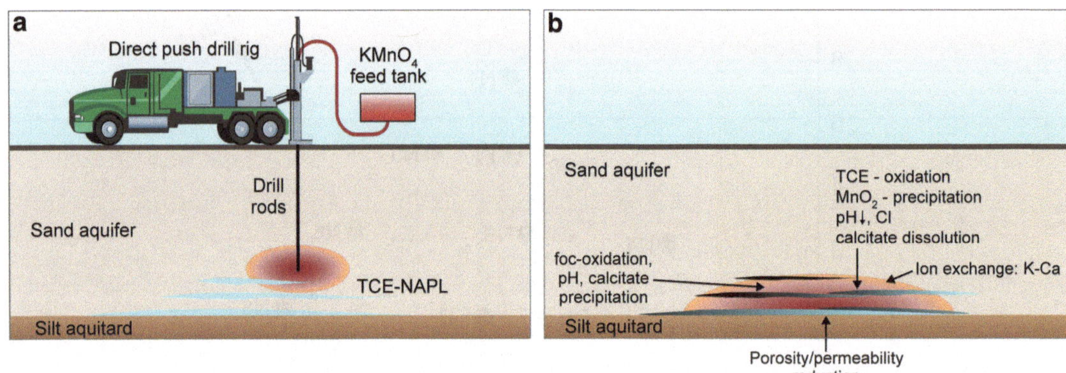

Figure 4.7. Conceptual model of density-driven chlorinated solvent oxidation by permanganate: (a) injection phase; (b) treatment phase including a summary of relevant geochemical processes. From Henderson et al., 2009.

In addition to the complexities of density-driven flow, there are additional complexities caused by multiple geochemical processes, as indicated in Figure 4.7b. The main TCE oxidation, presented earlier as Equation 4.8, leads to precipitation of manganese oxides, increases in concentrations of potassium, chloride, and carbonate species, and a decrease in pH. These species may participate in secondary geochemical reactions; for example, increase in potassium may lead to cation exchange reactions which can displace other major cations from exchange sites on clay minerals. Also, naturally occurring organic matter and other mineral phases react with permanganate and hence compete for the delivered oxidant. Therefore, in order to more fully investigate the impact of geochemical processes upon the remediation, the hydrogeochemical model MIN3P was applied (Henderson et al., 2009; Mayer et al., 2002).

Table 4.4 summarizes the kinetic and equilibrium reactions included in the model. The aqueous complexes participating in the equilibrium speciation species (see Equations 4.16, 4.17) are listed in Table 4.5. Referring to the earlier presentation of the governing equations in Section 4.3, we see that there are ten aqueous components ($N = 10$ in Equation 4.11) and eleven aqueous complexes ($N_x = 11$ in Equation 4.16). Only four immobile components must be explicitly modeled ($\bar{N}=4$ in Equation 4.12) because the cation exchange and TCE sorption reactions are equilibrium controlled and hence the immobile species can be computed from the aqueous species using sorption and exchange isotherm equations.

Table 4.4. Summary of the Reactions Used in the Model of Henderson et al. (2009)

Kinetically-controlled reactions	Equilibrium-controlled reactions
Oxidation of TCE by permanganate (i.e., Equation 4.8)	Sorption of TCE
Oxidation of soil organic matter by permanganate	Cation exchange reactions
Auto-decomposition of permanganate in solution	Aqueous speciation reactions
Dissolution of TCE from the NAPL to aqueous phase	
Precipitation of manganese oxide	

Table 4.5. Summary of the Primary Aqueous Components, Secondary Aqueous Complexes, and Immobile Species Used in the Model of Henderson et al. (2009)

Primary aqueous components (C_i)	Aqueous complexes (x_j)	Immobile species ($\bar{c}_i$)
Potassium (K^+)	OH^-	**Mineral phases**
Sodium (Na^+)	$MgOH^+$	Calcite
Hydrogen (H^+)	$MgCO_3^o$	Magnesium oxide
Calcium (Ca^{+2})	$MgHCO_3^+$	**Cation exchanger phase**
Magnesium (Mg^{+2})	$CaOH^+$	Potassium
Permanganate (MnO_4^-)	$CaHCO_3^+$	Sodium, calcium
Chloride (Cl^-)	$CaCO_3^o$	Magnesium
Carbonate (CO_3^{-2})	$NaCO_3^-$	
TCE (C_2HCl_3)	$NaHCO_3^o$	Soil organic matter
Oxygen ($O_{2\,(aq)}$)	HCO_3^-	TCE as trapped DNAPL
	H_2CO_3	TCE sorbed onto soil

MIN3P uses the finite volume technique to discretize the governing equations of density-dependent flow and multi-component reactive transport. A three-dimensional telescoping mesh was used to have higher spatial resolution in the center of the model domain, the location of the $KMnO_4$ injection zone. Grid dimensions ranged from 0.2 m to 5.5 m (0.6–18 ft) horizontally, and from 0.05 m to 0.25 m (0.16–0.8 ft) vertically. Many hydrogeological and reaction parameter values need to be specified in order to run the model. Values for parameters such as porosity, hydraulic conductivity, soil organic carbon content, and cation exchange capacity are reasonably well constrained from site characterization activities. Based upon the results of sensitivity analyses, the reaction rate parameter for TCE oxidation (Equation 4.8) was fixed to the literature value of Yan and Schwartz (1999), and the rate parameters for the remaining kinetic reactions listed in Table 4.4 were adjusted by trial and error in order to reproduce the observed spatial and temporal trends in TCE and other species. Figure 4.8 shows a comparison

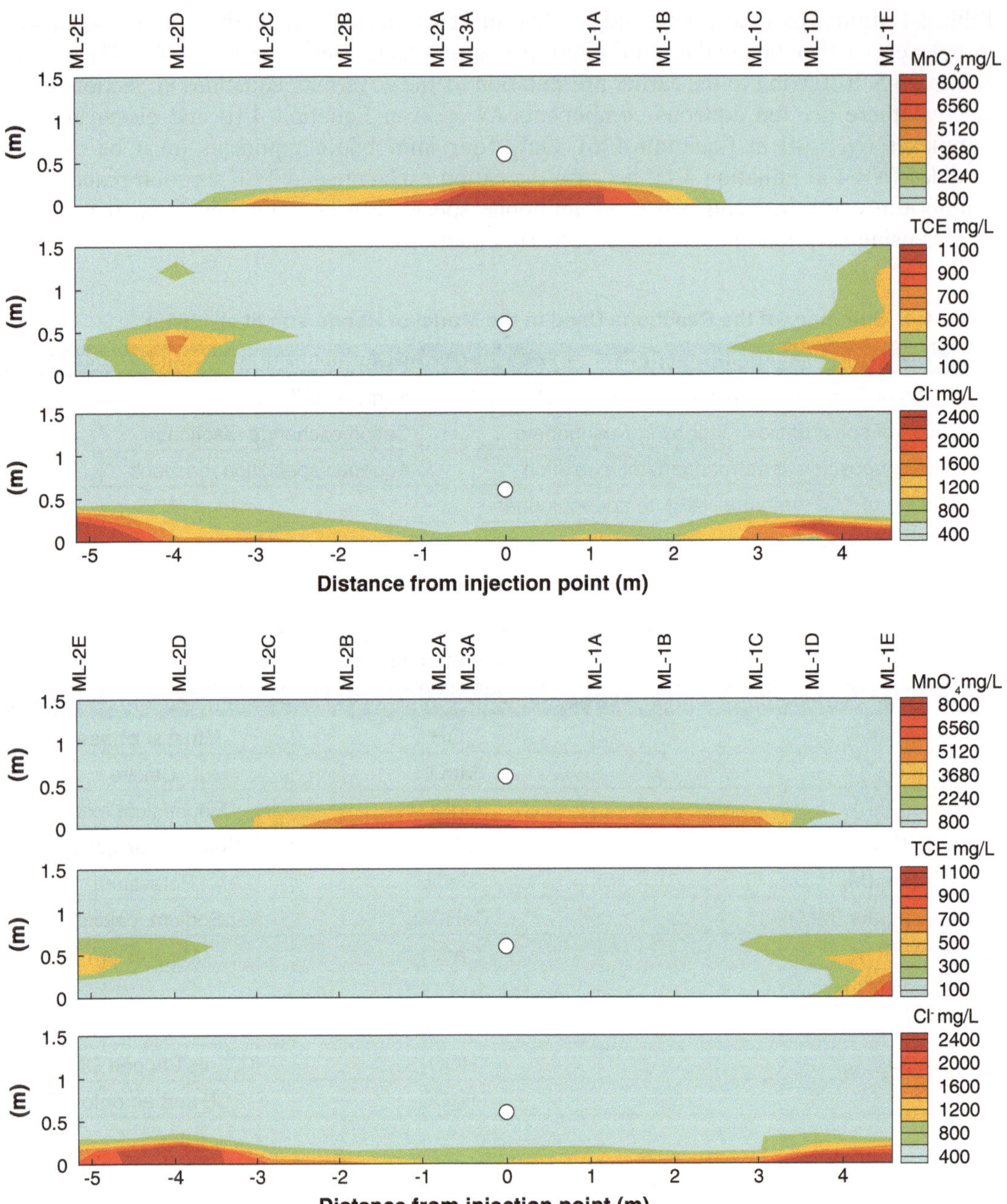

Figure 4.8. Permanganate, TCE, and chloride concentrations 55 days following injection of potassium permanganate. *Top* – observed; *Bottom* – simulated. The open circle denotes the injection point. From Henderson et al., 2009.

between the observed and simulated permanganate, TCE, and chloride concentrations 55 days following injection of the oxidant. These results, along with additional transient patterns from the individual multi-level samplers included in Henderson et al. (2009), indicate that the model is capable of reproducing the main trends in the spreading of permanganate, depletion of dissolved TCE, and generation of chloride.

The simulation model was also used to evaluate the overall remediation efficiency and to conduct additional investigations into the controlling physical and geochemical processes. It was found that approximately 35% of the injected permanganate was consumed by the soil organic matter oxidation, even though the aquifer material is sandy with relatively low organic carbon content. Additional simulations and sensitivity analyses revealed the key role played by density-driven flow patterns in mixing the permanganate reactant with the TCE contaminant. This mixing mechanism proved to be more significant than dispersion, due to the unique injection conditions and also perhaps due to the fact that flow rates diminish greatly following injection because the treatment zone is isolated by sheet pile.

4.7 SUMMARY AND CONCLUSIONS

This chapter has presented an overview of hydrogeochemical models – mathematical and numerical simulation models that couple physical transport and mass-transfer processes with geochemical and biochemical reactions. These models are becoming increasingly important tools for assessment and design of *in situ* remediation systems, because many challenging groundwater contamination problems involve mixtures of organic and inorganic pollutants, and because novel remediation strategies often require delivery of chemicals that react with the pollutants as well as with other dissolved substances and aquifer minerals. Therefore, it is imperative to model the interactions among multiple chemical species in order to assess natural attenuation processes or to properly design chemical delivery systems for *in situ* remediation. These models can be computationally intensive and relatively complicated to set up, but, as surveyed in Table 4.3, there are several widely available codes which are supported by graphical user interfaces that facilitate model building.

The enhanced capability of hydrogeochemical models requires the user to specify values for all the transport and reaction parameters. The case studies summarized in Section 4.6 highlight some of the challenges in applying these models for site-specific prediction. It is widely accepted that good site characterization is a crucial part of any remediation operation. This is required to develop the site conceptual model and to yield sufficient data on piezometric head and groundwater discharge to estimate hydraulic parameters; however, there is rarely sufficient data to estimate reaction parameters. Moreover, reactant delivery and mixing in the subsurface is highly dependent upon small-scale spatial variability of permeability which cannot be estimated with a high degree of certainty. The classical way to quantify these processes is through a dispersion parameter, but recent research summarized at the end of Section 4.5 suggests that this can over-estimate mixing and reaction, thus leading to overly optimistic prediction of remediation effectiveness.

Recognizing these practical challenges, we should not use hydrogeochemical models for site-specific predictions, but rather as a tool to refine and improve our underlying conceptual model of the coupled transport and biogeochemical processes. Hydrogeochemical models serve as a framework to integrate field data with knowledge from laboratory studies and the literature. Model processes and parameters should be continually adjusted and refined by comparing simulation forecasts with field observations. Therefore, during remediation it is imperative to collect field data on multiple chemical species, not only on the contaminant of concern. Hydrogeochemical models provide a tool for systematic evaluation of the main controlling processes, and, as demonstrated by all of the case studies, are invaluable for conducting quantitative sensitivity analyses. Alternative remediation designs (e.g., location and operation of wells for hydraulic control, reactant amount, delivery, and pulsing) can be evaluated, tested and compared. While the detailed predictions from these simulations should not be trusted, overall trends and relative comparisons of designs can be accepted with confidence.

ACKNOWLEDGEMENT

Financial support was provided by the U.S. Department of Energy through a SciDAC-2 project "Modeling Multiscale-Multiphase-Multicomponent Subsurface Reactive Flows using Advanced Computing." I am grateful to Prabhakar Clement, who provided valuable advice in developing Table 4.3.

REFERENCES

Allison JD, Brown DS, Gradac KJN. 1991. MINTEQA2/PRODEFA2, A Geochemical Assessment Model for Environmental Systems: Version 3.0 User's Manual. EPA/600/3-91/021. U.S. Environmental Protection Agency (USEPA), Office of Research and Development, Washington DC, USA, 106 p.

Anderson MA, Woessner WW. 1992. Applied Groundwater Modeling, Simulation of Flow and Advective Transport. Academic Press Inc., San Diego, CA, USA. 391 p.

Ball WP, Roberts PV. 1991. Long term sorption of halogenated organic chemicals by aquifer materials. 2. Intraparticle diffusion. Environ Sci Technol 25:1237–1249.

Barry DA, Prommer H, Miller CT, Engesgaard P, Brun A, Zheng C. 2002. Modelling the fate of oxidisable organic contaminants in groundwater. Adv Water Resour 25:945–983.

Bauer RD, Rolle M, Bauer S, Eberhardt C, Grathwohl P, Kolditz O, Meckenstock RU, Griebler C. 2009. Enhanced biodegradation by hydraulic heterogeneities in petroleum hydrocarbon plumes. J Contam Hydrol 105:56–68.

Bear J. 1979. Hydraulics of Groundwater. McGraw-Hill, New York, NY, USA. 569 p.

Benner SG, Blowes DW, Molson JWH. 2001. Modeling preferential flow in reactive barriers: Implications for performance and design. Ground Water 39:371–379.

Bethke C. 2008. Geochemical and Biogeochemical Reaction Modeling. Oxford University Press, Cambridge, United Kingdom. 543 p.

Binning PJ, Celia MA. 2008. Pseudokinetics arising from the upscaling of geochemical equilibrium. Water Resour Res 44:W07410, doi:10.1029/2007WR006147.

Brauner J, Widdowson M. 2001. Numerical simulation of a natural attenuation experiment with a petroleum hydrocarbon NAPL source. Ground Water 39:939–52.

Brennan RA, Sanford RA, Werth CJ. 2006. Chitin and corncobs as electron donor sources for the reductive dechlorination of tetrachloroethene. Water Res 40:2125–2134.

Brown CL, Pope GA, Abriola LM, Sepehrnoori K. 1994. Simulation of surfactant-enhanced aquifer remediation. Water Resour Res 30:2959–2078.

Brun A, Engesgaard P. 2002. Modelling of transport and biogeochemical processes in pollution plumes: Literature review and model development. J Hydrol 256:211–227.

Carrera J, Alcolea A, Medina A, Hidalgo J, Slooten LJ. 2005. Inverse problem in hydrogeology. Hydrogeo J 13:206–222.

Chilakapati A, Williams M, Yabusaki S, Cole C, Szecsody J. 2000. Optimal design of an in situ Fe(II) barrier: Transport limited reoxidation. Environ Sci Technol 34:5215–5221.

Christ JA, Abriola LM. 2007. Modeling metabolic reductive dechlorination in dense non-aqueous phase liquid source-zones. Adv Water Resour 30:1547–1561.

Christ JA, Goltz MN, Huang J. 1999. Development and application of an analytical model to aid design and implementation of in situ remediation technologies. J Contam Hydrol 37:295–317.

Chu M, Kitanidis PK, McCarty PL. 2005. Modeling microbial reactions at the plume fringe subject to transverse mixing in porous media: When can the rates of microbial reaction be assumed to be instantaneous? Water Resour Res. 41:W06002.

Cirpka OA. 2002. Choice of dispersion coefficients in reactive transport calculations on smoothed fields. J Contam Hydrol 58:261–282.

Cirpka OA. 2005. Effects of sorption on transverse mixing in transient flows. J Contam Hydrol 78:207–229.

Cirpka OA, Kitanidis PK. 1999. An advective-dispersive stream tube approach for the transfer of conservative-tracer data to reactive transport. Water Resour Res 36:1209–2220.

Cirpka OA, Valocchi AJ. 2007. Two-dimensional concentration distribution for mixing-controlled bioreactive transport in steady state. Adv Water Resour 30:1668–1679.

Cirpka OA, Frind EO, Helming R. 1999. Numerical simulation of biodegradation controlled by transverse mixing. J Contam Hydrol 40:159–182.

Cirpka OA, Schwede RL, Luo J, Dentz M. 2008. Concentration statistics for mixing-controlled reactive transport in random heterogeneous media. J Contam Hydrol 98:61–74.

Clement TP, Sun Y, Hooker BS, Petersen JN. 1998. Modeling multispecies reactive transport in ground water. Ground Water Monit Remediat 18:79–92.

Clement TP, Johnson CD, Sun YW, Klecka GM, Bartlett C. 2000. Natural attenuation of chlorinated ethene compounds: model development and field-scale application at the Dover site. J Contam Hydrol 42:113–140.

Clement TP, Truex MJ, Lee P. 2002. A case study for demonstrating the application of U.S. EPA's monitored natural attenuation screening protocol at a hazardous waste site. J Contam Hydrol 59:133–162.

Cline SR, West OR, Siegrist RL, Holden WL. 1997. Performance of in Situ chemical oxidation field demonstrations at DOE sites. American Society of Civil Engineers (ASCE) Geotechnical Conference, Minneapolis, MN, USA, October 5, Special Publication 71:338–348.

Coulibaly KM, Long CM, Borden RC. 2006. Transport of edible oil emulsions in clayey sands: One-dimensional column results and model development. J Hydrol Eng 11:230–237.

Curtis GP. 2005. Documentation and Applications of the Reactive Geochemical Transport Model RATEQ, Draft Report for Comment. Rockville, MD: U.S. Nuclear Regulatory Commission. Report NUREG/CR-6871. 97 p.

Curtis GP, Davis JA, Naftz DL. 2006. Simulation of reactive transport of uranium(VI) in groundwater with variable chemical conditions. Water Resour Res 42:W04404, doi:10.1029/2005WR003979.

Davis JA, Yabusaki SB, Steefel CI, Zachara JM, Curtis GP, Redden GD, Criscenti LJ, Honeyman BD. 2004. Assessing conceptual models for subsurface reactive transport of inorganic contaminants. EOS, Transactions, American Geophysical Union 85:449–455.

de Marsily G, Combes P, Goblet P. 1992. Comment on "Ground water models cannot be validated," by Konikow LF, Bredehoeft JD. Adv Water Resour 15:367–369.

Doherty J. 2002. Manual for PEST, 5th ed. Watermark Numerical Computing, Brisbane, Queensland, Australia.

Doherty J. 2003. Ground water model calibration using pilot points and regularization. Ground Water 41:170–177.

Dybas MJ, Barcelona M, Bezborodnikov S, Davies S, Forney L, Heuer H, Kawka O, Mayotte T, Sepulveda-Torres L, Smalla K, Sneathen M, Tiedje J, Voice T, Wiggert DC, Witt ME, Criddle CS. 1998. Pilot-scale evaluation of bioaugmentation for in-situ remediation of a carbon tetrachloride-contaminated aquifer. Environ Sci Technol 32:3598–3611.

Evans PJ, Koenigsberg SS. 2001. A bioavailable ferric iron assay and relevance to reductive dechlorination. In Leeson A, Alleman BC, Alvarez PJ, Magar VS, eds, Bioaugmentation, Biobarriers, and Biogeochemistry. Battelle Press, Columbus, OH, USA, pp 167–174.

Faybishenko B, Hazen TC, Long PE, Brodie E, Conrad M, Hubbard S, Christensen J, Joyner D, Borglin S, Chakraborty R, Williams K, Peterson J, Chen J, Brown S, Tokunaga T, Wan J,

Firestone M, Newcomer D, Resch C, Cantrell K, Willett A, Koenigsberg S. 2008. In situ long-term reductive bioimmobilization of Cr(VI) in groundwater using hydrogen release compound. Environ Sci Technol 42:8478–8485.

Fang YL, Yeh GT, Burgos WD. 2003. A general paradigm to model reaction-based biogeochemical process in batch systems. Water Resour Res 39:1083–1108.

Fennell DE, Gossett JM. 1998. Modeling the production of and competition for hydrogen in a dechlorinating culture. Environ Sci Technol 32:2450–2460.

Fernàndez-Garcia D, Snchez-Vila X, Guadagnini A. 2008. Reaction rates and effective parameters in stratified aquifers. Adv Water Resour 31:1364–1376.

Fruchter JS, Cole CR, Williams MD, Vermeul VR, Amonette JE, Szecsody JE, Istok JD, Humphrey MD. 2000. Creation of a subsurface permeable treatment zone for aqueous chromate contamination using in situ redox manipulation. Ground Water Monit Remedit 20:66–77.

Gandhi RK, Hopkins GD, Goltz MN, Gorelick SM, McCarty PL. 2002a. Full-scale demonstration of in situ cometabolic biodegradation of trichloroethylene in groundwater 1. Dynamics of a recirculating well system. Water Resour Res 38:101–1016.

Gandhi RK, Hopkins GD, Goltz MN, Gorelick SM, McCarty PL. 2002b. Full-scale demonstration of in situ cometabolic biodegradation of trichloroethylene in groundwater 2. Comprehensive analysis of field data using reactive transport modeling. Water Resour Res 38:111–1119.

Gelhar LW, Welty C, Rehfeldt KR. 1992. A critical review of data on field-scale dispersion in aquifers. Water Resour Res 28:1955–1974.

Goltz MN, Bouwer EJ, Huang J. 2001. Transport issues and bioremediation modeling for the in situ aerobic co-metabolism of chlorinated solvents. Biodegradation 12:127–140.

Gutierrez-Neri M, Ham PAS, Schotting RJ, Lerner DN. 2009. Analytical modelling of fringe and core biodegradation in groundwater plumes. J Contam Hydrol 107:1–9.

Gwo JP, D'Azevedo EF, Frenzel H, Mayes M, Yeh GT, Jardine PM, Salvage KM, Hoffman FM. 2001. HBGC123D: A high-performance computer model of coupled hydrogeological and biogeochemical processes. Comput Geoscienc 27:1231–1242.

Ham PAS, Schotting RJ, Prommer H, Davis GB. 2004. Effects of hydrodynamic dispersion on plume lengths for instantaneous bimolecular reactions. Adv Water Resour 27:803–813.

Hammond GE, Valocchi AJ, Lichtner PC. 2005. Application of Jacobian-free Newton-Krylov with physics-based preconditioning to biogeochemical transport. Adv Water Resour 28:359–376.

Hassan AE. 2004a. A methodology for validating Numerical Ground Water Models. Ground Water 42:347–362.

Hassan AE. 2004b. Validation of Numerical ground water models used to guide decision making. Ground Water 42:277–290.

Hazen T, Tabak H. 2005. Developments in bioremediation of soils and sediments polluted with metals and radionuclides: 2. Field research on bioremediation of metals and radionuclides. Rev Environ Sci Biotechnol 4:157–183.

Henderson TH, Mayer KU, Parker BL, Al TA. 2009. Three-dimensional density-dependent flow and multicomponent reactive transport modeling of chlorinated solvent oxidation by potassium permanganate. J Contam Hydrol 106:195–211.

Hill MC, Tiedeman CR. 2007. Effective Groundwater Model Calibration, With Analysis of Data, Sensitivities, Predictions, and Uncertainty. Wiley, New York, NY, USA. 456 p.

Hyndman DW, Dybas MJ, Forney L, Heine R, Mayotte T, Phanikumar MS, Tatara G, Tiedje J, Voice T, Wallace R and others. 2000. Hydraulic characterization and design of a full-scale biocurtain. Ground Water 38:462–474.

Ibaraki M, Schwartz FW. 2001. Influence of natural heterogeneity on the efficiency of chemical floods in source zones. Ground Water 39:660–666.

Innovative Technology Summary Report. 2000. In Situ Redox Manipulation. U.S. Department of Energy Office of Environmental Management, Office of Science and Technology. Report DOE/EM-0499.

Istok JD, Amonette JE, Cole CR, Fruchter JS, Humphrey MD, Szecsody JE, Teel SS, Vermeul VR, Williams MD, Yabusaki SB. 1999. In situ redox manipulation by dithionite injection: Intermediate-scale laboratory experiments. Ground Water 37:884–889.

Istok JD, Senko JM, Krumholz LR, Watson D, Bogle MA, Peacock A, Chang YJ, White DC. 2004. In situ bioreduction of technetium and uranium in a nitrate-contaminated aquifer. Environ Sci Technol 38:468–475.

Janssen GMCM, Cirpka OA, Van Der Zee SEATM. 2006. Stochastic analysis of nonlinear biodegradation in regimes controlled by both chromatographic and dispersive mixing. Water Resour Res 42:W01417, doi:10.1029/2005WR004042.

Jeen SW, Mayer KU, Gillham RW, Blowes DW. 2007. Reactive transport modeling of trichloroethene treatment with declining reactivity of iron. Environ Sci Technol 41:1432–1438.

Jin M, Butler GW, Jackson RE, Mariner PE, Pickens JF, Pope GA, Brown CL, McKinney DC. 1997. Sensitivity models and design protocol for partitioning tracer tests in alluvial aquifers. Ground Water 35:964–972.

Jones NL, Clement TP, Hansen CM. 2006. A three-dimensional analytical tool for modeling reactive transport. Ground Water 44:613–617.

Jung Y, Coulibaly KM, Borden RC. 2006. Transport of edible oil emulsions in clayey sands: 3D sandbox results and model validation. J Hydrol Eng 11:238–244.

Kao CM, Chen YL, Chen SC, Yeh TY, Wu WS. 2003. Enhanced PCE dechlorination by biobarrier systems under different redox conditions. Water Res 37:4885–4894.

Kapoor V, Gelhar LW, Miralles-Wilhelm F. 1997. Bimolecular second order reactions in spatially varying flows: Segregation induced scale-dependent transformation rates. Water Resour Res 33:527–536.

Kitanidis PK. 1994. The concept of the dilution index. Water Resour Res 30:2011–26.

Knutson C, Valocchi A, Werth C. 2007. Comparison of continuum and pore-scale models of nutrient biodegradation under transverse mixing conditions. Adv Water Resour 30:1421–1431.

Konikow LF, Bredehoeft JD. 1992. Ground-water models cannot be validated. Adv Water Resour 15:75–83.

Kräutle S, Knabner P. 2005. A new numerical reduction scheme for fully coupled multicomponent transport-reaction problems in porous media. Water Resour Res 41:W09414, doi:10.1029/2004WR003624.

Kräutle S, Knabner P. 2007. A reduction scheme for coupled multicomponent transport-reaction problems in porous media: Generalization to problems with heterogeneous equilibrium reactions. Water Resour Res. 43:W03429, doi:10.1029/2005WR004465.

Lee ES, Schwartz FW. 2007. Characteristics and applications of controlled-release KMnO4 for groundwater remediation. Chemosphere 66:2058–2066.

Lee I-S, Bae J-H, Yang Y, McCarty PL. 2004. Simulated and experimental evaluation of factors affecting the rate and extent of reductive dehalogenation of chloroethenes with glucose. J Contam Hydrol 74:313–331.

Li XD, Schwartz FW. 2004a. DNAPL mass transfer and permeability reduction during in situ chemical oxidation with permanganate. Geophys Res Let 31:1–5.

Li XD, Schwartz FW. 2004b. DNAPL remediation with in situ chemical oxidation using potassium permanganatey. Part I. Mineralogy of Mn oxide and its dissolution in organic acids. J Contam Hydrol 68:39–53.

Lichtner PC. 1996. Continuum formulation of multicomponent-multiphase reactive transport. Rev Mineral 34:1–81.

Liedl R, Valocchi AJ, Dietrich P, Grathwohl P. 2005. Finiteness of steady state plumes. Water Resour Res 41:1–8.

Lovley DR, Baedecker MJ, Lonergan DJ, Cozzarelli IM, Phillips EJP, Siegel DI. 1989. Oxidation of aromatic contaminants coupled to microbial iron reduction. Nat 339:297–300.

Lu G, Clement TP, Zheng C, Wiedemeier TH. 1999. Natural attenuation of BTEX compounds: Model development and field-scale application. Ground Water 37:707–717.

Luo J, Cirpka OA, Fienen MN, Wu W-m, Mehlhorn TL, Carley J, Jardine PM, Criddle CS, Kitanidis PK. 2006. A parametric transfer function methodology for analyzing reactive transport in nonuniform flow. J Contam Hydrol 83:27–41.

Luo J, Weber F-A, Cirpka OA, Wu W-M, Nyman JL, Carley J, Jardine PM, Criddle CS, Kitanidis PK. 2007. Modeling in-situ uranium(VI) bioreduction by sulfate-reducing bacteria. J Contam Hydrol 92:129–148.

Luo J, Dentz M, Carrera J, Kitanidis P. 2008. Effective reaction parameters for mixing controlled reactions in heterogeneous media. Water Resour Res 44:W02416, doi:10.1029/2006WR005658.

Maurer M, Rittmann BE. 2004a. Formulation of the CBC-model for modelling the contaminants and footprints in natural attenuation of BTEX. Biodegradation 15:419–434.

Maurer M, Rittmann BE. 2004b. Modeling intrinsic bioremediation for interpret observable biogeochemical footprints of BTEX biodegradation: The need for fermentation and abiotic chemical processes. Biodegradation 15:405–417.

Mayer KU, Blowes DW, Frind EO. 2001. Reactive transport modeling of an in situ reactive barrier for the treatment of hexavalent chromium and trichloroethylene in groundwater. Water Resour Res 37:3091–3103.

Mayer KU, Frind EO, Blowes DW. 2002. Multicomponent reactive transport modeling in variably saturated porous media using a generalized formulation for kinetically controlled reactions. Water Resour Res 38:1174–1194.

McCarty PL. 1997. Breathing with chlorinated solvents. Sci 276:1521–1522.

McCarty PL, Goltz MN, Hopkins GD, Dolan ME, Allan JP, Kawakami BT, Carrothers TJ. 1998. Full-scale evaluation of in situ cometabolic degradation of trichloroethylene in groundwater through toluene injection. Environ Sci Technol 32:88–100.

McLaughlin D, Townley LR. 1996. A reassessment of the groundwater inverse problem. Water Resour Res 32:1131–1161.

Molins S, Carrera JS, Ayora C, Saaltink MW. 2004. A formulation for decoupling components in reactive transport problems. Water Resour Res 40:W10301, doi:10.1029/2003WR002970.

Moore C, Doherty J. 2006. The cost of uniqueness in groundwater model calibration. Adv Water Resour 29:605–623.

NRC (National Research Council). 2000. Natural Attenuation for Groundwater Remediation. Water Science and Technology Board. National Academy Press, Washington DC, USA.

NRC. 2005. Contaminants in the Subsurface: Source Zone Assessment and Remediation. Water Science and Technology Board. National Academy Press, Washington DC, USA.

Neuman SP. 1990. Universal scaling of hydraulic conductivities and dispersivities in geologic media. Water Resour Res 26:1749–1758.

Neuman SP. 2003. Maximum likelihood Bayesian averaging of uncertain model predictions. Stochastic Env Res Risk Assess 17:291–305.

Oostrom M, Wietsma TW, Covert MA, Vermeul VR. 2007. Zero-valent iron emplacement in permeable porous media using polymer additions. Ground Water Monit Remediat 27:122–130.

Oya S, Valocchi AJ. 1997. Characterization of traveling waves and analytical estimation of pollutant removal in one-dimensional subsurface bioremediation modeling. Water Resour Res 33:1117–1127.

Parker JC, Park E. 2004. Modeling field-scale dense nonaqueous phase dissolution kinetics in heterogeneous aquifers. Water Resour Res 40:W05109, doi:10.1029/2003WR002807.

Parkhurst DL, Appelo CAJ. 1999. User's Guide to PHREEQC (version 2) – A Computer Program for Speciation, Batch-reaction, One-dimensional Transport, and Inverse Geochemical Calculations. Report 99–4259. U.S. Geological Survey (USGS), Water-Resources Investigations, Washington DC, USA.

Parkhurst DL, Thorstenson DC, Plummer LN. 1980. PHREEQE–A Computer Program for Geochemical Calculations. Report 80–96. USGS, Water Resources Investigations, Washington DC, USA. 195 p.

Parkhurst DL, Kipp KL, Engesgaard P, Charlton SR. 2004. PHAST – A Program for Simulating Ground-Water Flow, Solute Transport, and Multicomponent Geochemical Reactions. Report 6-A8. USGS, Washington DC, USA.

Patterson RR, Fendorf S, Fendorf M. 1997. Reduction of hexavalent chromium by amorphous iron sulfide. Environ Sci Technol 31:2039–2044.

Pavlostathis SG, Prytula MT, Yeh DH. 2003. Potential and limitations of microbial reductive dechlorination for bioremediation applications. Water Air Soil Pollut Focus 3:117–129.

Phanikumar MS, Hyndman DW, Zhao X, Dybas MJ. 2005. A three-dimensional model of microbial transport and biodegradation at the Schoolcraft, Michigan, site. Water Resour Res 41:1–17.

Poeter E, Anderson D. 2005. Multimodel ranking and inference in ground water modeling. Ground Water 43:597–605.

Poeter EP, Hill MC. 1999. UCODE: A computer code for universal inverse modeling. Comput Geosci 25:457–462.

Prommer H, Davis GB, Barry DA. 1999. Geochemical changes during biodegradation of petroleum hydrocarbons: Field investigations and biogeochemical modeling. Org Geochem 30:423–435.

Prommer H, Barry DA, Davis GB. 2000. Numerical modeling for design and evaluation of groundwater remediation schemes. Ecol Model 128:181–195.

Prommer H, Barry DA, Davis GB. 2002. Modeling of physical and reactive processes during biodegradation of a hydrocarbon plume under transient groundwater flow conditions. J Contam Hydrol 59:113–131.

Prommer H, Barry DA, Zheng C. 2003. MODFLOW/MT3DMS based reactive multicomponent transport modeling. Ground Water 41:247–257.

Prommer H, Tuxen N, Bjerg PL. 2006. Fringe-controlled natural attenuation of phenoxy acids in a landfill plume: Integration of field-scale processes by reactive transport modeling. Environ Sci Technol 40:4732–4738.

Quezada CR, Clement TP, Lee K-K. 2004. Generalized solution to multi-dimensional multi-species transport equations coupled with a first-order reaction network involving distinct retardation factors. Adv Water Resour 27:507–520.

Rittmann BE. 1993. The significance of biofilms in porous media. Water Resour Res 29:2195–2208.

Rittmann BE, McCarty PL. 2001. Environmental Biotechnology. McGraw Hill, Boston, MA, USA. 754 p.

Roberts PV, Hopkins GD, Mackay DM, Semprini L. 1990. A field evaluation of in-situ biodegradation of chlorinated ethenes: Part 1, methodology and field site characterization. Ground Water 28:591–604.

Robinson BA, Viswanathan HS, Valocchi AJ. 2000. Efficient numerical techniques for modeling multicomponent ground-water transport based upon simultaneous solution of strongly coupled subsets of chemical components. Adv Water Resour 23:307–324.

Saenton S, Illangasekare T. 2007. Upscaling of mass transfer rate coefficient for the numerical simulation of dense nonaqueous phase liquid dissolution in heterogeneous aquifers. Water Resour Res 43:W02428, doi:10.1029/2005WR004274.

Salvage KM, Yeh GT. 1998. Development and application of a numerical model of kinetic and equilibrium microbiological and geochemical reactions (BIOKEMOD). J Hydrol 209:27–52.

Schafer W, Therrien R. 1995. Simulating transport and removal of xylene during remediation of a sandy aquifer. J Contam Hydrol 19:205–236.

Scheibe TD, Fang Y, Murray CJ, Roden EE, Chen J, Chien Y-J, Brooks SC, Hubbard SS. 2006. Transport and biogeochemical reaction of metals in a physically and chemically heterogeneous aquifer. Geosphere 2:220–235.

Schirmer M, Molson JW, Frind EO, Barker JF. 2000. Biodegradation modeling of a dissolved gasoline plume applying independent laboratory and field parameters. J Contam Hydrol 46:339–374.

Schroth MH, Oostrom M, Wietsma TW, Istok JD. 2001. In-situ oxidation of trichloroethene by permanganate: effects on porous medium hydraulic properties. J Contam Hydrol 50:79–98.

Semprini L, McCarty PL. 1991. Comparison between model simulations and field results for in-situ biorestoration of chlorinated aliphatics: Part 1. Biostimulation of methanotrophic bacteria. Ground Water 29:365–374.

Semprini L, Dolan ME, Mathias MA, Hopkins GD, McCarty PL. 2007. Laboratory, field, and modeling studies of bioaugmentation of butane-utilizing microorganisms for the in situ cometabolic treatment of 1,1-dichloroethene, 1,1-dichloroethane, and 1,1,1-trichloroethane. Adv Water Resour 30:1528–1546.

Simunek J, Valocchi AJ. 2002. Geochemical Transport. In Topp C, Dane J, eds, Methods of Soil Analysis, Part I, Physical Methods. Soil Science Society of America, Madison, WI, USA, Chapter 6.9.

Srinivasan V, Clement TP, Lee KK. 2007. Domenico solution – Is it valid? Ground Water 45:136–146.

Steefel CI, MacQuarrie KTB. 1996. Approaches to modeling of reactive transport in porous media. Rev Mineral 34:82–129.

Sun Y, Petersen JN, Clement TP. 1999. Analytical solutions for multiple species reactive transport in multiple dimensions. J Contam Hydrol 35:429–440.

Szecsody JE, Fruchter JS, Sklarew DS, Evans JC. 2000. In Situ Redox Manipulation of Subsurface Sediments from Fort Lewis, Washington: Iron Reduction and TCE Dechlorination Mechanisms. Report PNNL-13178. Pacific Northwest National Laboratory, Richland, WA, USA.

Szecsody JE, Fruchter JS, Phillips JL, Rockhold ML, Vermeul VR, Williams MD, Devary BJ, Liu Y. 2005. Effect of Geochemical and Physical Heterogeneity on the Hanford 100 D Area In Situ Redox Manipulation Barrier Longevity. Report PNNL-15499 Rev. 1. Pacific Northwest National Laboratory, Richland, WA, USA.

Tebes-Stevens CL, Valocchi AJ. 2000. Calculation of reaction parameter sensitivity coefficients in multicomponent subsurface transport models. Adv Water Resour 23:591–611.

Tebes-Stevens C, Valocchi AJ, Van Briesen JM, Rittmann BE. 1998. Multicomponent transport with coupled geochemical and microbiological reactions: Model description and example simulations. J Hydrol 209:8–26.

Tonkin M, Doherty J, Moore C. 2007. Efficient nonlinear predictive error variance for highly parameterized models. Water Resour Res 43:W07429, doi:10.1029/2006WR005348.

Tuxen N, Albrechtsen H-J, Bjerg PL. 2006. Identification of a reactive degradation zone at a landfill leachate plume fringe using high resolution sampling and incubation techniques. J Contam Hydrol 85:179–194.

Valocchi AJ, Malmstead M. 1992. Accuracy of operator splitting for advection-dispersion-reaction problems. Water Resour Res 28:1471–1476.

van der Lee J, De Windt L. 2001. Present state and future directions of modeling of geochemistry in hydrogeological systems. J Contam Hydrol 47:265–282.

Vermeul VR, Bjornstad BN, Murray CJ, Newcomer DR, Rockhold ML, Szecsody JE, Williams MD, Xie Y. 2004. In Situ Redox Manipulation Permeable Reactive Barrier Emplacement: Final Report Frontier Hard Chrome Superfund Site, Vancouver, WA. Report PNWD-3361. Battelle Pacific Northwest Division Richland, WA, USA.

Viswanathan HS, Robinson BA, Valocchi AJ, Triay IR. 1998. A reactive transport model of neptunium migration from the potential repository at Yucca Mountain. J Hydrol 209:251–280.

Waddill DW, Widdowson MA. 2000. SEAM3D: A Numerical Model for Three-Dimensional Solute Transport and Sequential Electron Acceptor-Based Bioremediation in Groundwater. Report ERDC/EL-TR-00-18. U.S. Army Environmental Research Development Center, Vicksburg, MS, USA.

Wang S, Jaffe PR, Li G, Wang SW, Rabitz HA. 2003. Simulating bioremediation of uranium-contaminated aquifers; uncertainty assessment of model parameters. J Contam Hydrol 64:283–307.

Westall JC, Zachary JL, Morel FMM. 1976. MINEQL: A computer program for the calculation of chemical equilibrium composition of aqueous systems. Report 18. Department of Civil Engineering, Massachusetts Institute of Technology, Cambridge, MA, USA.

White MD, Oostrom M. 1998. Modeling surfactant-enhanced nonaqueous-phase liquid remediation of porous media. Soil Sci 163:931–940.

White MD, Oostrom M. 2006. STOMP Subsurface Transport Over Multiple Phases, Version 4.0, User's Guide. Pacific Northwest National Laboratory, Richland, WA, USA.

Wiedemeier TH, Rifai HS, Newell CJ, Wilson JT. 1999. Natural Attenuation of Fuels and Chlorinated Solvents in the Subsurface. John Wiley & Sons, New York, NY, USA.

Xu T, Pruess K. 2001. Modeling multiphase non-isothermal fluid flow and reactive geochemical transport in variably saturated fractured rocks: 1. Methodology. Am J Sci 301:16–33.

Xu T, Sonnenthal E, Spycher N, Preuss K. 2004. TOUGHREACT User's Guide: A Simulation Program for Non-Isothermal Multiphase Reactive Geochemical Transport in Variable Saturated Geologic Media. Report Report LBNL-55460. Lawrence Berkeley National Laboratory, Berkeley, CA, USA. 192 p.

Xu T, Apps JA, Pruess K. 2005. Mineral sequestration of carbon dioxide in a sandstone-shale system. Chem Geol 217:295–318.

Xu T, Sonnenthal E, Spycher N, Pruess K. 2006. TOUGHREACT – A simulation program for non-isothermal multiphase reactive geochemical transport in variably saturated geologic media: Applications to geothermal injectivity and CO2 geological sequestration. Comput Geosci 32:145–165.

Yabusaki S. 2001. Multicomponent reactive transport in an in situ zero-valent iron cell. Environ Sci Technol 35:1493–1503.

Yan YE, Schwartz FW. 1999. Oxidative degradation and kinetics of chlorinated ethylenes by potassium permanganate. J Contam Hydrol 37:343–365.

Yeh G-T, Salvage KM, Gwo JP, Zachara JM, and Szecsody JE. 1998. HydroBioGeoChem: A Coupled Model of Hydrologic Transport and Mixed Biogeochemical Kinetic/Equilibrium Reactions in Saturated-Unsaturated Media. Oak Ridge National Laboratory, Oak Ridge, TN, USA.

Yeh GT, Ward DS. 1980. FEMWATER: A finite-element model of water flow through saturated-unsaturated porous media. ORNL-5567, Environmental Sciences Division, Oak Ridge National Laboratory, Oak Ridge, TN, USA. 162 p.

Yeh GT, Tripathi VS. 1990. HYDROGEOCHEM: A Coupled Model of Hydrologic Transport and Geochemical Equilibrium in Reactive Multicomponent Systems. Environmental Sciences Division, Oak Ridge National Laboratory, Oak Ridge, TN, USA.

Yeh GT, Tripathi VS. 1991. A model for simulating transport of reactive multispecies components: Model development and demonstration. Water Resour Res 27:3075–3094.

Yeh GT, Sun JS, Jardine PM, Burgos WD, Fang Y, Li M-H. 2004. HYDROGEOCHEM 5.0: A Three-Dimensional Model of Coupled Fluid Flow, Thermal Transport, and HYDROGEOCHEMical Transport Through Variably Saturated Conditions – Version 5.0. Report ORNL/TM-2004/107. Oak Ridge National Laboratory, Oak Ridge, TN, USA.

Zhang F, Yeh G-T, Parker JC, Brooks SC, Pace MN, Kim Y-J, Jardine PM, Watson DB. 2007. A reaction-based paradigm to model reactive chemical transport in groundwater with general kinetic and equilibrium reactions. J Contam Hydrol 92:10–32.

Zhang H, Schwartz FW. 2000. Simulating the in situ oxidative treatment of chlorinated ethylenes by potassium permanganate. Water Resour Res 36:3031–3042.

Zheng C, Wang PP. 1999. MT3DMS: A Modular Three-Dimensional Multispecies Model for Simulation of Advection, Dispersion and Chemical Reactions of Contaminants in Groundwater Systems; Documentation and User's Guide. Report Contract Report SERDP-99-1. U.S. Army Engineer Research and Development Center, Vicksburg, MS, USA.

Zyvoloski G, Kwicklis E, Eddebbarh AA, Arnold B, Faunt C, Robinson BA. 2003. The site-scale saturated zone flow model for Yucca Mountain: Calibration of different conceptual models and their impact on flow paths. J Contam Hydrol 62–63:731–750.

CHAPTER 5

TRAVEL-TIME BASED REACTIVE TRANSPORT MODELING FOR *IN SITU* SUBSURFACE REACTOR

Jian Luo[1]

[1]School of Civil and Environmental Engineering, Georgia Institute of Technology, Atlanta, GA, USA

5.1 INTRODUCTION

Engineered *in situ* bioremediation of contaminated groundwater often uses multiple-well systems consisting of extraction and injection wells to create subsurface *in situ* reactors acting as treatment zones (McCarty et al., 1998; Christ et al., 1999; Hyndman et al., 2000; Cirpka and Kitanidis, 2001; Cunningham and Reinhard, 2002; Gandhi et al., 2002; Luo and Kitanidis, 2004; Luo et al., 2006b; Wu et al., 2006a, b). Substrates and nutrients are added into injection wells and delivered to the subsurface treatment zones to stimulate indigenous microbial activity as well as abiotic reactions for contaminant removal. The role of these systems in chemical delivery and mixing are discussed in detail in Chapter 6.

A simple example of such a treatment system is the flow field created by an injection-extraction well pair with equal pumping rate (see Figure 5.1) in an aquifer where the flow is primarily horizontal and there originally is uniform regional flow (i.e., the velocity is the same everywhere). The well pair modifies the flow by dividing the domain into three zones of primary interest: a capture zone, a recirculation zone, and a release zone. Outside of these zones flows water from upstream that is not mixed with water from the wells. The recirculation and release zones function as *in situ* treatment reactors for groundwater remediation. Contaminated groundwater from the capture zone is withdrawn by the extraction well and then, mixed with additives, is returned through the injection well into the recirculation and release zones, where reactions occur. In order to design such *in situ* reactor systems and evaluate their performance, efficient and practical modeling tools are needed for simulating subsurface flow and bioreactive transport. Modeling flow and transport is described primarily in Chapter 4. This chapter focuses on a specialized modeling approach that is well suited for simulating reactive transport in *in situ* reactors created by injection and extraction wells.

Several factors make it challenging to model transport and reactions in recirculation systems of interest:

1. Flow fields created by injection and extraction wells are highly non-uniform;
2. Characterization of the hydraulic, chemical, and biological properties of aquifers is costly for heterogeneous media; and
3. Reactions are often nonlinear and involve multiple species.

The optimal design and monitoring of system operations requires a reliable mathematical model of transport and reactions that is computationally efficient and can be calibrated based on data that can be obtained in practice at a reasonable cost. One must properly account for mixing of water from the capture zone with water from the recirculation zone and the addition of substrates in the wells, the nonuniformity of flow and reaction rates within reaction zones,

P.K. Kitanidis and P.L. McCarty (eds.), *Delivery and Mixing in the Subsurface: Processes and Design Principles for In Situ Remediation*, doi: 10.1007/978-1-4614-2239-6_5,

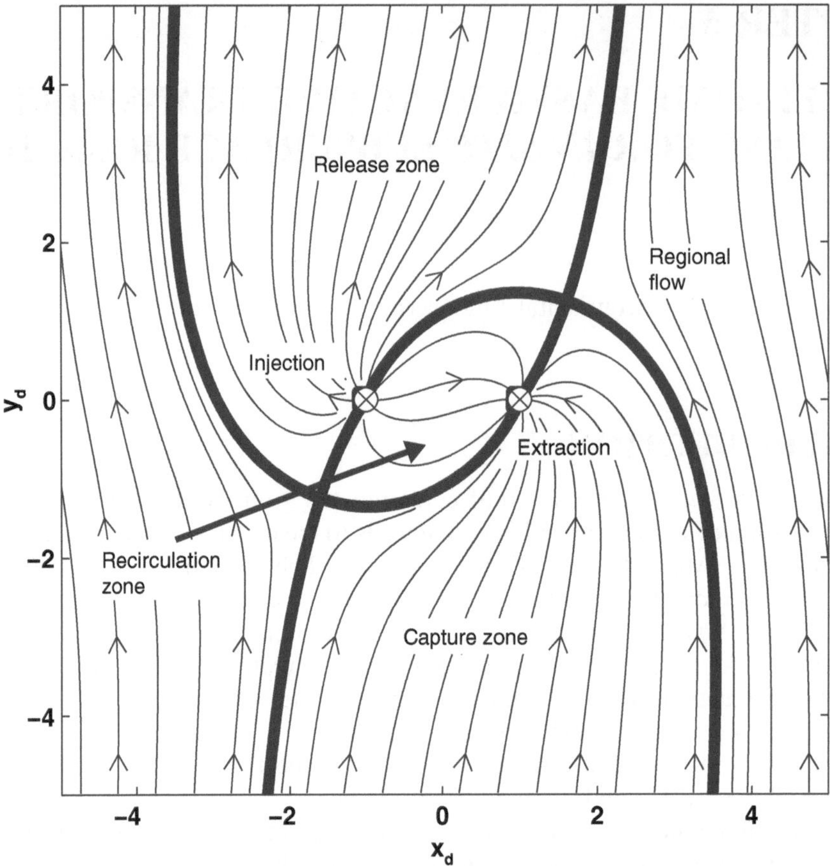

Figure 5.1. Plan view of a flow field created by an injection-extraction well pair in the presence of uniform regional flow. The *solid lines* are streamlines with *arrows* indicating the flow direction. The *solid thick lines* are the separation streamlines passing through stagnation points. The coordinates are normalized by the half separation distance d between the wells. The injection and extraction wells have the same flow rates. Modified from Luo and Kitanidis, 2004.

and other factors that make it difficult, if not impossible, to apply simple analytical techniques, such as one-dimensional (1-D) flow with uniform coefficients.

In addition, during a long-term groundwater remediation project, flow fields may be influenced by many processes, such as biomass accumulation, gas production, solids precipitation, etc., which may require to re-characterize the aquifer. These complexities make it difficult to employ the most common numerical transport models, which require a detailed description of spatially variable parameters, flow velocity, and concentration. Instead, one must consider alternative methods for simulating transport problems in such reactor systems that can avoid the massive effort required for a highly resolved multi-dimensional model, particularly in the planning and designing phase of an *in situ* bioremediation project.

This chapter describes a practical and efficient approach that can circumvent the challenges mentioned above and is well suited for simulating engineered remediation systems involving multiple injection and extraction wells. This approach stems from the residence-time theory developed for reactor design in chemical engineering. Similar approaches have been developed for hydrogeology, including a stochastic-convective approach (e.g., Simmons, 1982; van der Zee and van Riemsdijk, 1987; Shapiro and Cvetkovic, 1988; Simmons et al., 1995), a streamline-based methodology (Crane and Blunt, 1999), and a travel-time based model (Cirpka and

Kitanidis, 2001; Robinson and Viswanathan, 2003; Luo and Cirpka, 2008). All such approaches have been developed based on what is known as a "travel-time domain" instead of the conventional multi-dimensional spatial domain. That is, flow and transport variables are described as functions not of their location in space but their time of travel from a starting point (typically an injection well).

This chapter is organized as follows: Section 5.2 briefly introduces the residence-time theory for a flow-through reactor system and the concept of complete segregation. Section 5.3 presents a travel-time based reactive transport model. Section 5.4 outlines the approaches for evaluating travel-time distributions. Section 5.5 presents a generic example of a travel-time based model with detailed solution algorithms. Section 5.6 discusses various extensions of the travel-time based model. Finally, Section 5.7 gives a summary.

5.2 RESIDENCE-TIME THEORY

"*Residence-time theory deals with particles that enter, flow through, and leave a system*" (Nauman and Buffham, 1983). Consider a simple closed reactor system with only one inlet and one outlet and constant flow rate (see Figure 5.2). Suppose that a number of conservative-tracer particles are introduced as a single unit impulse (i.e., injected all together and having unit total mass) with the influent flow into the reactor system via the inlet and are collected with the effluent flow at the outlet. Residence time (or "age") is defined as the period of time a particle spends in the reactor. It is realistic to expect variability in residence times, described through a distribution, instead of a single residence time for all the particles. Let us focus on a

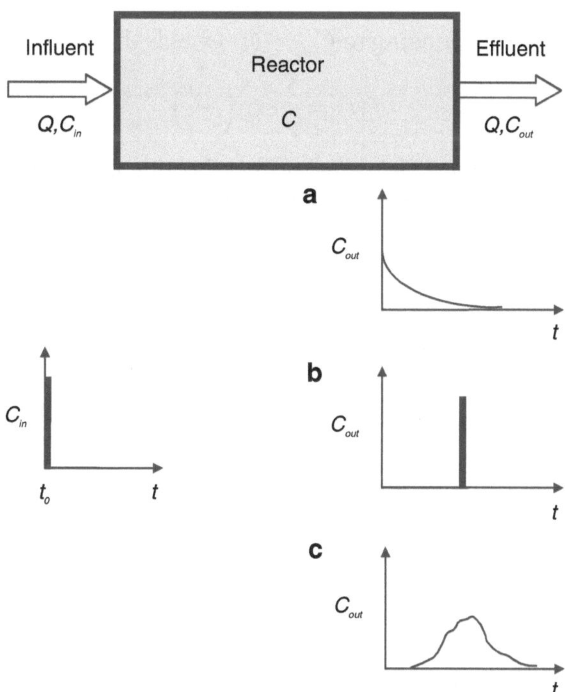

Figure 5.2. Schematic of a closed reactor system with one inlet and outlet. The input at the inlet is a delta function. Response functions are collected at the outlet for different reactors: (a) continuous-flow stirred-tank reactor (CSTR); (b) plug-flow reactor (PFR); and (c) an advection-dispersion reactor.

steady-state (i.e., time invariant) flow system. The probability that the residence time takes a value in the small interval from t to $t + dt$ is $f(t)dt$, where $f(t)$ is called the residence-time density function. In terms of transport of a conservative tracer, $f(t)$ may be interpreted as the system impulse response function or transfer function, i.e., the effluent concentration, C_{out}, corresponding to a concentrated input of unit load at the inlet (i.e., in mathematical terms,, $C_{in} = \delta(t)$, where δ denotes the delta function, $\delta(t) = 0$ except at $t = 0$ and $\int_{-\infty}^{\infty} \delta(\tau)d\tau = 1$).

Consider two basic reactor types: a continuous-flow stirred-tank reactor (CSTR) and a plug-flow reactor (PFR). A CSTR is known as a perfectly mixed reactor that achieves compositional uniformity throughout its volume. Thus, the mass balance of the system can be described by:

$$Q\Delta tC_{in} - Q\Delta tC_{out} = V\Delta C \qquad \text{(Eq. 5.1)}$$

where Q is the constant flow rate, V is the reactor volume, C is the uniform concentration in the reactor, and Δt is the time interval. Due to the compositional uniformity, the effluent concentration leaving a CSTR is identical to that at any point within the reactor, $C_{out} = C$. Substituting $C = C_{out}$ into Equation 5.1, dividing both sides by Δt, and adopting a differential (infinitesimally small) Δt, lead to the differential equation:

$$\frac{\mathrm{d}C_{out}}{\mathrm{d}t} = -\frac{1}{\bar{t}}(C_{out} - C_{in}) \qquad \text{(Eq. 5.2)}$$

where $\bar{t}$ is the average hydraulic residence time, which is a property of the system:

$$\bar{t} = \frac{V}{Q} \qquad \text{(Eq. 5.3)}$$

The solution of C_{out} corresponding to $C_{in} = \delta(t)$ yields the residence time density function:

$$f(t) = \frac{1}{\bar{t}}\exp\left(-\frac{t}{\bar{t}}\right) \qquad \text{(Eq. 5.4)}$$

Thus, a CSTR has an exponential distribution of residence times (see Figure 5.2a). (However, a reactor with an exponential distribution of residence times is not necessarily a CSTR, as pointed out by Nauman and Buffham, 1983). If the initial concentration is $C_{out}(0)$ and no new mass is added after t, $C_{in}(t) = 0$, then $C_{out}(t) = C_{out}(0)\exp\left(-\frac{t}{\bar{t}}\right)$, indicating that the concentration decays exponentially.

Unlike a CSTR, tracer particles introduced into a PFR move downstream in the reactor as a "plug". Every particle stays in the company of particles of the same age and there is no mixing between particles introduced earlier or later. Hence, in an ideal PFR, the residence time density function is given by:

$$f(t) = \delta(t - \bar{t}) \qquad \text{(Eq. 5.5)}$$

as shown in Figure 5.2b.

In an actual subsurface reactor of interest in applications, the result lies somewhat between a CSTR and a PFR as a conservative tracer undergoes both advection and dispersion. The residence-time density function may resemble an inverse-Gaussian density function, which is an adequate representation in some cases (see Figure 5.2c) (Cirpka and Kitanidis, 2001; Robinson and Viswanathan, 2003):

$$f(t) = \sqrt{\frac{\mu^3}{2\pi\sigma^2 t^3}}\exp\left(-\frac{\mu(\mu - t)^2}{2\sigma^2 t}\right) \qquad \text{(Eq. 5.6)}$$

where μ and σ^2 are the mean and variance of the residence times, respectively.

The residence-time distribution of a subsurface reactor system is not sufficient information to determine the actual degree of mixedness within an actual (i.e., non-ideal) reactor (Nauman and Buffham, 1983). Two particles may have the same age (meaning that they were introduced at the same time) but different life expectancy, (meaning one may stay in the reactor longer than the other). For example, all particles in a CSTR have the same distribution of life expectancy independently of age. In a PFR, the life expectancy of a particle is determined strictly by its age. Actual cases fall in between these two extremes.

Given a residence-time distribution, the approach that we will describe here to simulate transport assumes that tracer particles are advected through discrete, non-interacting stream-tubes. Thus, the original reactor is considered to be comprised of a number of PFRs in parallel, each of which experiences the same influent concentration, C_{in}, but is assigned with a specific residence time, τ, and a weighting probability, $f(\tau)\,\Delta\tau$ according to the residence-time density function (see Figure 5.3). The local effluent concentration of a single PFR is given by:

$$c_{out}(t) = \begin{cases} C_{in}(t-\tau), & t \geq \tau \\ 0, & t < \tau \end{cases} \qquad \text{(Eq. 5.7)}$$

where c_{out} is the local effluent concentration for a PFR with the residence time τ. Then, by weighting c_{out} with the probability density of the residence time τ, the averaged effluent breakthrough curve of a conservative tracer corresponding to a unit impulse injection exactly reproduces the density function of residence time:

$$C_{out}(t) = \int_0^\infty \delta(t-\tau) f(\tau) \mathrm{d}\tau = f(t) \qquad \text{(Eq. 5.8)}$$

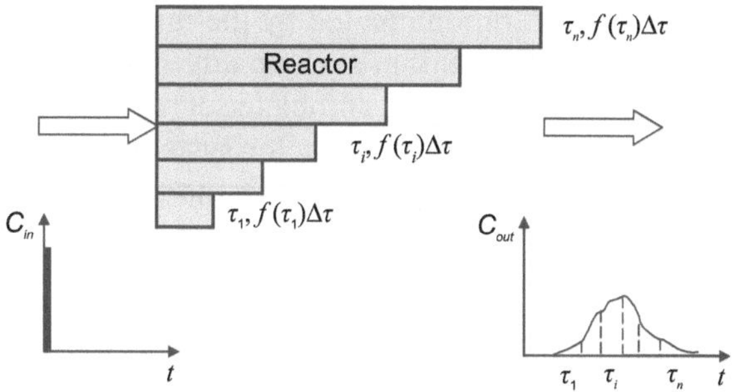

Figure 5.3. The conceptualization of the complete-segregation reactor model. The reactor consists of a number of individual PFRs in parallel, each of which is assigned with a specific residence time and probability to reproduce the residence-time distribution.

Such an approach is known in the chemical-engineering literature as complete segregation (Nauman and Buffham, 1983). Other approaches, such as maximum mixedness, may also be available to describe the extent of mixing given with a residence-time distribution (Zwietering, 1959; Nauman and Buffham, 1983). Robinson and Viswanathan (2003) used both the approach of complete segregation and maximum mixedness as the bounding models for reactive systems.

In an alternative complete-segregation approach, the reactor can be modeled as a single PFR with side exits (see Figure 5.4). The side exits are arranged in a manner to reproduce the residence-time distribution. The advantage of this model is that one only needs to solve a

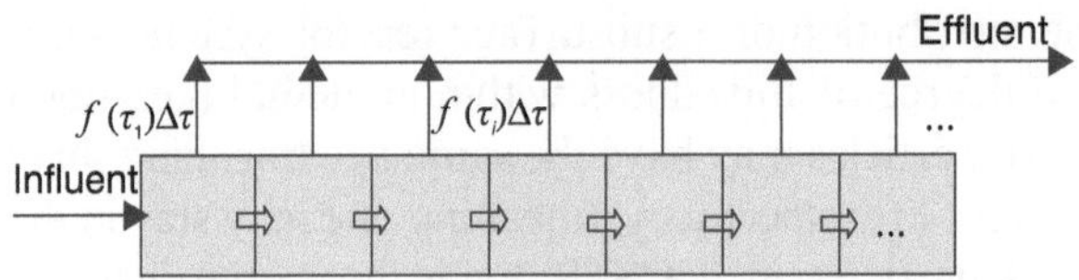

Figure 5.4. Complete segregation by a PFR with side exits. The probability of the side exit is assigned to replicate the residence-time distribution. Modified from Nauman and Buffham, 1983.

transport problem in a single PFR rather than transport problems in multiple parallel PFRs (see Figure 5.3). Within the complete segregation framework, mixing of tracer particles occurs only at the inlet and outlet, while mixing within the reactor is neglected. Thus, for multi-species reactive transport, all reactants must be mixed at the inlet. Otherwise, the reactants would not mix within the reactor, and no reaction would occur. As will be discussed in the following sections, this approach is well suited to simulate the subsurface reactor systems created by injection and extraction wells.

5.3 TRAVEL-TIME BASED REACTIVE TRANSPORT

Compare the recirculation zone in Figure 5.1 and the conceptualization of complete segregation shown in Figure 5.3. The injection well is the inlet, the recirculation zone is the reactor, and the extraction well is the outlet. The streamtubes (i.e., channels between streamlines) within the recirculation zone can be considered as the parallel PFRs. The assumption that all mixing takes place at the inlet and outlet is reasonable for the subsurface reactor system because the residence-time distribution within the recirculation zone is usually a reflection of the highly nonuniform flow configuration created by the injection and extraction wells rather than actual mixing; mixing is usually weak in such an advection-dominated field (Cirpka and Kitanidis, 2001; Luo et al., 2007a). Therefore, complete segregation appears to be a reasonable approximation in simulating such a subsurface reactor system. In groundwater hydrogeology, the travel-time based models and stochastic-convective models are similar to the approach of complete segregation (Simmons, 1982; Simmons et al., 1995; Ginn et al., 1995; Cirpka and Kitanidis, 2001). One may also consider the weights of the parallel PFRs as the flux ratios of streamtubes and the effluent concentration is the flux-averaged concentration (Simmons et al., 1995; Luo and Cirpka, 2008).

Similarly to conservative transport, reactive transport is simulated on the travel-time domain instead of the spatial domain. One may transform the spatial coordinate of a point to the travel time required to reach this point along a streamline path:

$$\tau(\mathrm{x}) = \int_{\mathrm{x}_{in}}^{\mathrm{x}} \frac{\mathrm{d}\xi}{\|v(\xi)\|} \quad \text{(Eq. 5.9)}$$

where ξ is the coordinates along a streamline path. The travel time is zero at the injection point, $\mathbf{x}_{in}$ where every streamtube begins, and is maximum for each streamtube at the extraction point. The travel-time distribution from $\mathbf{x}_{in}$ to $\mathbf{x}$ is discretized into a number of discrete, non-overlapping and non-interacting streamtubes. Advective-reactive transport is simulated in each streamtube:

$$\frac{\partial c_i}{\partial t} + \nabla \cdot (vc_i) = r_i(c_1, c_2, \cdots) \quad \text{(Eq. 5.10)}$$

where the subscript represents species i, c is the concentration, v is the spatially variable velocity, and r is the reaction rate, which may result from both equilibrium and kinetic reactions. Substituting Equation 5.9 into Equation 5.10 yields the travel-time based advective-reactive transport equation (e.g., Simmons et al., 1995; Ginn et al., 1995; Crane and Blunt, 1999; Chilakapati and Yabusaki, 1999; Cirpka and Kitanidis, 2001):

$$\frac{\partial c_i}{\partial t} + \frac{\partial c_i}{\partial \tau} = r_i(c_1, c_2, \cdots) \quad \text{(Eq. 5.11)}$$

By solving Equation 5.11, or rather a system of equations – one for each species, the solution c_i is then computed as a function of actual time and, instead of spatial coordinates, travel time. Predictions of the effluent concentration breakthrough curve, C_i, at the outlet are obtained by integrating the concentration computed at each travel time over the entire range of travel times:

$$C_i(t) = \int_0^{\infty} c_i(t, \tau) f(\tau) \mathrm{d}\tau \quad \text{(Eq. 5.12)}$$

Travel-time based models can be viewed as macroscopic models that can simulate transport in highly nonuniform flow fields or highly heterogeneous media better than macroscopic models in spatial domains because travel-time distributions are much more versatile as modeling tools than mechanistic transport models (e.g., Selroos and Cvetkovic, 1992; Berkowitz and Scher, 1997, 1998; Berkowitz et al., 2000, 2006; Di Donato et al., 2003; Di Donato and Blunt, 2004; Fiori et al., 2006). In addition, a major advantage of simulating transport in the travel-time domain is that transport described by Equation 5.11 becomes one-dimensional with a uniform "velocity" (i.e., the coefficient that multiplies $\frac{\partial c_i}{\partial \tau}$ is 1) whereas traditional spatial models, Equation 5.10, describe transport in a multi-dimensional domain with potentially highly variable velocity (Simmons et al., 1995; Crane and Blunt, 1999; Cirpka and Kitanidis, 2001). The numerical solution of a travel-time system can be achieved much more efficiently and accurately. Furthermore, the travel-time distributions needed for the approach may be conveniently evaluated by conducting conservative tracer tests, and the aquifer heterogeneities do not need to be characterized in an explicit form (Simmons, 1982; Simmons et al., 1995). This is a major advantage when it comes to practical implementation. Thus, travel-time based models are well suited to simulate breakthrough curves at locations where conservative tracer data are available and the travel-time distributions are estimated. Contrarily, the travel-time based model may not be as efficient as a spatially-based model for predicting concentrations at locations with unknown travel-time distributions.

5.4 ESTIMATION OF TRAVEL-TIME DISTRIBUTION

Travel-time distributions are essential for developing travel-time based models. As mentioned above, travel-time distributions may be conveniently evaluated by conducting conservative-tracer tests. In subsurface media without significant tracer retention in stagnation zones, the transfer function corresponding to a unit impulse input may be interpreted as the travel-time distribution, described by Equation 5.8. For other input functions, the concentration breakthrough curve of a conservative tracer is described by:

$$C_{out}(t) = \int_0^{\infty} C_{in}(t - \tau) f(\tau)\, d\tau \quad \text{(Eq. 5.13)}$$

Given the concentration profile measured in the injection well, C_{in}, and the breakthrough curve in the extraction well or observation well, C_{out}, the travel-time distribution, f, can be estimated through mathematical techniques known as deconvolution. In the special case that the tracer is continuously injected with a constant concentration, i.e., C_{in} is constant, we have the following relationship:

$$f = \frac{1}{C_{in}} \frac{dC_{out}}{dt} \quad \text{(Eq. 5.14)}$$

Both parametric and nonparametric methods are available for estimating the travel-time distribution (Fienen et al., 2006; Luo et al., 2006a). The method based on parametric distributions such as gamma distribution (Luo et al., 2006a) and inverse Gaussian distribution (Cirpka and Kitanidis, 2001) is quite efficient but may not be versatile enough to describe multimodal, heavily-tailing travel-time distributions. By contrast, nonparametric methods are more flexible because they do not predetermine the shape of the travel-time distribution, but are computational more expensive (Fienen et al., 2006).

Numerical methods based on spatial models can be used to evaluate travel-time distributions by simulating the flow field and conducting streamline-tracing or particle-tracking (e.g., Pollock, 1988; Zheng and Bennett, 2002; Fienen et al., 2006). However, the spatial distribution of hydraulic conductivity field is required. In a preliminary designing phase, by assuming two-dimensional homogeneous media, the streamline-tracing technique based on the complex potential theory is efficient and well suited to evaluate the travel-time distribution in the recirculation zone created by injection-extraction wells (Strack, 1989; Fienen et al., 2005).

5.5 AN ILLUSTRATIVE EXAMPLE

In this section, we will illustrate by example the procedures to set up a travel-time based model for simulating bioreactive transport in an *in situ* reactor created by injection and extraction wells. The reaction system follows the bioreactive kinetics presented by Semprini and McCarty (1991). More complicated cases can be found in Cirpka and Kitanidis (2001), Weber (2002), and Luo et al. (2007a). Consider the two-dimensional dipole flow field shown in Figure 5.1. All the water extracted from one well is re-injected into another. The line connecting the two wells is perpendicular to the regional flow direction. The domain is initially contaminated by compound A with a uniform concentration distribution. Substrate S is continuously added into the water in the injection well to force a constant injection concentration, $S_b = 100$mg/L. Indigenous biomass, X, is stimulated in the subsurface for the removal of A. Table 5.1 summarizes the hydraulic and reaction parameters. The procedure to build a travel-time based model for evaluating the system performance consists of the following steps:

Table 5.1. Hydraulic and Reactive Parameters for the Illustrative Example

Hydraulic parameters	Value
Half separation distance between wells, d [m]	1
Well pumping rate per unit aquifer thickness, Q_w [m^2/day]	12.56
Regional flow rate per unit aquifer thickness, q_r [m/day]	1
Effective porosity, n_e [–]	0.4
Regional flow orientation to the positive x-axis, α	$\pi/2$
Reaction parameters	**Value**
Maximum specific growth rate, $\hat{\mu}$ [1/d]	1
Half saturation constant of electron donor, K_S [mg$_S$/L]	0.5
Half saturation constant of electron acceptor, K_A [mg$_A$/L]	0.1

(*continued*)

Table 5.1 (continued)

Reaction parameters	Value
Yield coefficient of electron donor, Y_S [mg_{VSS}/mg_S]	0.1
Yield coefficient of electron acceptor, Y_A [mg_{VSS}/mg_A]	0.2
Endogenous-decay coefficient, b [/d]	0.05
Initial biomass concentration, X_0 [mg_{VSS}/L]	1×10^{-3}
Initial concentration of electron donor, S_0 [mg_S/L]	0
Initial concentration of electron acceptor, A_0 [mg_A/L]	100

1. Evaluation of the travel-time distribution. The travel-time distribution $f(\tau)$ (see Figure 5.5a) from the injection well to the extraction well is evaluated by a semi-analytical streamline tracing scheme (Luo and Kitanidis, 2004):

$$\tau = \frac{1 - \pi F \cot(\pi F)}{\sin^2(\pi F)} \qquad \text{(Eq. 5.15)}$$

where F is the cumulative distribution function (cdf). For a field case, a conservative-tracer test without tracer recirculation is a more convenient and accurate way to obtain the travel-time distribution. Suppose that a certain amount of conservative tracer is instantaneously released at the injection well and a concentration breakthrough curve is measured at the extraction well. The travel-time distribution, $f(\tau)$, is given by:

$$f(\tau) = \frac{1}{M} C_{out}(t) \qquad \text{(Eq. 5.16)}$$

where M is the total mass of captured tracer by the extraction well. For tests with tracer recirculation, deconvolution is usually needed to obtain the travel-time distribution.

2. Discretization of the travel-time distribution. $f(\tau)$ is discretized into a number of bins with a constant travel-time interval $\Delta\tau$. Each travel-time bin, indexed by i, is assigned a probability, $p_i = f(i\Delta\tau)\,\Delta\tau$ (see Figure 5.5b). Note that for this system the extracted water in the extraction well consists of the recirculated water and the captured regional flow. That is, not all the streamlines starting at the injection well will arrive at the extraction well. The proportion of recirculated water to the well pumping rate is $P_r = 0.47$. For example, suppose that 100 particles are captured by the extraction well. 47 are from the injection well through the recirculation zone, and the other 53 from the capture zone. By virtue of flow balance, among 100 fluid particles released in the injection well, 47 travel in the recirculation zone to the extraction well, and 53 move to the downgradient areas.
3. 1-D reactive transport simulation on the travel-time domain. The one-dimensional advective-reactive transport is governed by:

$$\frac{\partial S}{\partial t} + \frac{\partial S}{\partial \tau} = r_S \qquad \text{(Eq. 5.17)}$$

$$\frac{\partial A}{\partial t} + \frac{\partial A}{\partial \tau} = r_A \qquad \text{(Eq. 5.18)}$$

$$\frac{dX}{dt} = r_X \qquad \text{(Eq. 5.19)}$$

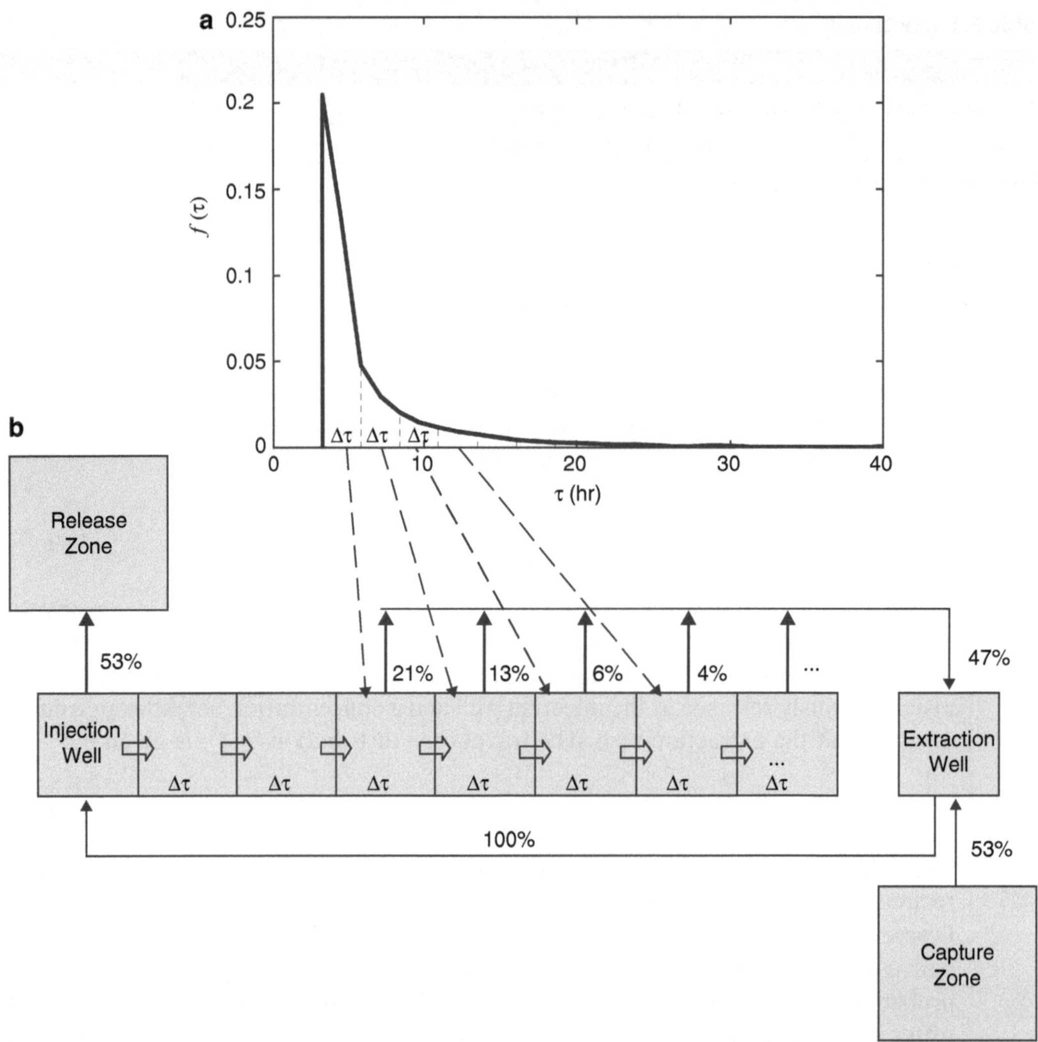

Figure 5.5. Travel-time discretization to build a travel-time based model for the generic example. (a) The travel-time distribution evaluated by streamline tracing; (b) the 1-D advective-reactive travel-time model with side exits for each travel-time element. The probability is assigned to replicate the travel-time distribution.

where the reaction rates r_S, r_A, and r_X are described by the multiplicative Monod model:

$$r_S = -\frac{\hat{\mu}}{Y_S}\frac{S}{K_S + S}\frac{A}{K_A + A}X \tag{Eq. 5.20}$$

$$r_A = -\frac{\hat{\mu}}{Y_A}\frac{S}{K_S + S}\frac{A}{K_A + A}X \tag{Eq. 5.21}$$

$$r_X = \hat{\mu}\frac{S}{K_S + S}\frac{A}{K_A + A}X - bX \tag{Eq. 5.22}$$

Because the initial biochemical conditions are uniform, only one advective-reactive transport model needs to be solved by applying the methodology of a PFR with side exits shown by Figure 5.4. The travel-time domain is discretized into a finite number of elements, each of which represents a travel-time interval $\Delta\tau$, the same as the discretization of the travel-time distribution (Figure 5.5b). Correspondingly, the side exit of each travel-time element is assigned with the probability, $p_i = f(i\Delta\tau)\,\Delta\tau$. Operator-splitting schemes can be applied to solve the equations (Valocchi and Malmstead, 1992). The time step is chosen to be $\Delta t = \Delta\tau$. For each time step, Equations (5.17, 5.18, 5.19) are solved for each travel-time element. Then, the obtained aqueous concentrations, S and A, are shifted one element downstream for the next-step calculation. The biomass concentration is assumed to be immobile. Robinson and Viswanathan (2003) introduced another method to construct the 1-D model with side exits. Figure 5.6 shows the concentration distributions over the travel-time domain at different time steps.

The first element is the injection well, where the concentration S is forced to be constant and A is equal to the concentration in the extracted water from the extraction well:

$$S(t, \tau = 0) = S_{in}(t) = S_b \quad \text{(Eq. 5.23)}$$

$$A(t, \tau = 0) = A_{in}(t) = A_{out}(t) \quad \text{(Eq. 5.24)}$$

The extracted water is the mixture of all the effluent from the side exits and the captured water from the uniform regional flow (Figure 5.5b). Thus, the concentrations in the extracted water are given by:

$$S_{out}(t) = \int_0^{\infty} f(\tau)\,S(t,\tau)\,\mathrm{d}\tau + \left(1 - \int_0^{\infty} f(\tau)\,\mathrm{d}\tau\right) S_0 = \sum_{i=1}^{N_\tau} p_i S(t, i\Delta\tau) + (1 - P_r)\,S_0 \quad \text{(Eq. 5.25)}$$

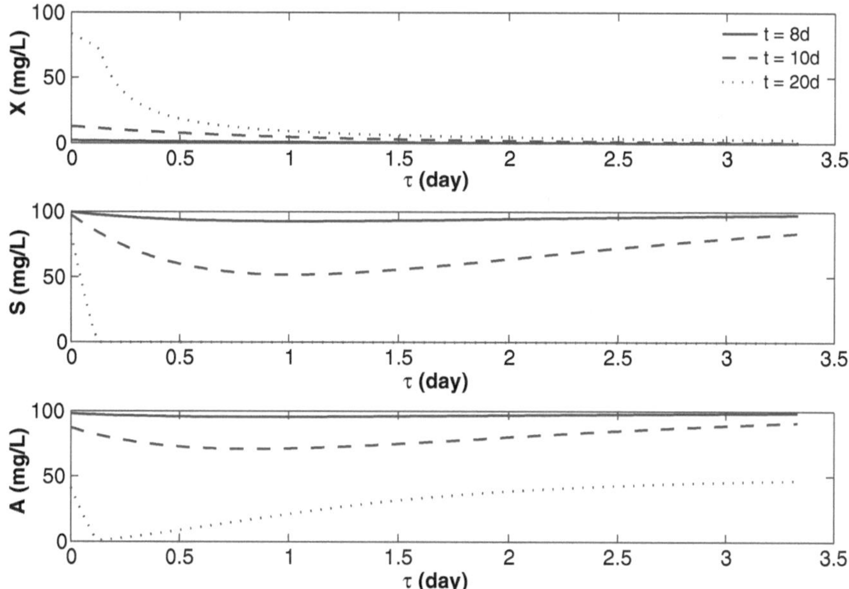

Figure 5.6. The concentration distributions of biomass X, substrate S, and species A over the travel-time domain at different times.

$$A_{out}(t) = \int_0^\infty f(\tau)A(t,\tau)\,\mathrm{d}\tau + \left(1 - \int_0^\infty f(\tau)\,\mathrm{d}\tau\right)A_0 = \sum_{i=1}^{N_\tau} p_i A(t, i\Delta\tau) + (1 - P_r)A_0 \qquad \text{(Eq. 5.26)}$$

Figure 5.7 shows the concentration breakthrough curves in the extraction well. The required mass of S injected into the injection well to force the constant injection concentration is then given by:

$$m_S(t) = Q_w(S_b - S_{out}) \qquad \text{(Eq. 5.27)}$$

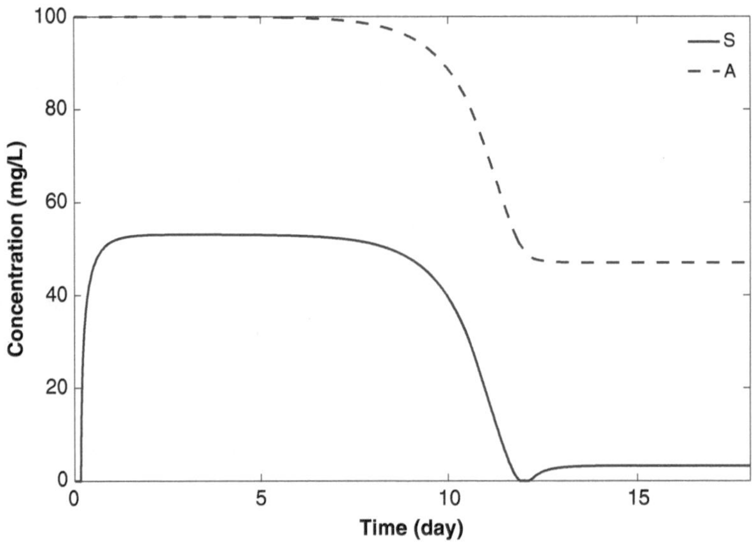

Figure 5.7. Concentration breakthrough curves of substrate S and species A at the extraction well.

5.6 DISCUSSION AND EXTENSIONS

5.6.1 Spatial Mapping

In general, travel-time models can only be applied to evaluate the concentration breakthrough curves at locations where the travel-time distributions are available. Only when the spatial distribution of travel times is available can the spatial concentration distribution be evaluated. The spatial travel-time distribution relates the spatial coordinates with the travel times, which may be available if tracer tests are sampled by a dense monitoring network or a numerical streamline-tracing is applicable. The following summarizes the procedures for spatial concentration mapping:

1. Evaluation of the spatial travel-time distribution. Figure 5.8 shows the spatial distribution of the isochrones (i.e., lines of equal travel time) in the recirculation and release zone evaluated by a streamline-tracing approach.
2. 1-D reactive transport simulation on the travel-time domain. This step has been described above.
3. Spatial mapping of 1-D travel-time based solution onto the spatial domain. At a certain time t, species concentration distributions over the travel-time domain (see Figure 5.6)

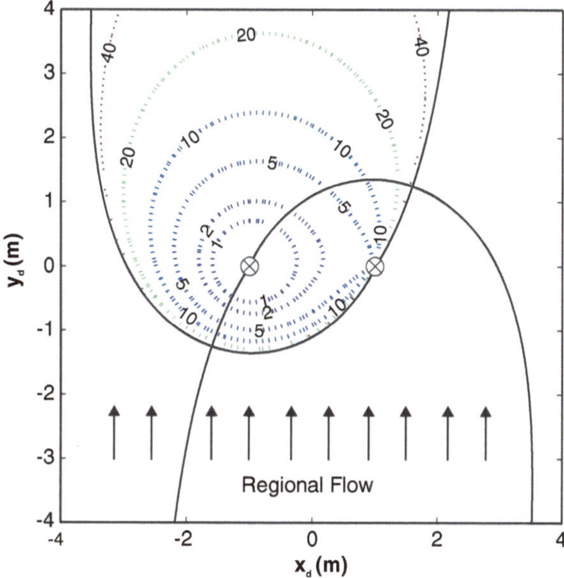

Figure 5.8. Spatial travel-time distribution in the recirculation zone and the release zone. The *dashed lines* are isochrone lines (hours after release), the *thick solid lines* are the separation streamlines delineating the capture zone, the recirculation zone, and the release zone, and the *arrows* indicate the regional flow. The simulation is accomplished by a streamline-tracing technique based on the mathematically modeled flow field.

can be mapped onto the spatial coordinates to visualize the spatial transport model (Cirpka and Kitanidis, 2001; Weber, 2002) (Figure 5.9). Because only one reactive transport is solved in this case, all spatial locations that have a same travel time, e.g., the isochrones in Figure 5.8 have the same species concentrations at each time moment. In this case, the release zone also functions as a treatment zone because the well-mixed injected substrates and contaminants can stimulate the biomass growth in the release zone for contaminant removal.

5.6.2 Multiple-Reactor System

The example above considers only one pair of the injection and extraction wells. In field applications, multiple injection and extraction wells may be installed for groundwater remediation (e.g., Gandhi et al., 2002; Wu et al., 2006a, b). The travel-time approach can be conveniently extended to simulate reactive transport problems in such a multiple-reactor system. For example, Figure 5.10 shows a three-well system consisting of an upgradient extraction well and two downgradient injection wells. Such a well system creates one capture zone, two recirculation zones, and two release zones.

The travel-time model is conceptualized in Figure 5.11. Two travel-time distributions are evaluated and discretized for the two recirculation zones. For each injection well, we need to develop and solve a one-dimensional travel-time based reactive transport model. Of course, if the injection concentration profiles in the two injection wells are identical and the initial biochemical conditions are uniform, one may only need to solve one travel-time based model. In addition, if the system is symmetrical (see Figure 5.10a), i.e., the travel-time

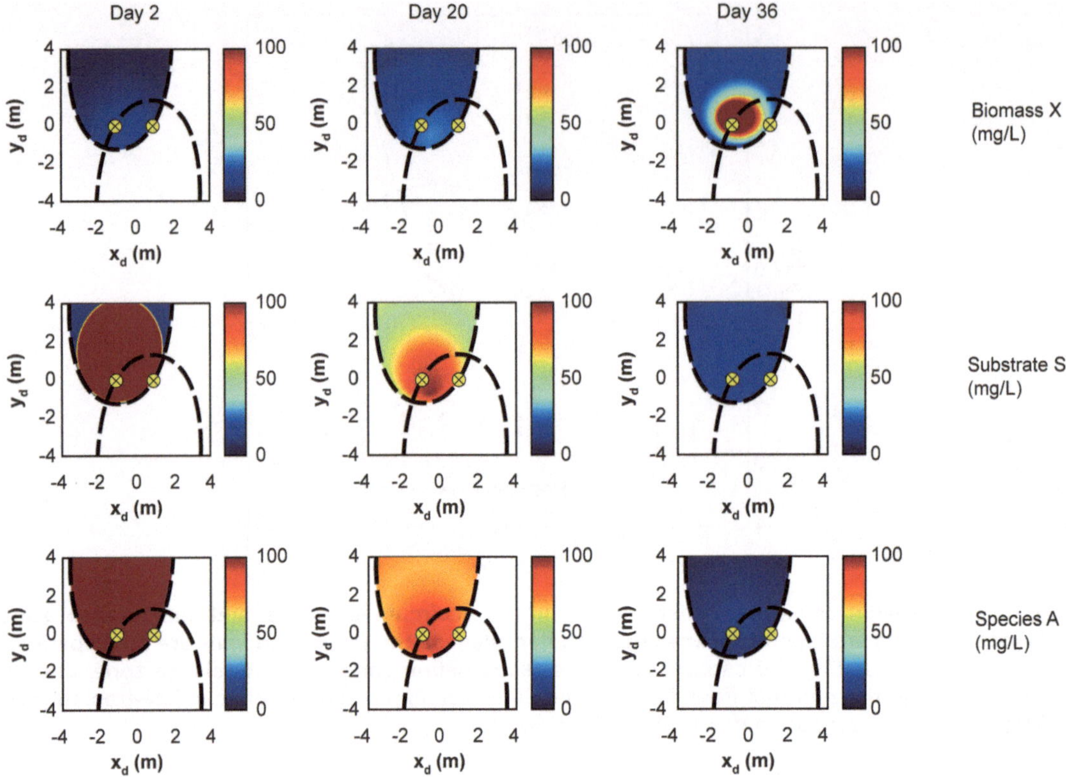

Figure 5.9. Spatial distributions of species concentrations at different time moments. The *first row* is the biomass X, the *second* is the substrate S, and the *bottom row* is the species A.

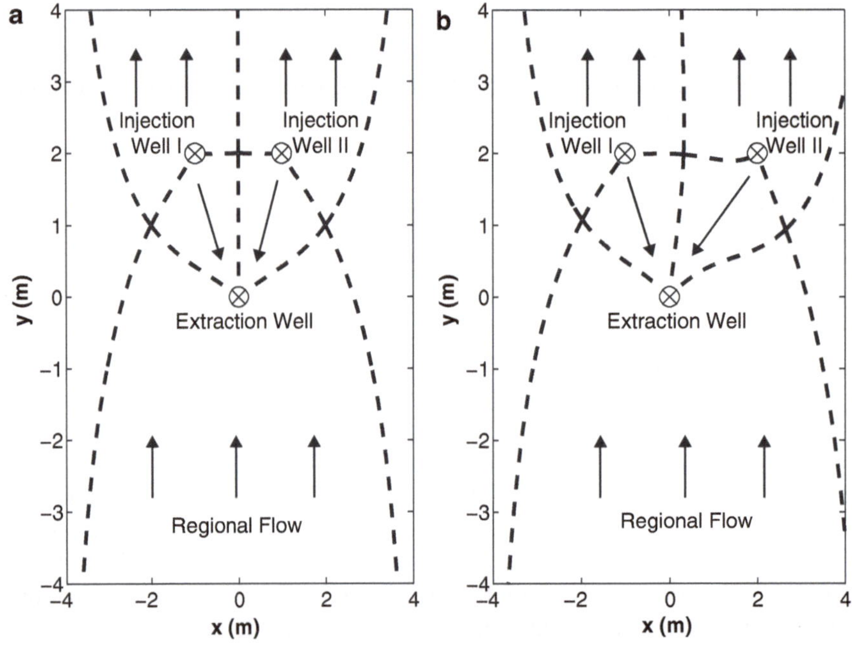

Figure 5.10. A three-well system consisting of an upgradient extraction well and two downgradient injection wells. (a) A symmetric system; (b) a nonsymmetric system.

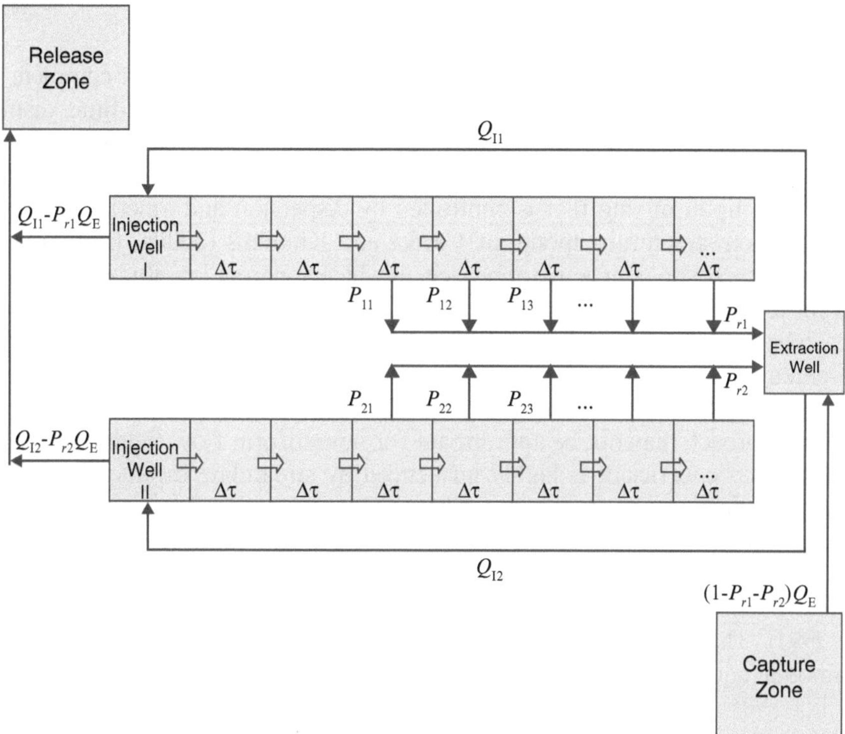

Figure 5.11. The travel-time based model for the three-well system.

distributions are identical from the two injection wells to the extraction, then the probabilities associated with the side exits of the travel-time elements are identical, i.e., $P_{1i} = P_{2i}$, where P_{1i} and P_{2i} are the probabilities of the *i*th travel-time element for the injection wells I and II, respectively. In field applications, tracer tests with multiple tracer injection may be used to identify the travel-time distributions at different monitoring wells or from different injection wells (Reimus et al., 2003).

Figure 5.11 illustrates the general case, in which the water compositions in the two injection wells may not be the same and the system may not be symmetrical (see Figure 5.10b). Two advective-reactive travel-time based transport problems are solved, and the concentration in the extraction well is the weighted concentration of all the streamtubes in the two recirculation zones and the regional flow, which can be evaluated as:

$$C_{out}(t) = \sum_i p_{1i} c(t, i\Delta\tau) + \sum_j p_{2j} c(t, j\Delta\tau) + (1 - P_{r1} - P_{r2})\, C_0 \qquad \text{(Eq. 5.28)}$$

where *i* and *j* are the indexes of the travel-time elements, C_0 is the contaminant concentration in the regional flow, P_{r1} and P_{r2} are the ratios of recirculated flow to the injection flow rates in injection well I and II, respectively. Furthermore, to visualize the spatial concentration distributions, the spatial locations must be associated with a specific reactive transport, e.g., to map concentrations in the recirculation zone and the release zone associated with injection well I, the solution to the advective-reactive transport from injection well I should be used.

5.6.3 Mixing Within Reactor

The travel-time based model presented above is based on the concept of complete segregation which is the simplest approach for transport simulations using travel-time distributions by assuming all mixing occurs within the wells and neglecting local mixing within the reactor altogether. One must modify the approach to simulate bimolecular reactions that involve compounds that react upon mixing that is controlled by dispersion and kinetic mass transfer. The advective-dispersion streamtube model of Cirpka and Kitanidis (2000a, b) may be seen as an extension of the stochastic-convective approach, in which apparent longitudinal dispersion is used to account for local mixing, i.e., Equation 5.11 includes an additional longitudinal dispersion term. The approach has been successfully applied to bioreactive transport by Janssen et al. (2006). Ginn (2001) presented a similar approach allowing longitudinal dispersion in a single streamtube, which was applied to simulate an intermediate-scale biodegradation experiment (Ginn et al., 2001). However, this approach may not be appropriate for nonuniform flow fields because the derived apparent dispersion coefficient is highly influenced by streamline distances. In addition, flow fields created by injection and extraction wells are often advection dominated.

The travel-time model with kinetic mass transfer neglects local dispersion, but uses the mass exchange between the mobile and immobile phase to account for mixing resulting from both local dispersion and kinetic mass transfer (Shapiro and Cvetkovic, 1990; Simic and Destouni, 1991). This model is well suited to describe transport of sorbing compounds in mobile-immobile aquifers. The travel-time based transport equation is now:

$$\frac{\partial c_i}{\partial t} + \frac{\partial c_i}{\partial \tau} = r_i + F(c_i, \bar{c}_i) \qquad \text{(Eq. 5.29)}$$

in which $\bar{c}_i$ is the concentration of species i in the immobile phase and F represents the mass transfer flux between the mobile and immobile phase. The general solution procedure is similar to the illustrative example presented above, i.e., evaluating the travel-time distribution, solving the reactive-transport model, Equation 5.29, and calculating the averaged concentration breakthrough curves. However, an immobile phase needs to be added to each travel-time element (see Figure 5.12). Reactions and kinetic mass transfer between mobile and immobile zones are

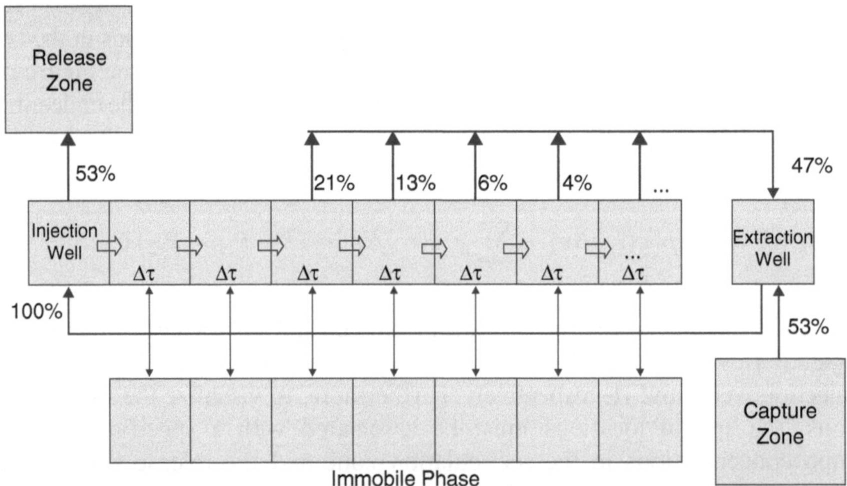

Figure 5.12. The travel-time based model with kinetic mass transfer. An immobile phase is connected to each travel-time element in the mobile phase, and mass transfer occurs between the corresponding mobile and immobile elements.

calculated for each time step. The advective displacement of aqueous species concentrations is only applied in the mobile phase, and the extracted water in the extraction well is also from the mobile phase.

The travel-time based models accounting for local mixing within the reactor involve two key issues that need to be resolved: the conceptualization and quantification of effective mixing mechanisms and the estimation of the travel-time distribution. When dispersion or mass transfer is considered, the measured breakthrough curve of a conservative tracer test with respect to a unit impulse injection cannot be interpreted as the travel-time distribution and directly applied to build a travel-time model because it results from all transport processes. For example, in a mobile-immobile travel-time domain, the residence time of a tracer particle within the reactor is the sum of the travel time in mobile zones and the retention time in stagnant zones. Thus, the travel-time distribution reflecting the particle residence in mobile zones where reactions occur, instead of the overall residence-time distribution, should be used to set up the model. In a mobile-immobile medium, the breakthrough curve of a conservative tracer with a unit impulse input is given by:

$$\tilde{c} = \tilde{f}\left(s + s\tilde{\phi}\right) \qquad \text{(Eq. 5.30)}$$

where c is the breakthrough curve, f is the travel-time distribution, ϕ is a memory function describing mass transfer processes between the mobile and immobile phases (Carrera et al., 1998), $\tilde{c}, \tilde{f}$, $\tilde{\phi}$ are the Laplace transforms of c, f, and ϕ, and s is the coordinate in the Laplace domain. Thus, the breakthrough curve, subject to both transport in the mobile phase and retention in the immobile phase, is a function of the travel-time distribution resulting from transport only in the mobile phase in the Laplace domain (Villermaux, 1987; Sardin et al., 1991).

There are two approaches to build a travel-time based model with kinetic mass transfer:

1. First approximating the travel-time distribution, and then determining the mass transfer model and parameters (Rubin and Ezzeddine, 1997; Bellin and Rubin, 2004). Local-scale (point-like) breakthrough curves need to be measured to estimate the travel-time distribution. This approach may be unsuitable to characterize multi-modal, asymmetric travel-time distributions because it requires an extremely large number of measurement points.
2. First estimating apparent local mass transfer parameters, and then determining the travel-time distribution. Similar to the first approach, local breakthrough curves need to be measured. However, the local information is used to estimate apparent local mixing parameters, not the travel-time distributions. Then, by fitting the concentration breakthrough curves, both parametric and nonparametric approaches are available for estimating travel-time distributions (Cirpka and Kitanidis, 2000b; Ginn, 2001; Luo et al., 2006a; Fienen et al., 2006). A detailed discussion of travel-time based models including mixing descriptions within reactors is presented by Luo and Cirpka (2008).

5.6.4 Chemical Heterogeneities

The procedure presented above to build a travel-time based model assumes homogeneous biogeochemical conditions and uniform initial concentration distributions. This assumption simplifies multi-dimensional transport problems to one 1-D advective-reactive transport problem on the travel-time domain. In order to extend the travel-time based model to incorporate spatial variability of biogeochemical parameters, all spatial parameters must be associated with travel times (Cvetkovic and Dagan, 1994; Cvetkovic et al., 1998; Sanchez-Vila and Rubin, 2003). Thus it requires correlations between travel times and reactive parameters, which may not be available in field applications. The streamline-method of Crane and Blunt (1999), among others, also considers one-dimensional advective-reactive transport. In contrast to the other methods, the streamline method aims to simulate the actual spatial distribution of concentration and thus requires determining the distribution of streamlines and computing the travel-time distribution along each of them. Obi and Blunt (2004) extended the streamline-method to model diffusion and dispersion in solute transport problems using an operator splitting technique where dispersive transport is solved on an underlying spatial grid. The main advantage of the streamline-method is that it reduces the computational effort by transforming a multi-dimensional problem to a number of one-dimensional problems. However, it cannot avoid the costly aquifer characterizations in order to accurately predict the streamline paths and incorporate the biochemical heterogeneities.

5.6.5 Reaction Rate Estimation

Travel-time based models are convenient ways to estimate effective (i.e., lumped-parameter) first-order reaction rate coefficients based on concentration breakthrough curves. By assuming reactions only occur in the mobile zones, first-order reactions are fully controlled by the travel times in the mobile zones, but are independent of specific transport and mixing mechanisms because a first-order reaction is spontaneous and affects each molecules separately and independently of the degree of mixedness (Nauman and Buffham, 1983). Although the estimated reaction rate coefficients are lumped parameters, they still provide useful information for describing the fate and transport of subsurface contaminants during the remediation operations.

Consider that a conservative tracer and a reactant undergo the same transport processes. In a steady-state flow system without retention in immobile zones, the travel-time distribution is considered as the impulse response function or the transfer function, and the concentration breakthrough curve with respect to an arbitrary input function can be described by linear convolution:

$$C_{out}(t) = \int_0^t C_{in}(t-\tau)f(\tau)d\tau \qquad \text{(Eq. 5.31)}$$

Similarly, the reactant concentration breakthrough curve subject to a first-order reaction is given by:

$$C^*_{out}(t) = \int_0^t C^*_{in}(t-\tau)f(\tau)\exp(-k\tau)d\tau \qquad \text{(Eq. 5.32)}$$

where C^*_{out} and C^*_{in} are the output and input concentrations for the reactant, and k is the first-order reaction rate coefficient. Thus,

$$f^*(\tau) = f(\tau)\exp(-k\tau) \qquad \text{(Eq. 5.33)}$$

is the impulse response function of the reactant. To estimate the first-order reaction rate coefficient, one may evaluate the impulse response functions of the conservative tracer and the reactant by deconvoluting the breakthrough curves, and k is given by:

$$k = -\frac{d\ln(f^*/f)}{d\tau} \qquad \text{(Eq. 5.34)}$$

Applications of this approach can be found in Luo et al. (2006a) and Fienen et al. (2006). Gong et al. (2011) presented the travel-time based approach for estimating more complex reaction kinetics.

5.7 SUMMARY

Multiple injection-extraction well systems create highly non-uniform flow fields, in which the created recirculation and/or release zones can serve as the treatment zones for *in situ* groundwater remediation. Travel-time based models provide practical and efficient tools for simulating reactive transport in such systems. Complete segregation is the simplest travel-time based conceptualization, which assumes all mixing occurs at the inlet and outlet of the reactor system but there is no mixing at all within the reactor. By transforming multi-dimensional problems on spatial coordinates into one-dimensional problems on travel-time domains, the travel-time approach can minimize computational effort. In particular, when reaction parameters and initial concentrations are uniformly distributed, one only needs to solve as many reactive transport problems as the number of the injection wells. Travel-time based models can be extended to incorporate local mixing processes, such as dispersion and kinetic mass transfer. Travel-time distributions must be corrected to distinguish advection from local mixing processes. To describe biochemical heterogeneities, all spatial parameters must be associated with travel times. Spatial concentration distributions can be visualized provide with the spatial travel-time distribution which relates the spatial coordinates with travel times.

More importantly, the development of travel-time based models rely on travel-time distribution, which may be obtained by conducting tracer tests, and thereby, do not require the detailed spatial characterizations of aquifer heterogeneities explicitly. Over a long-term groundwater remediation project, flow fields may be influenced by bioreactions, such as biomass accumulation, gas production, and solids precipitation, etc., which raises the question of the applicability of the initial aquifer characterization and suggests that aquifer characterizations may need to be repeated during bioremediation (Luo et al., 2007b). Travel-time based models, which may be updated by tracer tests, are economical, flexible, and well suited to simulate long periods of remediation implemented in *in situ* reactors created by injection and extraction wells.

REFERENCES

Bellin A, Rubin Y. 2004. On the use of peak concentration arrival times for the inference of hydrogeological parameters. Water Resour Res 40:W07401.

Berkowitz B, Scher H. 1997. Anomalous transport in random fracture networks. Phys Rev Lett 79:4038–4041.

Berkowitz B, Scher H. 1998. Theory of anomalous chemical transport in random fracture networks. Phys Rev E 57:5858–5869.

Berkowitz B, Scher H, Silliman SE. 2000. Anomalous transport in laboratory-scale, heterogeneous porous media. Water Resour Res 36:149–158.

Berkowitz B, Cortis A, Dentz M, Scher H (2006) Modeling non-Fickian transport in geological formations as a continuous time random walk. Rev Geophys, 44:RG2003.

Carrera J, Sanchez-Vila X, Benet I, Medina A, Galarza G, Guimera J. 1998. On matrix diffusion: Formulations, solution methods and qualitative effects. Hydrogeol J 6:178–190.

Cirpka OA, Kitanidis PK. 2000a. An advective-dispersive streamtube approach for the transfer of conservative tracer data to reactive transport. Water Resour Res 36:1209–1220.

Cirpka OA, Kitanidis PK. 2000b. Characterization of mixing and dilution in heterogeneous aquifers by means of local temporal moments. Water Resour Res 36:1221–1236.

Cirpka OA, Kitanidis PK. 2001. Travel-time based model of bioremediation using circulation wells. Ground Water 39:422–432.

Chilakapati A, Yabusaki S. 1999. Nonlinear reactions and nonuniform flows. Water Resour Res 35:2427–2438.

Christ JA, Goltz MN, Huang JQ. 1999. Development and application of an analytical model to aid design and implementation of in situ remediation technologies. J Contam Hydrol 37:295–317.

Crane MJ, Blunt MJ. 1999. Streamline-based simulation of solute transport. Water Resour Res 35:3061–3078.

Cunningham JA, Reinhard M. 2002. Injection-extraction treatment well pairs: An alternative to permeable reactive barriers. Ground Water 40:599–607.

Cvetkovic VD, Dagan D. 1994. Transport of kinetically sorbing solute by steady random velocity in heterogeneous porous formations. J Fluid Mech 265:189–215.

Cvetkovic VD, Dagan G, Cheng H. 1998. Contaminant transport in aquifers with spatially variable hydraulic and sorption properties. Proc Royalty Soc London A 454:2173–2207.

Di Donato G, Blunt MJ. 2004. Streamline-based dual-porosity simulation of reactive transport and flow in fractured reservoirs. Water Resour Res 40:W04203.

Di Donato G, Obi EO, Blunt MJ. 2003. Anomalous transport in heterogeneous media demonstrated by streamline-based simulation. Geophys Res Lett 30:1608.

Fienen MN, Luo J, Kitanidis PK. 2005. Semi-analytical, homogeneous, anisotropic capture zone delineation. J Hydrol 312:39–50.

Fienen MN, Luo J, Kitanidis PK. 2006. A bayesian geostatistical transfer function approach to tracer test analysis. Water Resour Res 42:W07426.

Fiori A, Jankovic I, Dagan G. 2006. Modeling flow and transport in highly heterogeneous three-dimensional aquifers: Ergodicity, Gaussianity, and anomalous behavior – 2 Approximate semianalytical solution. Water Resour Res, 42:W06D13.

Gandhi RK, Hopkins GD, Goltz MN, Gorelick SM, McCarty PL. 2002. Full-scale demonstration of in situ cometabolic biodegradation of trichloroethylene in groundwater, 1: Dynamics of a recirculating well system. Water Resour Res 38:10.1029/2001 WR000380.

Ginn TR. 2001. Stochastic-convective transport with nonlinear reactions and mixing: finite streamtube ensemble formulation for multicomponent reaction systems with intra-streamtube dispersion. J Contam Hydrol 47:1–28.

Ginn TR, Simmons CS, Wood BD. 1995. Stochastic-convective transport with nonlinear reaction: Biodegradation with microbial growth. Water Resour Res 31:2689–2700.

Ginn TR, Murphy EM, Chilakapati A, Seeboonruang U. 2001. Stochastic-convective transport with nonlinear reaction and mixing: Application to intermediate-scale experiments in aerobic biodegradation in saturated porous media. J Contam Hydrol 48:121–149.

Gong R, Lu C, Wu W-M, Carley J, Cheng H, Watson D, Criddle CS, Kitanidis PK, Jardine PM, Luo J. 2011, Estimating reaction rate coefficients within a travel-time modeling framework. Ground Water 49:209–218.

Hyndman DW, Dybas MJ, Forney L, Heine R, Mayotte T, Phanikumar MS, Tatara G, Tiedje J, Voice T, Wallace R, Wiggert D, Zhao X, Criddle CS. 2000. Hydraulic characterization and design of a full-scale biocurtain. Ground Water 38:462–474.

Janssen G, Cirpka OA, van der Zee SEATM. 2006. Stochastic analysis of nonlinear biodegradation in regimes controlled by both chromatographic and dispersive mixing. Water Resour Res 42:W01417.

Luo J, Cirpka OA. 2008. Travel-time based descriptions of transport and mixing in heterogeneous domains. Water Resour Res 44: W09407.

Luo J, Kitanidis PK. 2004. Fluid residence times within a recirculation zone created by an extraction-injection well pair. J Hydrol 295:149–162.

Luo J, Cirpka OA, Fienen MN, Wu WM, Mehlhorn TL, Carley J, Jardine PM, Criddle CS, Kitanidis PK. 2006a. A parametric transfer function methodology for analyzing reactive transport in nonuniform flow. J Contam Hydrol 83:27–41.

Luo J, Wu W-M, Fienen MN, Jardine PM, Mehlhorn TL, Watson DB, Cirpka OA, Criddle CS, Kitanidis PK. 2006b. A nested-cell approach for in situ remediation. Ground Water 44:266–274.

Luo J, Weber F-A, Cirpka OA, Wu W-M, Carley J, Nyman J, Jardine P, Criddle CS, Kitanidis PK. 2007a. Modeling in-situ U(VI) bioreduction by sulfate-reducing bacteria. J Contam Hydrol 92:129–148.

Luo J, Wu W-M, Carley J, Ruan C, Gu B, Jardines PM, Criddle CS, Kitanidis PK. 2007b. Hydraulic performance analysis of a multiple injection-extraction well system. J Hydrol 336:294–302.

McCarty PL, Goltz MN, Hopkins GD, Dolan ME, Allan JP, Kawakami BT, Carrothers TJ. 1998. Full-scale evaluation of in situ cometabolic degradation of trichloroethylene in groundwater through toluene injection. Environ Sci Technol 32:88–100.

Nauman EB, Buffham BA. 1983. Mixing in Continuous Flow System. John Wiley and Sons, Inc., New York, NY, USA.

Obi EO, Blunt MJ. 2004. Streamline-based simulation of advective-dispersive solute transport. Adv Water Resour 27:913–924.

Pollock DW. 1988. Semianalytical computation of path lines for finite-difference models. Ground Water 26:743–750.

Reimus PW, Haga MJ, Adams AI, Callahan TJ, Turin HJ, Counce DA. 2003. Testing and parameterizing a conceptual solute transport model in saturated fractured tuff using sorbing and nonsorbing tracers in cross-hole tracer tests. J Contam Hydrol 62:613–636.

Robinson BA, Viswanathan HS. 2003. Application of the theory of micromixing to groundwater reactive transport models. Water Resour Res 39:1313.

Rubin Y, Ezzeddine S. 1997. The travel times of solutes at the Cape Cod tracer experiment: Data analysis, and structural parameter inference: Theory and unconditional numerical simulations. Water Resour Res 26:691–701.

Sanchez-Vila X, Rubin Y. 2003. Travel time moments for sorbing solutes in heterogeneous domains under nonuniform flow conditions. Water Resour Res 39:1086.

Sardin M, Schweich D, Leij FJ, van Genuchten MT. 1991. Modeling the nonequilibrium transport of linearly interacting solutes in porous media: A review. Water Resour Res 27:2287–2307.

Selroos JO, Cvetkovic V. 1992. Modeling solute advection coupled with sorption kinetics in heterogenous formations. Water Resour Res 28:1271–1278.

Semprini L, McCarty PL. 1991. Comparison between model simulation on field results for in-situ biorestoration of chlorinated aliphatics: Part 1. Biostimulation of the methanotrophic bacteria. Ground Water 29:365–374.

Shapiro AM, Cvetkovic VD. 1988. Stochastic-analysis of solute arrival time in heterogeneous porous-media. Water Resour Res 24:1711–1718.

Shapiro AM, Cvetkovic VD. 1990. Mass arrival of sorptive solute in heterogeneous porous-media. Water Resour Res 26:2057–2067.

Simic E, Destouni G. 1991. Water and solute residence times in a catchment: Stochastic-mechanistic model interpretation of 180 transport. Water Resour Res 35:2109–2119.

Simmons CS. 1982. A stochastic-convective transport representation of dispersion in one-dimensional porous-media systems. Water Resour Res 18:1193–1214.

Simmons CS, Ginn T, Wood B. 1995. Stochastic-convective transport with nonlinear reaction: Mathematical framework. Water Resour Res 31:2675–2688.

Strack ODL. 1989. Groundwater Mechanics. Prentice-Hall, Englewood Cliffs, NJ, USA.

Valocchi AJ, Malmstead M. 1992. Accuracy of operator splitting for advection-dispersion -reaction problems. Water Resour Res 28: 1471–1476.

van der Zee SEATM, van Riemsdijk WH. 1987. Transport of reactive solutes in spatially variable soil systems. Water Resour Res 23:2059–2069.

Villermaux J. 1987. Chemical-engineering approach to dynamic modeling of linear chromatography – A flexible method for representing complex phenomena from simple concepts. J Chromatogr 406:11–26.

Weber F-A. 2002. Reactive transport modeling: stimulation of microbial uranium(VI) reduction for remediation of a contaminated aquifer. MS Thesis. University of Stuttgart, Germany.

Wu W, Carley J, Fienen MN, Mehlhorn T, Lowe K, Nyman J, Luo J, Gentile ME, Rajan R, Wagner D, Hickey RF, Gu B, Watson D, Cirpka OA, Kitanidis PK, Jardine PM, Criddle CS. 2006a. Pilot-scale bioremedation of uranium in a highly contaminated aquifer I: Conditioning of a treatment zone. Environ Sci Technol 40:3978–3985.

Wu W, Carley J, Gentry T, Ginder-Vogel MA, Fienen M, Mehlhorn T, Yan H, Carroll S, Nyman J, Luo J, Gentile ME, Fields MW, Hickey RF, Watson D, Cirpka OA, Fendorf S, Zhou J, Kitanidis P, Jardine PM, Criddle CS. 2006b. Pilot-scale bioremediation of uranium in a highly contaminated aquifer II: Reduction of U(VI) and geochemical control of U(VI) bioavailability. Environ Sci Technol 40:3986–3995.

Zheng C, Bennett G. 2002. Applied Contaminant Transport Modeling, Theory and Practice. John Wiley and Sons, Inc., New York, NY, USA.

Zwietering TN. 1959. The degree of mixing in continuous flow systems. Chem Eng Sci 11:1–15.

CHAPTER 6

RECIRCULATION SYSTEMS

Mark N. Goltz[1] and John A. Christ[2]

[1]Air Force Institute of Technology, Wright Patterson Air Force Base, OH, USA
[2]U.S. Air Force Academy, Colorado Springs, CO, USA

6.1 INTRODUCTION

In this chapter we discuss how recirculation systems can be engineered to achieve mixing in order to facilitate *in situ* remediation. As noted in Chapters 1 and 5, recirculation systems have a number of advantages. First, as active systems, they can be designed to maximize the probability of contaminant control and capture, even under changing hydrological conditions. Second, since the systems use wells, the recirculation zone(s) can be established at depths that can't be impacted by other systems (e.g., permeable reactive barriers). Third, as will be discussed subsequently, mixing of the amendment and the contaminant occurs in an engineered reactor, either in-well or aboveground, and is therefore relatively complete. Fourth, for systems that rely upon biodegradation, it has been shown that recirculation systems can be used to establish *in situ* biological treatment zones that are effective in biodegrading the target contaminant, even when initially biodegradation activity is sparse (Hoelen et al., 2006). Finally, as net loss of water is minimized when recirculation is used, the systems are very useful in regions where water needs to be conserved. They can be designed to confine a source of contamination and treat it there, or to act similar to a barrier wall by removing contaminants in a passing plume to prevent downgradient contamination. Of course, as active systems that use wells, there are a number of attendant disadvantages to recirculation; the most obvious being the operation, maintenance, and monitoring costs associated with a "pump-and-treat" system (the fact that the treatment happens to occur *in situ* notwithstanding). Additionally, as systems which rely on pumping, the water which is captured and amended will come preferentially from the most permeable zones of the aquifer. Contaminated water that is resident in low permeability regions may not be treated.

In the coming chapter, we discuss these advantages and disadvantages in more detail, as well as present principles that may be used to design recirculation systems and optimize their performance. Finally, we present case studies that demonstrate a range of recirculation system field applications.

6.2 TYPES OF RECIRCULATION SYSTEMS

There are two basic types of recirculation systems that are used to mix an amendment with contaminated groundwater: (1) systems that extract water from the subsurface and mix the amendment into the water aboveground, and (2) systems that add the amendment into the water as it flows through wells, without the need to pump groundwater to the surface. Both types of systems have certain limitations. Since the systems rely on operation of pumping wells, the aquifer must be sufficiently permeable (hydraulic conductivity $>10^{-4}$ centimeters per second [cm/s]) to permit the wells to pump water at the designed flow rates. Plugging/fouling of the injection or extraction well screens may also impact long-term system operability

P.K. Kitanidis and P.L. McCarty (eds.), *Delivery and Mixing in the Subsurface: Processes and Design Principles for In Situ Remediation*, doi: 10.1007/978-1-4614-2239-6_6, © Springer Science+Business Media New York 2012

(USEPA, 1995). Since these systems rely upon injection of water containing contaminants (and chemical amendments) into an aquifer, the injection well screens must be located within the zone of contamination to avoid contaminating previously uncontaminated aquifer zones.

While aboveground mixing and/or treatment are easier to engineer and allow for more straightforward process control, there are cost and regulatory advantages to be gained by in-well subsurface mixing and treatment. In the following sections, two types of systems are presented; injection-extraction systems, where mixing for treatment is done aboveground, and groundwater circulation well and tandem recirculating well systems, where the mixing is accomplished in-well.

6.2.1 Injection-Extraction

Injection-extraction wells have been used to add amendments into groundwater to achieve *in situ* bioremediation at least since 1971, when the Raymond process (see Figure 6.1) was developed (Suthersan, 1997). This system is designed to confine a contaminant source and treat it in place. As shown in Figure 6.1, contaminated water here is extracted from the subsurface downgradient from the source, amendments that promote biological or chemical contaminant degradation are mixed above ground with the extracted contaminated water, and the water with amendments is then reinjected upgradient.

As will be discussed more fully in subsequent sections, hydraulic control for contaminant containment as well as treatment is one possible objective when designing an injection-extraction system. In order to assure hydraulic control, injection-extraction systems are typically designed in a line-drive layout with the extraction wells downgradient of the contamination and

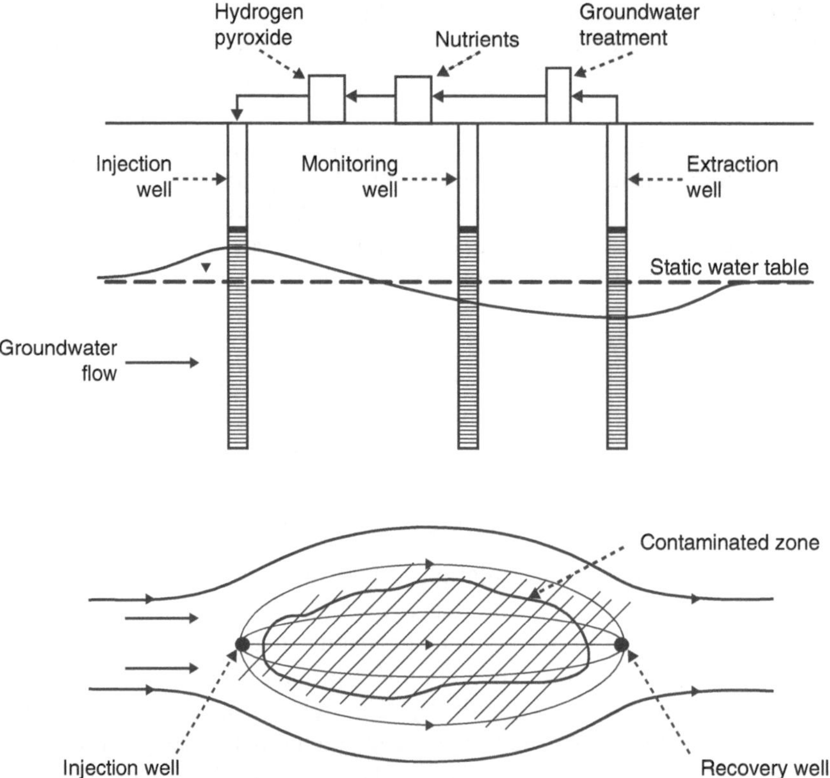

Figure 6.1. Description of the Raymond process (Suthersan, 1997).

the injection wells. To help ensure hydraulic control, injection rates are generally less than extraction rates. Alternative designs may use injection wells along the contamination plume centerline with the extraction wells at the plume boundaries, or use an upgradient infiltration gallery to distribute chemicals into the contaminated zone with downgradient extraction wells (USEPA, 1995). Injection-extraction well pairs have also been proposed and used to establish a barrier downgradient of a contaminant plume (Christ et al., 1999; Dybas et al., 2002). An alternative design to treat a passing plume or a source uses horizontal wells for the injection and extraction wells (Hazen et al., 1996).

Perhaps the main disadvantage of injection-extraction systems, which extract contaminated water to the surface, is the requirement to manage the extracted water. Since less water is generally injected here than extracted, there is a need to treat and dispose of the excess water. Additionally, in some jurisdictions with stringent regulatory guidance, reinjection may not even be an option, necessitating the treatment of all extracted water, thus excluding the possibility for using this type of system.

6.2.2 Groundwater Circulation Wells (GCWs)

GCWs are an alternative recirculation system that is sometimes used. GCWs are multi-screened wells, with the screens vertically separated and isolated by in-well packers. Water is extracted from the aquifer through one screen and injected into the aquifer through the other. This injection-extraction scheme establishes a vertical circulation zone in the aquifer, as water flows from the injection screen to the extraction screen. The wells may be operated in either upflow or downflow modes (Johnson and Simon, 2007). Figure 6.2 illustrates a GCW system operated in a downflow mode designed to mix methane and oxygen with TCE contaminated groundwater for cometabolic biodegradation.

GCWs have often been used to treat groundwater contamination since their introduction in the early 1990s (Herrling and Stamm, 1991). Most frequently, the wells have been used as "*in situ* air strippers." In one version using this approach air is injected into water as it enters the well through the lower screen. The decreased density of the air-water mixture causes the water to rise through the well by "air-lift pumping." As the air and contaminated water rise, volatile

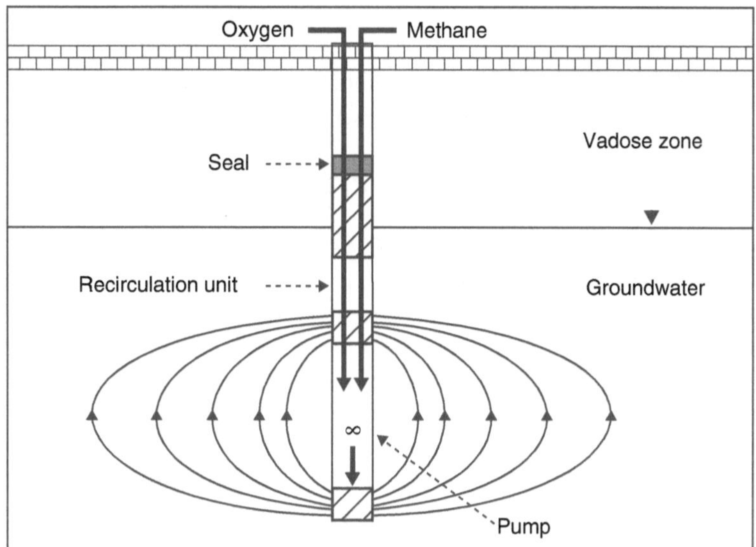

Figure 6.2. Groundwater circulation well (GCW) (McCarty and Semprini, 1994).

contaminants partition to the gas phase. The water then reenters the groundwater through the upper screen, while the gas phase, which now contains gaseous contaminant, is treated aboveground such as in an activated carbon adsorber. Some commercial implementations of GCWs, all of which are based on the above-described principles, are the NoVOCs™, Unterdruck-Verdampfer-Brunnen (UVB™), Density Driven Convection (DDC) and C-Sparge™ systems. While NoVOCs™ and UVB™ are designed to rely on in-well air stripping alone to treat volatile contaminants, GCWs have also been used as "*in situ* mixers" to amend chemicals into contaminated groundwater flowing through the wells. GCWs have been used to add oxidizing agents, electron donors, and surfactants into contaminated groundwater (Johnson and Simon, 2007). The DDC system stimulates aerobic biodegradation through the injection and mixing of oxygen into the contaminated water, and the C-Sparge™ system employs chemical oxidation resulting from injection of ozone and air (USEPA, 1998).

The main advantage of GCWs is that their use obviates the need to extract contaminated water from the subsurface. This can result in significant cost savings, especially when the treated water is deep. Additionally, since contaminated water is not brought to the ground surface, regulatory requirements (particularly regarding the need for discharge permits) are much less stringent. Since GCWs induce vertical groundwater flow, water may also flow through low permeability layers, thereby flushing contamination that might otherwise be bypassed if flow were horizontal. Operation of GCWs in an upflow mode creates water table mounding rather than drawdown. Thus, there are no isolated and untreated zones of contamination above a cone of depression such as may result during operation of a conventional extraction well (ESTCP, 1997).

The main disadvantage of GCWs is that establishment of the circulation cell between the injection and extraction screens, which is crucial to operation of the technology, relies on inducing vertical flow. If the aquifer is anisotropic (as a rule of thumb, if the ratio of horizontal to vertical conductivity is greater than 10) the weakness of the vertical flow will significantly decrease system effectiveness (ESTCP, 1999). As aquifers that have anisotropy ratios greater than 10 are quite common, this limits the technology's applicability at many sites. Conversely, if the aquifer is relatively isotropic (horizontal to vertical hydraulic conductivity ratio is less than three) GCWs may also be ineffective, as groundwater will short-circuit from the injection to the extraction well screens. Short-circuiting may result in a small GCW radius of influence, as well as an injected water residence time that is insufficient to promote a desired *in situ* reaction (ESTCP, 1999). Compounding this problem are the difficulties in determining (1) the anisotropy ratio in order to evaluate whether the GCW technology will be effective at a particular site, and (2) the extent of recirculation established during GCW operation (ESTCP, 1999).

6.2.3 Tandem Recirculating Wells (TRWs)

To address the problems associated with GCWs, i.e., establishment of recirculation zones in anisotropic aquifers, TRWs were developed. TRWs were first applied to stimulate *in situ* bioremediation of trichloroethene-contaminated groundwater at a site on Edwards Air Force Base (AFB), California (McCarty et al., 1998). Like GCWs, TRWs are configured so that groundwater need not be extracted from the subsurface. However, in contrast to GCWs, which rely upon vertical flow of groundwater from the injection well screen to the extraction screen, TRWs rely primarily on horizontal flow.

As shown in Figure 6.3a, the TRW system employs two dual-screened treatments wells. One well is operated in an upflow mode, where contaminated water is extracted from the aquifer through the lower screen and flows upward through the well where it is injected back into the aquifer through the upper screen, while the second well operates in a downflow mode,

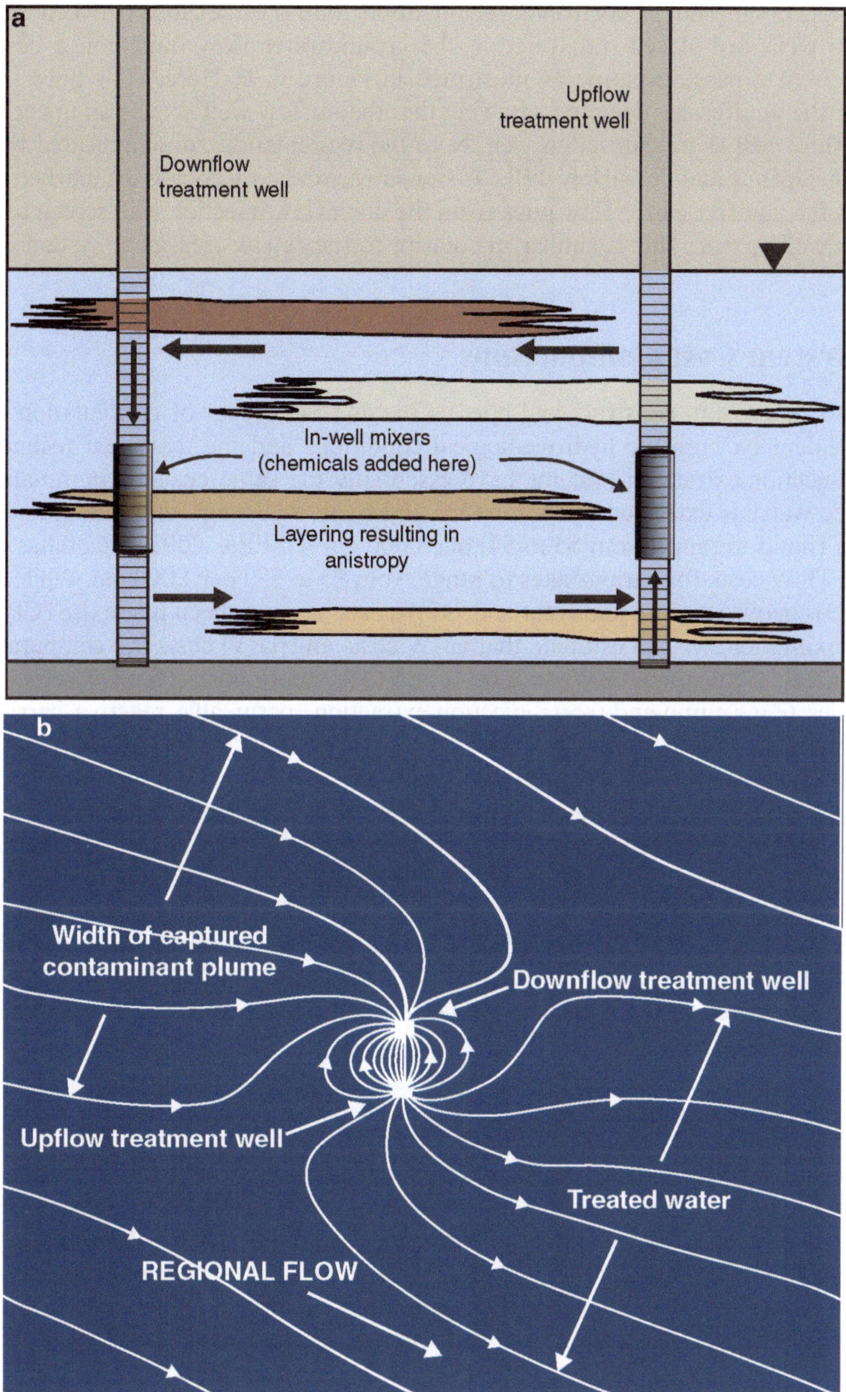

Figure 6.3. Tandem recirculating wells (TRWs): (a) cross-section, (b) plan view showing flow patterns in upper aquifer portion.

extracting contaminated water through the upper screen and injecting it through the lower screen. As the water flows through the wells, in-well mixers are used to add chemicals. When the chemically-amended groundwater is injected back into the aquifer, under typical anisotropic conditions, the water will flow horizontally toward the adjacent wells' extraction screen

(Christ et al., 1999). Thus, a horizontal recirculation zone is established between the two wells. If the two wells are placed transverse to the groundwater flow direction, a barrier system results to treat a passing plume as illustrated in Figure 6.3b. Here, flow lines in the upper portion of the aquifer are depicted, showing that the upflow well acts as an injection well and the downflow well as an extraction well. Note the recirculation zone, indicated by flow lines between the upflow and downflow wells. A similar recirculation zone is established in the lower portion of the aquifer, where flow goes from the downflow injection well screen to the upflow extraction well screen. This is similar in concept to the *in situ* treatment system discussed in Chapter 5.

6.2.4 System Cost Comparisons

Of course, costs to construct and operate the different types of recirculation systems are very dependent on both the hydrogeological conditions and the treatment technology being utilized. Injection-extraction systems have essentially the same costs as pump-and-treat systems, since water is extracted to the surface and treated aboveground. Pump-and-treat costs have been found to range from $5 to $41 per 1,000 gal (USEPA, 2001). Mandalas et al. (1998) estimated TRW costs for various sites to range from $6 to $80 per 1,000 gal. While meaningful comparisons between pump-and-treat and GCW costs have not been made (ESTCP, 1999), as a first approximation, we can estimate that GCW costs and TRW costs are comparable.

In general, GCWs and TRWs will be found to be more cost effective versus competing technologies (e.g., pump-and-treat, injection-extraction, permeable reactive barriers (PRBs)) for deeper plumes, as the costs of pumping contaminated water to the surface (or, in the case of PRBs, constructing very deep trenches) are avoided.

As with GCWs, the main advantage of TRWs over injection-extraction is that there's no need to extract contaminated groundwater from the subsurface. And because TRWs, like injection-extraction, rely on horizontal flow between wells, the technology is applicable at more sites than GCWs. As a rule of thumb, an anisotropy ratio of 10:1 or greater is required for TRWs to induce adequate horizontal flow without excessive recirculation between screens within a single well. This level of anisotropy is readily realized at many sites (Christ et al., 1999). Also, a good seal is required in the bore hole between screens to prevent such excessive single-well recirculation.

6.3 DESIGN PRINCIPLES

Having reviewed the three most common recirculation system applications, this section investigates the environmental and engineering factors that must be considered when (1) determining whether a recirculation system is appropriate for use at a particular site (i.e., technology screening) and (2) designing a recirculation system. We also present examples of models, which incorporate these environmental and engineering factors, and that may be used for technology screening and design. We begin with a discussion of how the remediation goal must be considered when selecting and designing a recirculation system.

6.3.1 Effect of Remediation Goal

In general, recirculation systems are applied to achieve one of two goals: contaminant source isolation and destruction or contaminant plume containment. System design is dependent on the remediation goal. Figure 6.1 depicts a typical source isolation and destruction application. The strategy is to isolate a volume of contaminated aquifer, and mix a compound into that volume to stimulate contaminant destruction. Here, the extraction well is placed downgradient and the injection well upgradient of the contaminant location. Regional flow then passes around the isolated area. Luo et al. (2007a) used a double-cell nested design originally proposed by Wilson (1984) to isolate a volume of aquifer contaminated with uranium (VI) for biotreatment. The design used two injection wells upgradient of the treatment zone, and two extraction wells downgradient, in order to establish inner and outer treatment cells. Goltz et al. (2005) used a design with three TRWs (an upgradient TRW pumping in an upflow mode coupled with two downgradient TRWs pumping in a downflow mode) to isolate a trichloroethene (TCE) source area for treatment.

In contrast, plume containment involves establishment of a treatment fence/barrier. In contrast to a source destruction application, an isolated treatment zone is not created. Instead, the system is open – contaminated water enters the treatment zone from upgradient, an amendment is mixed into the water to stimulate contaminant destruction, and treated water flows downgradient. Establishment of the treatment fence/barrier can be accomplished either actively or semi-passively. An active fence involves continuous pumping of the recirculating wells. Examples of this type of system design may be seen in Mandalas et al. (1998) and Christ et al. (1999). Figure 6.3 shows a typical application of this strategy. A semi-passive approach has the recirculating wells only operating periodically. This approach was used to establish a biobarrier by adding nutrients, electron donor, a strong base, and microorganisms at a carbon tetrachloride-contaminated site in Schoolcraft, Michigan (Dybas et al., 2002). After amendment of the microorganisms to the subsurface, it was only necessary to actively inject chemicals and extract water for 6 h per week. Other semi-passive designs might involve continuous pumping and *in situ* treatment for relatively long periods (perhaps weeks, depending on hydrogeology), followed by a lengthy rest period, during which contaminated water would flow into the treatment zone, to be treated during the subsequent pumping period.

6.3.2 Environmental Factors to Consider in Design

6.3.2.1 Physical

Hydrogeology

Hydrogeological factors are important in helping to determine whether a recirculation system is appropriate for application at a site, and if it is, how to design such a system. These factors include such basic hydrogeological data as depth to the water table, aquifer thickness, and hydraulic conductivity.

Hydraulic Conductivity

For any system that relies on pumping, the most critical physical factor to consider is hydraulic conductivity. As a rule of thumb, hydraulic conductivities greater than 10^{-4} cm/s are required to ensure adequate circulation of water (USEPA, 1995). There are a number of commonly used slug and pump tests that may be employed to estimate hydraulic conductivity at a site (Domenico and Schwartz, 1998). However, even more important than knowing the overall volume-averaged hydraulic conductivity at a site (as would be measured by a pump test or tracer test), is knowing how hydraulic conductivity varies vertically, as well as knowing the level of hydraulic conductivity anisotropy. Vertical variations in hydraulic conductivity are important to measure, in order to decide where well screens should be installed. In partially screened recirculation systems (e.g., GCW and TRW), installing a screen in a low permeability zone would lead to ineffective operation. One qualitative way of estimating hydraulic conductivity variations with depth is through interpretation of well logs. For quantitative analysis, a number of direct-push methods (cone penetrometry, Geoprobe) have been applied to measure vertical variations in hydraulic conductivity (Butler et al., 2002; Cho et al., 2000; Scaturo and Widdowson, 1997). Another quantitative method uses pump tests in conjunction with borehole flowmeters, which measure flows at different depths within the pumping well, to infer vertical variation in hydraulic conductivity (Boman et al., 1997). An innovative quantitative method developed by Huang et al. (2004) uses a genetic algorithm to interpret the results of a conventional step pump test in order to estimate the hydraulic conductivity of individual layers. These estimates have been successfully used to determine the placement of TRW well screens.

As discussed earlier, hydraulic conductivity anisotropy is an important parameter to determine in order to decide whether GCWs and/or TRWs are applicable at a site. Given estimates of the hydraulic conductivity in horizontal layers, it is possible to obtain an estimate of anisotropy, since an aquifer that is modeled as several isotropic layers of differing hydraulic conductivities may be transformed so as to be modeled as a single homogeneous, anisotropic layer (e.g., Domenico and Schwartz, 1998). In addition, dipole flow tests have been developed and used to directly measure anisotropy (Kabala, 1993; Zlotnik and Ledder, 1996).

Regional Hydraulic Head and Flow Field

The hydraulic head field is important to ascertain, since the heads, in conjunction with the hydraulic conductivities, determine the direction and magnitude of the regional flow. This information, which may be obtained in a relatively straightforward manner using conventional techniques, is required in order to properly locate the recirculating and monitoring wells.

Dispersion

As mixing in a recirculation system occurs either aboveground or in an in-well reactor, the dispersion process (quantified by the longitudinal, transverse, and vertical dispersivities), which describes dissolved compound spreading and mixing in a porous medium is not crucial to system design. A number of recirculation system applications have been successfully simulated using models that neglected dispersion (e.g., McCarty et al., 1998; Luo et al., 2007a; Cunningham et al., 2004). As the conventional method for quantifying dispersivity is to conduct relatively costly tracer tests (Pickens and Grisak, 1981), it is recommended that dispersivity may be assumed to be negligible, when designing a recirculation system.

6.3.2.2 Chemical/Biological

Groundwater Chemistry

As is discussed in the section on reaction kinetics below, knowing the groundwater chemistry (pH, redox conditions, concentrations of electron acceptors like oxygen, nitrate, sulfate, key microorganism presence, etc.) is crucial to determining the rate and extent of the chemical or biological reactions that are being engineered through use of the recirculation system. The contaminant plume location, and concentrations of contaminant and co-contaminants are also critical for the effective design of a recirculation system.

Sorption

The portion of a contaminant or chemical amendment that is sorbed is typically assumed to be unavailable to participate directly in biologically-mediated reactions (e.g., McCarty and Semprini, 1994). In addition, the transport of a sorbing compound would be retarded, due to the fact that the compound in the sorbed phase is immobile. Thus, in order to design a recirculation system, it is important to characterize the distribution of chemicals (both contaminants and amendments) between the sorbed and aqueous phases.

Normally, reversible, equilibrium sorption is assumed, and a sorption isotherm, which describes the relation between sorbed and aqueous concentrations, is developed using sorption batch tests, where various concentrations of the target compound are equilibrated between water and soil. At the low concentrations that are typically found in contaminated groundwater, linear isotherms are applicable, though Freundlich or Langmuir isotherms may also be appropriate to describe the relation between sorbed and dissolved chemical (Fetter, 1999). Sorption parameters may be obtained by fitting the hypothesized model to the isotherm data. An adequate understanding of sorption may lead to a better prediction of whether contaminants will separate, thereby decreasing reaction rates, or better mix, thereby increasing reaction kinetics (Oya and Valocchi, 1998).

Reaction Kinetics

The motivation for adding an amendment to contaminated groundwater is to initiate some chemical or biochemical reaction that will destroy the contaminant. In assessing the feasibility of a recirculation system, or designing the system, it is therefore crucial to understand and quantify the kinetics and stoichiometry of the reaction involved in target contaminant destruction. As discussed in Chapter 2, there are a number of model formulations that may be used to simulate reaction kinetics (e.g., first-order, second order, Monod, dual-Monod). Determining the appropriate model to apply is dependent on several factors. Perhaps a primary factor is the model's purpose. If the model is used only to screen whether a particular approach will work at a site, simple first-order kinetics may be adequate. However, if the purpose of the model analysis is to design the remediation system, and to determine an amendment dosing schedule, a model that's founded on the biochemical process of interest would be more suitable. In general, though, Occam's razor should be used when selecting a model – simpler models are preferred over complex ones. Similarly, determination of model parameter values should depend upon the model's purpose. For screening purposes, literature values of kinetic parameters may be appropriate, while for design, parameter values that are based on laboratory analyses that use soil and water from the actual site may be needed. Laboratory analyses normally involve recreating, as best as possible in a laboratory microcosm, the biogeochemical conditions (pH, redox, etc.) found in the field, as these conditions will often control the rate and extent of reaction. Addition of the target contaminant, along with the amendment, to the microcosm, may be monitored over time to quantify the reaction kinetics (e.g., Jenal-Wanner and McCarty, 1997; Hatzinger, 2002).

6.3.3 Engineering Factors to Consider in Design

Using the relevant environmental data obtained in accordance with the above discussion, design decisions regarding system construction and operation can be made. The use of these environmental data to make engineering design decisions is discussed below.

6.3.3.1 Construction

System Type

The decision on which system design to deploy will largely depend on remediation objectives and the basic hydrogeologic conditions encountered. Assuming hydraulic conductivity is sufficiently high to support a pumping system, the following considerations come into play.

Based on economic considerations, a deep contaminated zone lends itself to application of either a GCW or TRW system, where water need not be pumped to the ground surface, while shallow contamination can in most cases be treated most cost effectively using an injection-extraction well system. A thick contaminated zone also lends itself to treatment by a GCW or TRW system, while a thin zone of contamination is best treated by injection-extraction. Note that to use any recirculation system, all well screens should be adjacent to contaminated zones, to avoid circulating contaminated water into previously uncontaminated areas. As described previously, sites with anisotropy ratios greater than 10:1, lend themselves to systems that rely on horizontal flow (injection-extraction or TRW systems), while sites less than 10:1 encourage vertical recirculation cells (GCWs). For relatively isotropic systems (horizontal to vertical hydraulic conductivity ratio is less than 3:1) that require a minimum residence time for circulating water in the aquifer (for instance, in order to allow sufficient time for an *in situ* reaction to occur), GCWs may also be ineffective (ESTCP, 1999).

Well Screen Locations

Injection and extraction well screen locations for given remediation objectives are best determined through flow modeling. The placement of the extraction and injection wells, whether upgradient, downgradient, or cross gradient, depends upon remediation objectives. That is, as discussed earlier, whether the goal of the remediation is source isolation and treatment or plume containment will control decisions on well screen location. Combining objectives with environmental inputs of hydraulic conductivity (considering anisotropy), aquifer characteristics (e.g., aquifer thickness) and the regional hydraulic gradient with engineered inputs such as the injection and extraction well screen separation distances, screen lengths, and well pumping rates, flow models may be used to determine such things as capture zone widths, fluid residence times, and the fraction of recirculated water. A number of analytical flow models, which may be used as screening tools, have been developed for injection-extraction systems (Christ et al., 1999; Luo and Kitanidis, 2004; Dacosta and Bennett, 1960; Cunningham and Reinhard, 2002), GCW (Philip and Walter, 1992), and TRW (Huang and Goltz, 2005; Cirpka and Kitandis, 2001; Cunningham et al., 2004) systems. These models, under the simplifying assumptions of steady-state flow in a homogeneous and isotropic medium, can be used as tools to prepare preliminary designs of recirculating well systems. Recently, more sophisticated models have been presented that simulate the steady-state flow field for an injection-extraction system in a two-dimensional, homogeneous, anisotropic medium (Fienen et al., 2005), the flow field for a GCW in a heterogeneous, anisotropic medium (Elmore and DeAngelis, 2004), and the flow field for a TRW in a three-dimensional, homogeneous, anisotropic medium (Huang and Goltz, 2005). Use of models to locate wells and well screens is discussed in the section on modeling applications, below.

Mixing Mechanism

For GCW and TRW systems, in-well mixers are required. Static or inline mixers, which are commonly used in industrial applications, have been effective in mixing chemical amendments (both dissolved and gaseous) into contaminated water, as the water flows through the treatment wells (McCarty et al., 1998; Cunningham et al., 2004; Goltz et al., 2005; Hoelen et al., 2006). Hollow-fiber membrane diffusers have also been applied to achieve in-well mixing of gases (Agarwal et al., 2005). Of course, injection-extraction systems can make use of conventional industrial mixers, as mixing is done aboveground. It is even possible to have in-well mixing combined with treatment, utilizing fast catalytic reactions (e.g., Davie et al., 2008).

6.3.3.2 Operation

Water Flow Rates in Wells

As discussed in the previous section, flow modeling may be used to determine water flow rates in the treatment wells at the same time the modeling is used to determine the well screen locations. Of course, specification of flow rates largely depends upon the aquifer hydraulic conductivity, as determined by pumping (or other) tests. Also as discussed earlier, for given flow rates and well screen locations, modeling may be used to estimate capture zone widths, fraction of recirculated water, and fluid residence times, all crucial design elements. Use of models to determine well flow rates is illustrated in the section on modeling applications.

Chemical Amendment

Selection of the chemical to add to the contaminated water is, of course, dependent on the chemical or biological degradation process that is intended to be induced (see Chapter 2). The large majority of recirculating well applications have been to stimulate *in situ* bioremediation. Table 6.1 lists the target contaminant, amended chemical, system type, and degradation process for some field applications of recirculating wells.

For relatively well understood processes (e.g., oxidation of fuel hydrocarbons) there is no need to conduct microcosm studies prior to selecting a chemical amendment for a recirculating well system. However, for processes that are less well understood (e.g., perchlorate reduction, trichloroethene cometabolism) or where the requisite microorganisms may not be present (e.g., methyl tertiary-butyl ether [MTBE] oxidation, reductive dechlorination of chlorinated ethenes)

Table 6.1. Contaminants and Chemical Amendments Used for Recirculating Well Field Applications

Contaminant	Amendment	Degradation process	Type of recirculating well system	Microcosm study conducted?	Reference
Trichloroethene	Toluene, oxygen gas, hydrogen peroxide	Biological – aerobic cometabolism	TRW	Yes	McCarty et al. (1998), Jenal-Wanner and McCarty (1997)
Perchlorate	Citrate	Biological – reduction	TRW	Yes	Hatzinger et al. (2009)
TCE and perchlorate	Acetate and, subsequently, lactate[a]	Biological – reduction	Injection-extraction	Yes	ESTCP (2004)
Carbon tetrachloride	Acetate, phosphate, alkali	Biological – reduction	Injection-extraction	Yes	Dybas et al. (2002)
Gasoline (benzene, toluene, ethylbenzene, xylenes)	Air	Biological – oxidation	GCW	No	ESTCP (1999), USEPA (1998)
Cis-dichloroethene	Propionate	Biological – reduction	TRW	Yes	Hoelen et al. (2006)
Perchloroethene	Ozone and air	Chemical – oxidation	GCW	No	USEPA (1998)

[a] Bioaugmentation of KB-1 dehalorespiring microbial culture also accomplished, but not through use of the recirculating well system

microcosm studies may be important to determine whether and how fast a degradation reaction will occur, and what concentration of amendment to add. Bioaugmentation might be used for aquifers where needed organisms are not naturally present. Table 6.1 outlines whether microcosm studies were conducted in support of the design of a particular field application.

Chemical Addition Schedule

Once appropriate chemical amendments have been chosen, there are several parameters that must be selected to specify a chemical addition schedule. The parameters are (1) daily mass loading of chemical (or alternatively, time-averaged chemical concentration), (2) injected chemical concentration, and (3) injection schedule (e.g., hours injected per day, minutes injected per hour). Modeling is useful in determining a chemical addition schedule. System design models, as discussed below, can be used to account for the multiple factors that affect the demand for the amendment (e.g., contaminant concentration in the treatment zone, reaction kinetic parameters, concentration of electron donors and/or acceptors, hydrology).

For a given time-averaged concentration or mass loading, options for the chemical addition schedule range from continuously adding a relatively low concentration, to adding short pulses of very high concentrations. As will be discussed below, short pulses of high concentrations are a strategy that has been used in an effort to control biofouling at the injection well screens (McCarty et al., 1998; Goltz et al., 2001).

Biofouling Control

If recirculating wells are applied to stimulate biodegradation, the system designer must be concerned with the potential that the chemicals that are being added to sustain *in situ* microbial growth and activity will also result in microbial attachment, growth, and blocking of the well screens, filter pack, or aquifer in the immediate vicinity of the injection well (i.e., biofouling). This biofouling may limit or prevent the proper functioning of the well (ESTCP, 2005a). ESTCP (2005a) published an excellent review of biofouling control methods for application in recirculating wells. The following is a very brief summary of the contents of the review. For a practitioner interested in designing or operating a recirculating well system to stimulate biodegradation, biofouling must be considered, and the ESTCP biofouling review is a good resource to have at hand.

In general, biofouling may be controlled by either (1) well rehabilitation, after fouling has resulted in either increased pressure or decreased flow at the well, or (2) control measures that attempt to prevent biofouling from occurring (ESTCP, 2005a). Well rehabilitation may consist of physical methods such as surging with overpumping or through the use of surge blocks, jetting, or manual brushing. Chemical rehabilitation methods may involve shock chlorination, followed by acid treatment. Other methods may involve injection of hot water, cryogenic carbon dioxide, or combinations of physical, chemical, and thermal processes (ESTCP, 2005a).

In order to limit or eliminate the need for expensive well rehabilitation, biofouling preventive controls may be applied. The ideal biofouling control (1) prevents biofouling at the injection well, but doesn't inhibit the desired biodegradation reactions in the aquifer, (2) is cost effective, (3) is safe and easy to apply, (4) does not adversely impact groundwater geochemistry, and (5) is acceptable to regulators (ESTCP, 2005a). Chemicals used for biofouling prevention include oxidizing agents like hydrogen peroxide, chlorine and chlorine dioxide, nonoxidizing biocides, and chelating and dispersing agents (e.g., enzymes and surfactants). Other approaches may involve using physical controls (e.g., periodic brushing or jetting), or applying more "exotic" methods like ultraviolet (UV) radiation, ultrasound, or impressed currents to control biogrowth (ESTCP, 2005a).

Operational procedures and well construction methods may also be used to minimize biofouling. A number of projects have relied on the use of short pulses of high concentrations of nutrients to minimize the time that nutrients are available to microorganisms at or near the injection well screens (McCarty et al., 1998; Goltz et al., 2001; Hoelen et al., 2006; Hatzinger et al., 2009). Construction methods include use of wide well-screen slots or larger filter pack material in order to reduce bioclogging potential. Application of nonfouling coatings on well screens has also been considered (ESTCP, 2005a).

6.3.3.3 Regulatory Considerations

All recirculating well systems involve addition of some chemical compound or microorganism into the subsurface, and regulations clearly play an important role in determining what compounds are acceptable, and what restrictions apply to the additions. Note that the regulations for injecting microorganisms (other than genetically engineered microorganisms (GEMs)) in order to effect bioaugmentation are the same as for injecting chemicals (ESTCP, 2005b). GEMs may be considered "new" organisms, so injection of GEMs would also be regulated under the Toxic Substances Control Act (Gentry et al., 2004).

For GCW and TRW systems, the regulatory issue is limited to the injection of the amendment, while for injection-extraction systems, there is regulatory concern both with injection of the amendment and with reinjection of the contaminated groundwater. Thus, for injection-extraction systems, there is an additional regulatory barrier that must be overcome.

That said, the main regulatory barrier to reinjection of contaminated groundwater, Section 3020 of the Resource Conservation and Recovery Act (RCRA), which states that "...contaminated groundwater must be treated to substantially reduce hazardous constituents prior to reinjection", has been interpreted by the U.S. Environmental Protection Agency (USEPA) to allow reinjection of contaminated groundwater as part of an *in situ* recirculation treatment system (ITRC, 2002). This interpretation applies to site remediation conducted under the Comprehensive Environmental Response, Compensation, and Liability Act (CERCLA) or under RCRA. Unfortunately, the fact that the USEPA's interpretation does not apply to non-CERCLA and non-RCRA sites is a significant regulatory obstacle to implementation of injection-extraction systems (ITRC, 2002).

The other statute that is relevant to recirculating well systems is the Safe Drinking Water Act, which was used to establish the Underground Injection Control (UIC) program. The UIC program states that injection cannot potentially cause a violation of the primary drinking water standards nor have potentially adverse health effects. Most states have interpreted this as requiring an injection permit (ITRC, 2002). The injection permit may require establishment of a containment area using extraction or monitoring wells, and/or demonstration of containment through modeling. It is also likely that results of monitoring during the remediation would need to be reported to the state, and the compound being injected would have to be analyzed by a state-approved laboratory as permit conditions (ITRC, 2002).

To summarize, the ease or difficulty of obtaining regulatory approval to operate a recirculating well system depends on both the type of system (injection-extraction versus GCW/TRW) and the regulatory environment (CERCLA/RCRA versus non-CERCLA/RCRA), with the least stringent rules applying to CERCLA/RCRA sites where groundwater is not pumped to the surface (i.e., GCW/TRW system). Obviously, the compound being added is also a significant factor. Nonhazardous substances such as lactate or oxygen would be approved as amendments much more readily than potentially hazardous amendments like toluene, phenol, hydrogen peroxide, and nitrate. Also, it should be noted that seemingly harmless substances may be of regulatory concern because of potential reactions that may occur to produce harmful daughter products or undesired geochemical changes. For example, bromide, which was used as a tracer in a TRW implementation, was of concern because of the potential production of bromate (Goltz et al., 1998). Addition of electron donors such as lactate or ethanol to implement *in situ* perchlorate bioremediation was also of regulatory concern because of the potential that the resulting strong aquifer reducing conditions could cause manganese and iron mobilization (USEPA, 2005a). Additionally, some amendments, which may be harmless, may have impurities that may be problematic and cause regulatory concern. For instance, sodium hypochlorite (bleach) which could be used to control biofouling, has been shown to contain perchlorate as an impurity (Renner, 2004). Taken together, successful implementation of a recirculation system will require continual collaboration with the site regulatory body throughout the design, construction, and operation processes.

6.3.4 Modeling Applications

6.3.4.1 Screening Models

Screening models are tools that may be useful in determining the applicability of a remediation technology at a particular site, as well as during the preliminary design process. In the case of recirculation systems, a number of screening models have been developed. Typically, screening models are analytical or semi-analytical models that solve the flow equations for a recirculation system under simplifying assumptions (e.g., assuming homogeneous and isotropic hydraulic conductivity). For specified well locations, pumping rates, and hydrogeologic conditions, the models will output information on the fraction of recirculation between the injection and extraction well screens, the travel times between screens, and the extent of the capture zones. The travel-time approach discussed in Chapter 5 is an example of this type of modeling. Other screening models have been developed that couple flow codes with simple models of the chemical/biological destruction processes. Models such as these provide a potential technology user with an estimate of contaminant concentrations (e.g., upgradient and downgradient of the recirculation system) and perhaps even a preliminary design and cost estimate (Mandalas et al., 1998). Table 6.2 lists some of the screening models that have appeared in the literature for injection-extraction, GCW, and TRW systems.

6.3.4.2 System Design Models

System design models are normally numerical codes that simulate the fate and transport of the chemical compounds of interest (the target contaminant(s), amendments, and degradation daughter products) during operation of a recirculating well system. These models couple simulations of the groundwater flow field induced by operation of a recirculating well system with a model of transport, which accounts for the physical (advection/dispersion), and biochemical (sorption, biodegradation, redox reactions, etc.) processes that affect the fate and transport of the compounds of interest.

In addition to being applied to design recirculating well systems, these system design models are used to help site managers gain an understanding of system operations and to guide adjustment of operating parameters in order to optimize system performance. Use of system design models to optimize recirculation system performance is discussed further in the section on system optimization. Table 6.3 lists some of the system design models that have appeared in the literature for injection-extraction, GCW, and TRW systems.

Table 6.2. Screening Models Used for Recirculation Systems

Type of recirculating well system	Type of model	Processes modeled	Model output	Reference
TRW	Analytical	Flow, aerobic cometabolic biodegradation	System design parameters and cost needed to achieve required performance (downgradient vs. upgradient concentrations)	Mandalas et al. (1998)
		Flow	Plume capture zone width and fraction of water recirculated	Cunningham et al. (2004)
		Flow, aerobic cometabolic biodegradation	Travel times, substrate and contaminant concentrations, biomass	Cirpka and Kitanidis (2001)
		Flow, Pd-catalyzed hydrodehalogenation	Ratio of down gradient to up gradient contaminant concentrations	Stoppel and Goltz (2003)
Injection-extraction	Analytical	Flow	Plume capture zone width	Cunningham and Reinhard (2002)
	Semi-analytical	Flow, first-order degradation	Contaminant concentrations over time	Tenney et al. (2004)
GCW	Not specified	Flow	Circulation cell width	ESTCP (1999)

Table 6.3. System Design Models Used for Recirculation Systems

Type of recirculating well system	Type of model	Processes modeled	Model application	Reference
TRW	Numerical	Flow, aerobic cometabolic biodegradation	Interpretation of field experimental results	Gandhi et al. (2002a, b)
		Flow, biological reduction of perchlorate	System design	Parr et al. (2003)
Injection-extraction	Numerical	Flow, aerobic cometabolic biodegradation	System design and optimization	Lang et al. (1997)
		Flow, nonreactive transport	System design and optimization	Hyndman et al. (2000)
		Flow (travel-time approach), U(VI) bioreduction	Interpretation of field experimental results	Luo et al. (2007a)

(*continued*)

Table 6.3. (continued)

Type of recirculating well system	Type of model	Processes modeled	Model application	Reference
Injection-extraction (horizontal wells)	Numerical	Flow, aerobic cometabolic biodegradation	Interpretation of field experimental results	Travis and Rosenberg (1997)
GCW	Numerical	Flow, advective/ dispersive transport, removal by volatilization	Interpretation of laboratory experimental results	Pinto et al. (1997)
		Flow (in unsaturated and saturated zones), advective/ dispersive transport	Interpretation of laboratory experimental results	Stallard et al. (1996)

6.3.5 Example Designs

In the following sections, simplified examples are presented that demonstrate how environmental factors (e.g., hydrogeological conditions, reaction chemistry) and remediation objectives impact the design of recirculation systems.

6.3.5.1 Effect of Physical Factors

Hydraulic conductivity (magnitude and degree of anisotropy), magnitude and direction of regional groundwater flow, aquifer thickness, and aquifer heterogeneity are all physical factors that impact recirculation system design. Making a number of simplifying assumptions (two-dimensional steady flow in a homogenous, confined aquifer with uniform thickness; regional flow with a constant Darcy velocity), Christ et al. (1999) derived an analytical solution that could be used to calculate the flow in an extraction well that originated from an adjacent injection well in a line of multiple injection/extraction well pairs. Other similar models have also been developed (e.g., Shan, 1999; Zhan, 1999; Fienen et al., 2005).

As demonstrated below, these models can be used to determine the overall capture zone and treatment efficiency for a line of injection/extraction well pairs or TRWs as a function of engineered (e.g., well pumping rates, distance between wells) and environmental (e.g., aquifer thickness, hydraulic conductivity, regional gradient) parameters. Figure 6.4 depicts the influence of these parameters on system recirculation and capture zone width for a typical two-well injection-extraction system. To simplify the analysis the environmental factors are consolidated into a modified transmissivity (T'), which quantifies the strength of the background regional flow, and the engineered parameters are consolidated into a modified flow rate (Q'), which quantifies the strength of the recirculation system:

$$T' = KBi \text{ and } Q' = \frac{Q}{D} \qquad \text{(Eq. 6.1)}$$

where K is the hydraulic conductivity (L/T), B is the aquifer thickness (L), i is the regional gradient (L/L), Q is the in-well flow rate (L^3/T) (assumed to be equal in both wells), and D is the distance between wells (L). The direction of regional flow with respect to the well pair orientation is indicated by (α). For regional flow in the positive x-direction, (α) is the angle

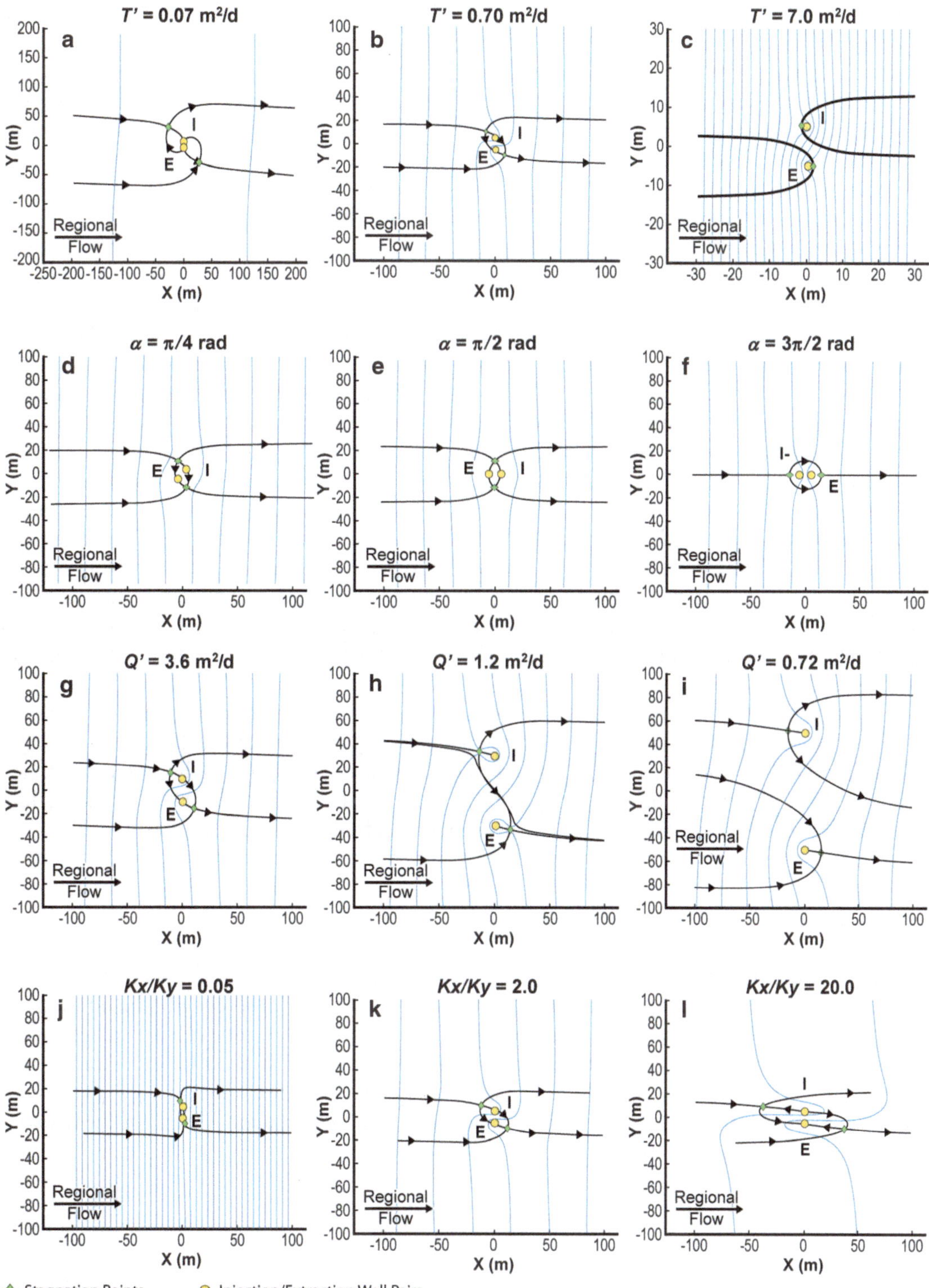

Figure 6.4. Plan view of equipotential and streamlines for an injection-extraction well pair as a function of: (a–c) modified transmissivities (T'), (d–f) regional flow directions, (g–i) modified flow rates (Q'), and (j–l) levels of anisotropy (*Kx/Ky*). *Streamlines* shown indicate the boundaries of the injection-extraction regions. The model parameters, direction of regional flow, and orientation of the injection and extraction well are as depicted in the figure. All figures were generated following the methods of Fienen et al., 2005.

measured clockwise from the positive y-axis to a line running from the extraction well to the injection well. Anisotropy in the x-y plane, defined as the ratio of horizontal to vertical hydraulic conductivity (K_x/K_y), is used to approximate the effects of heterogeneity following the methods of Fienen et al. (2005). All model parameters were chosen to be consistent with typical recirculation system applications (see case studies section). Figure 6.4a–c demonstrates how the strength of the background regional flow (T') strongly influences the magnitude of recirculation for a given value of Q'. As the strength of regional flow increases, the fraction of water recirculated between the injection and extraction well decreases, until ultimately no water recirculates between the two wells (Figure 6.4c), which decreases overall treatment efficiency. The same conclusion may be drawn as the strength of the recirculation system (Q') declines relative to the background regional flow (Figure 6.4g–i). These comparisons illustrate that for given aquifer characteristics and distance between wells, the extent of recirculation is a function of pumping rate, Q. The direction of regional flow may also have a significant influence on the fraction of water recirculated and the capture zone width. For example, Figure 6.4e depicts a scenario with the injection well down gradient of the extraction well, resulting in minimal recirculation, while Figure 6.4f depicts a scenario with the injection well up gradient of the extraction well. Although T' and Q' are the same for both examples, simply changing the orientation of the wells has shifted the recirculation system from minimal recirculation to nearly 100% recirculation, as depicted by the near-zero capture zone width (Figure 6.4f). Subsurface heterogeneity may also play an important role in the fraction of water recirculated. Even relatively minor differences in the hydraulic conductivity may lead to significant differences in the fraction of water recirculated (Figure 6.4j–l), suggesting characterization of the site will be an important factor for accurate prediction of technology performance.

A simple example illustrates the major features of an injection-extraction system for plume containment. Assume an aquifer has a depth (B) of 10 m and a Darcy velocity q of 0.1 m/day. Assume also that for each pass through a treatment system (as discussed in previous sections, the system can be either in-well, *in situ*, or above ground) 90% of the influent contaminant is removed. We define the single-pass treatment efficiency (η) as

$$\eta = 1 - \frac{C_{eff}}{C_{\inf}} \qquad \text{(Eq. 6.2)}$$

where C_{inf} is the contaminant concentration in the influent water entering the treatment system and C_{eff} is the effluent contaminant concentration. Thus, in our example, $\eta = 90\%$.

If the extraction and injection wells are sufficiently far apart, there would be no recirculation between the wells. What then would be the width of the capture zone (CZW) and the overall treatment efficiency of the extraction-injection system ($\eta_{overall}$) if the rate of extraction and injection were both 0.05 m^3/min? We define the overall treatment efficiency ($\eta_{overall}$) in terms of the contaminant concentrations up gradient (C_{up}) and down gradient (C_{down}) of the injection-extraction system as follows:

$$\eta_{overall} = 1 - \frac{C_{down}}{C_{up}} \qquad \text{(Eq. 6.3)}$$

Each day 0.05 cubic meters per minute (m^3/min) (72 m^3/day) would be pumped, so that the capture zone width would equal (72 m^3/day)/(10 m $\times$ 0.1 m/day) or 72 m. The overall treatment efficiency would equal the single-pass treatment efficiency of 90%. Now, let's assume the wells were placed sufficiently close together so that 90% of the water pumped in the extraction well originated in the injection well. We define this fraction as the recirculation ratio (I). In this case, the water captured from up gradient is only 10% of what had been captured in the previous scenario, and now the $CZW = 7.2$ m. The overall treatment efficiency, however, increases. Mass balance indicates that for a two-well injection-extraction system with a single-pass treatment efficiency of η and a recirculation ratio I, the overall treatment efficiency would then equal:

$$\eta_{overall} = \frac{\eta}{1 - I(1-\eta)} \quad \text{(Eq. 6.4)}$$

or for the given example, where $\eta = \mathrm{I} = 0.90$, the overall treatment efficiency would be 99% and the down gradient concentration would be 1% of the up gradient concentration.

Christ et al. (1999) derived an analytical solution for $\eta_{overall}$ for a line of N injection/extraction wells ($N/2$ injection wells and $N/2$ extraction wells) which might be used as a plume containment barrier. If each well is pumping at a rate Q, Christ et al. (1999) defined a "total" recirculation fraction (I_T) as the flow through all extraction wells that originated in injection wells normalized by the flow through a single extraction well (i.e., Q). As an aside, note that the recirculation ratio, I, in each well is different, depending on the well's location in the line of wells. An average recirculation ratio, I_{avg}, can be defined as the total recirculation fraction (I_T) divided by the number of extraction wells ($N/2$). Christ et al. (1999) showed that:

$$I_T = \frac{N}{2} - \frac{\psi_1 - \psi_2}{Q/B} = \frac{N}{2} I_{avg} \quad \text{(Eq. 6.5)}$$

where ψ_1 is the stream function evaluated at the stagnation point associated with the extraction well at one end of the line of wells, and ψ_2 is the stream function evaluated at the stagnation point associated with the injection well at the other end of the line of wells. Complex potential theory may be used to compute the stream functions (ψ_1 and ψ_2) as a function of engineered parameters (e.g. orientation of the line of wells with respect to the regional flow direction (α), well pumping rate (Q)), environmental parameters (e.g. aquifer hydraulic conductivity (K) and thickness (B)), and total number of wells (N) (Christ et al., 1999). Knowing I_T, calculation of the width of the up gradient groundwater captured by the line of wells (CZW) is straightforward:

$$CZW = \frac{Q}{T'}\left[\frac{N}{2} - I_T\right] = \frac{Q}{T'}\frac{N}{2}\left[1 - I_{avg}\right] \quad \text{(Eq. 6.6)}$$

and mass balance can be used to calculate $\eta_{overall}$

$$\eta_{overall} = \frac{\eta}{1 - \frac{2I_T}{N}(1-\eta)} = \frac{\eta}{1 - I_{avg}(1-\eta)} \quad \text{(Eq. 6.7)}$$

Thus, for given environmental conditions, up gradient contaminant concentrations, and treatment efficiency, Equations 6.5, 6.6, and 6.7 can be used to produce a simplified design for a system that will meet specified treatment goals.

Figure 6.5 shows how the capture zone width and the fraction of water recirculated behave as a function of environmental parameters (Figure 6.5a), direction of regional flow (Figure 6.5b) and engineered parameters (Figure 6.5c). As the strength of the background regional flow (T') increases, the fraction of water recirculated and the capture zone width decrease (Figure 6.5a). Generally, a reduction in recirculation results in an increase in capture zone width. In this case, however, the influence of the recirculation system is reduced as it becomes less significant relative to the background regional flow. The opposite trend is true for the case where modified transmissivity (T') is held constant, and modified flow rate (Q') is increased (Figure 6.5c). Among the most important factors influencing recirculation, which is often overlooked, is the direction

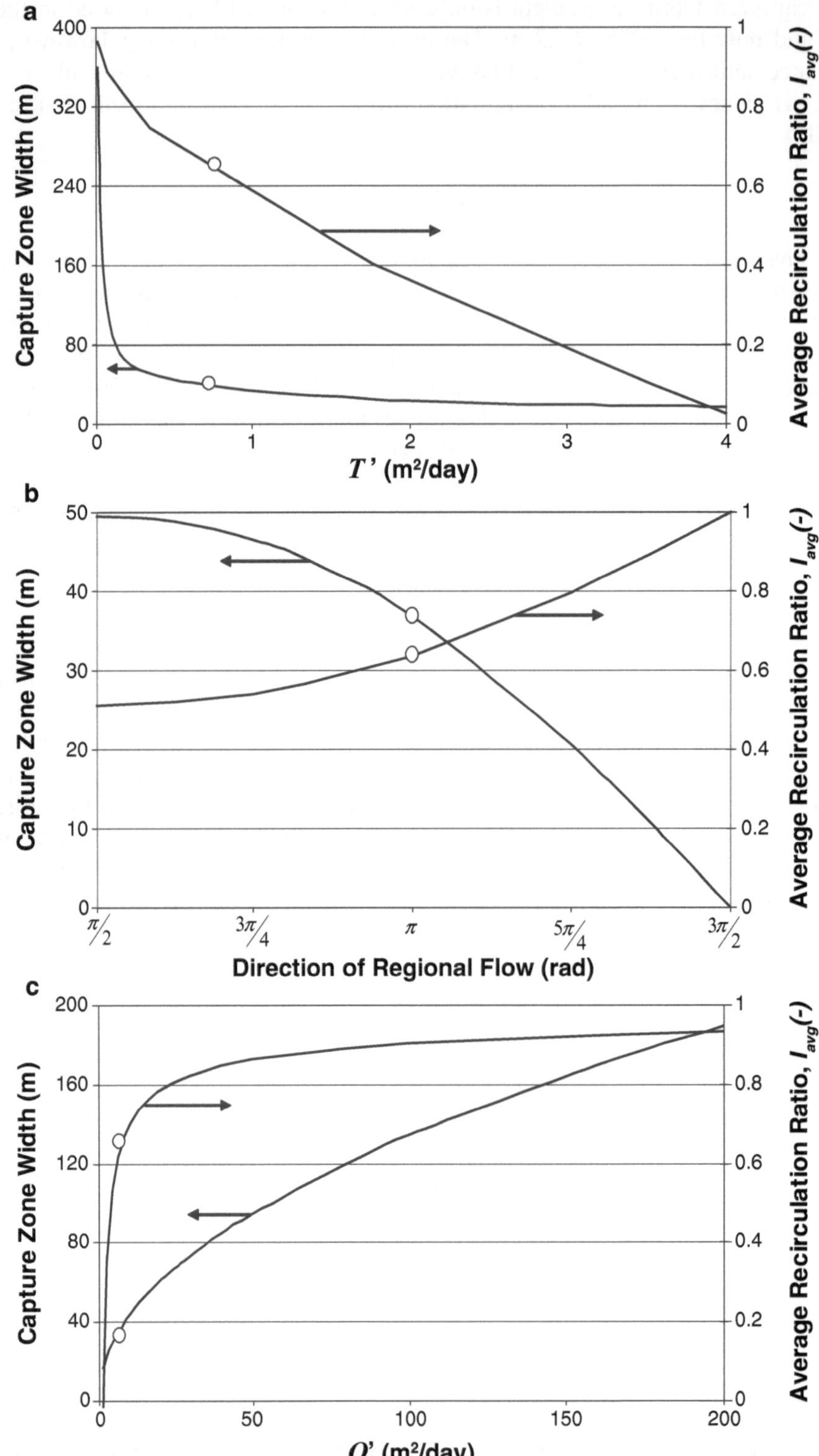

Figure 6.5. Capture zone width and fraction of water recirculated as a function of (a) modified transmissivity (T'), (b) direction of regional flow (α), and (c) modified in-well flow rate (Q'). Baseline parameters are $T' = 0.70$ m^2/day, $Q' = 7.2$ m^2/day and $\alpha = 0$ or π. *Circles* indicate the operating location of the baseline figure, Figure 6.4b.

of regional flow. As Figure 6.5b suggests, well alignment relative to background regional flow can have a significant influence on the fraction of water recirculated and captured. This is especially important when one considers the potential for seasonal variations in background regional flow. These results imply that most recirculation systems will fall within a range of operation throughout the year as the strength and direction of background regional flow vary with the seasons.

6.3.5.2 Effect of Chemical/Biological Factors

Physical, chemical, and biological factors will typically control the efficacy of the treatment technology selected for implementation using the recirculation system. *In situ* bioremediation, in-well vapor stripping, and in-well metal-catalyst reactors are all technologies that have been proposed for application in a recirculating well system (Wu et al., 2008). Application of these technologies requires a firm understanding of reaction kinetics, characteristic times, and operational life span.

Reaction models, such as described in Chapter 2, can be used to estimate the treatment efficiency. Treatment efficiency may be a function of contaminant concentration, concentration of amendments (electron donors, acceptors, and nutrients), aquifer geochemistry, concentration of microorganisms, and residence time in reactive zones. In the simplest case, first-order plug-flow contaminant degradation might be assumed, resulting in the following expression for single-pass treatment efficiency (η):

$$\eta = 1 - e^{-kt_r} \qquad \text{(Eq. 6.8)}$$

where k (T^{-1}) is the first-order rate constant and t_r (T) is the residence time in the reactive zone. Stoppel and Goltz (2003) used a first-order model to calculate treatment efficiency in an in-well catalytic reactor that was a component of a TRW system. Stoppel and Goltz (2003) presented dimensionless curves that demonstrated how a given overall reduction in contaminant concentration (C_{down}/C_{up}) could be achieved through different combinations of treatment efficiency (quantified as the product of k and t_r) and system hydraulics (quantified as Q'/T').

A much more complex model of contaminant degradation in a reactive zone was presented by Luo et al. (2007a). This model was used to simulate uranium (VI) reduction by sulfate-reducing bacteria in a recirculation system, where ethanol was added as an electron donor. The model used Monod kinetics, and simulated reduction of U(VI) as a function of the concentration of sulfate-reducing bacteria (which is, in turn, a function of sulfate concentration), the concentration of electron donor (ethanol), the concentration of U(VI), the concentration of a competing electron acceptor (nitrate), and the concentration of denitrifying bacteria. Although considerably more complex than the simple first-order model, the principles are the same – biochemical reaction kinetics are used to establish treatment efficiency.

As described previously, if the single-pass treatment efficiency of the technology (η) is known, this treatment efficiency can be combined with the fraction of water recirculated (I_{avg}) to predict the overall treatment efficiency of the system. As the fraction of water recirculated or the single-pass treatment efficiency of the technology increases, the concentration down gradient of the recirculation system decreases. Figure 6.6 depicts C_{down}/C_{up} as a function of single-pass treatment efficiency (η) and recirculation (I_{avg}). As shown in the figure, practically, there is a minimum level of treatment required to attain a high level of contaminant removal (e.g., $C_{down}/C_{up} \leq 0.10$). As the fraction of water recirculated declines, the required level of treatment (η) increases eventually reaching the efficiency of the overall system ($1 - C_{down}/C_{up}$) when the fraction of recirculated water reaches zero. This figure suggests that investing in increasing technology single-pass treatment efficiency will pay greater dividends than

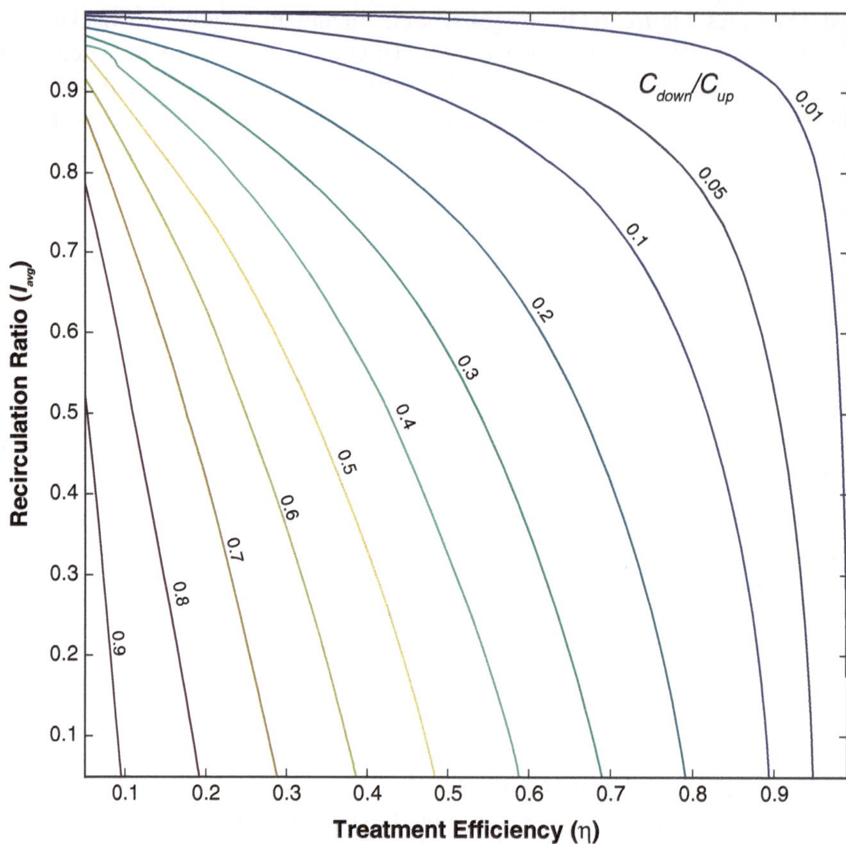

Figure 6.6. C_{down}/C_{up} as a function of the fraction of water recirculated (I_{avg}) and the treatment efficiency (η) of a hypothetical treatment technology (e.g., *in situ* bioremediation, in-well vapor stripping, in-well catalytic reactors).

increasing recirculation, especially when considering increasing recirculation will likely lead to increased operating costs, and decreased capture from up gradient. Additionally, this figure, along with other published screening models and type curves (e.g., Wu et al., 2008), provides a simple tool that may be used to examine alternative treatment technology-recirculation system combinations to select an effective technology and system design for a specific contaminant and site condition.

6.4 SYSTEM OPERATION AND MAINTENANCE (O&M)

6.4.1 Process and Performance Monitoring

Process monitoring is conducted to evaluate how well a system is operating in order that adjustments can be made to optimize performance (ESTCP, 2004). Process monitoring of recirculation systems should focus on two areas: physical (hydrologic) processes and chemical/biological processes.

Monitoring of hydrologic processes in a recirculation system is done to answer the questions: (1) what are the hydraulic gradients, (2) what are the flow rates in the recirculating wells, (3) to what extent is recirculation occurring, (4) what are the travel times between injection and extraction well screens, and (5) what is the extent of the capture zones? The quantitative answers to these questions should be compared with model simulations, and used for model calibration. Calibrated models can then be used to increase understanding of the system, as well as to simulate the system response to operational changes that are being considered to improve system performance. While the first two measurements, hydraulic gradients and flow, are relatively straightforward, the measurements needed to determine recirculation, travel times, and capture zones will normally require tracer testing. Although tracer tests are relatively expensive, they will frequently be necessary, except in the simplest of geologies. Tracer testing is critical, as it provides the best evidence of the establishment of groundwater circulation, which is the heart of recirculation system operation. In fact, the failure to prove that recirculation occurred during GCW testing "…has lead to a roadblock toward broad acceptance of GCW and forestalled endorsement by DoD" (ESTCP, 1999). Tracer tests are typically run when recirculation systems are operated (e.g., McCarty et al., 1998; Wu et al., 2006; Cunningham et al., 2004; Hyndman et al., 2000). A good discussion of the use of tracer testing to evaluate system performance may be found in Cunningham et al. (2004). Less clear-cut evidence of recirculation has been obtained through various other measurements (e.g., water level changes, transducer measurements, flow sensors) (ESTCP, 1999).

In addition to cost, another disadvantage of tracer testing is that it only provides a measurement at a single point in time (ESTCP, 1999). Changes in hydraulic conditions over time, either due to seasonal variations, changes in well flow rates, or the effect of the remediation system itself (upon hydraulic conductivities, for instance), would not be captured by a single test. Luo et al. (2007b) reported that two tracer tests conducted 2 months apart in an injection-extraction bioremediation system showed significantly different results, presumably due to the influence of biological growth that resulted from operation of the system.

In addition to determining the hydraulic gradients, an important use of hydraulic head measurements is to monitor for biofouling. An increase in the hydraulic head at an injection well is an indicator of biofouling. During two separate field evaluations of TRWs, one at Edwards AFB and the other at Moffett Field, California, McCarty et al. (1998) and Hoelen et al. (2006), respectively, observed a gradual increase in pumping heads over time, which was interpreted as being due to biofouling. In the Edwards AFB evaluation, well redevelopment was accomplished to return heads to close to their initial values.

Monitoring of chemical/biological processes is necessary to determine the rate and extent of destruction of the target contaminant. Essential to developing a monitoring plan is an understanding of the destruction kinetics and stoichiometry (ITRC, 2002 and Chapter 2). This understanding allows for identification of daughter products, which along with the target contaminant, chemical amendment, geochemical indicators (dissolved oxygen, pH, oxidation-reduction potential [ORP], etc.), and other compounds that may be of regulatory concern (e.g., dissolved metals), should be measured. As in any conventional chemical engineering process, these measurements indicate whether the destruction process is proceeding as designed. Details of process monitoring and development of a monitoring plan (monitoring well location, frequency of sampling, analytes to sample, etc.) for recirculation systems may be found in ESTCP (2004). Groundwater sampling at monitoring wells located up gradient and down gradient of the recirculation system, as well as within the treatment zone, may be used for process monitoring. Note that sampling at different depths is important; in particular, at the depths of the injection and extraction screens.

Performance monitoring is accomplished to evaluate how well the system is achieving remedial objectives (ESTCP, 2004). Lines of evidence for performance evaluation include spatial and/or temporal reduction in target contaminant concentrations and reduction of target contaminant mass (USEPA, 2000). Recent work also suggests that reduction in contaminant mass flux is another factor to consider when evaluating system performance (Einarson and Mackay, 2001; Soga et al., 2004; USEPA, 2005b). Groundwater sampling at monitoring wells located up gradient and down gradient of the recirculation system may be used to evaluate system performance.

6.4.2 System Optimization

The data obtained from process and performance monitoring should be used, in conjunction with models, to optimize system performance. Optimization of recirculating well systems can include strategies such as: (1) changing the chemical addition schedule to minimize cost while maximizing target contaminant destruction, (2) changing the treatment well pumping rates or even initiating "pulsed pumping" (i.e., turning off the treatment well pumps periodically), and (3) adjusting biofouling controls to minimize costs.

A system design model, such as those discussed earlier, is an extremely useful tool that will help managers determine how a proposed operational change may affect system performance. These system design models, which simulate fate and transport of the chemical compounds of interest (target contaminant, amendments, degradation daughter products), may be used in conjunction with optimization codes to help site managers decide what operational changes to implement to improve system performance. Knarr et al. (2003) used a system design model in conjunction with a genetic algorithm to show how a TRW system designed to remediate perchlorate-contaminated groundwater could be optimized to minimize operating cost while maximizing contaminant mass removal by adjusting the treatment well pumping rates and the chemical addition schedule. Garrett et al. (1999) did a similar analysis for a TRW system used to induce aerobic cometabolic bioremediation of TCE-contaminated groundwater. Minsker and Shoemaker (1998) determined optimal pumping rates, chemical injection schedules, and well locations for an injection-extraction system used to induce aerobic biodegradation.

6.5 CASE STUDIES

To demonstrate actual recirculation system performance, the following section presents case studies for an injection-extraction system, a GCW system, and a TRW system. Additional case studies showing applications of the various recirculating well systems may be found in the following references: ESTCP (1999), USEPA (1998), ESTCP (2004, 2005b), and Grindstaff (1998).

6.5.1 Injection-Extraction Application: Schoolcraft, Michigan Site

A biocurtain near the leading edge of a groundwater plume at a site in Schoolcraft, Michigan was established by using closely spaced (1 m) injection/extraction wells to add microorganisms, as well as acetate (an electron donor), alkali, and phosphorus into groundwater contaminated by carbon tetrachloride (CT) and nitrate. The purpose of the biocurtain was to create conditions amenable to the *in situ* biodegradation of the contaminants, in order that the contaminants would be reduced and not travel downgradient of the curtain (Dybas et al., 2002).

The CT plume was about 1.6 kilometers (km) long, 160 m wide, and extended between 9 and 27 m below ground surface (bgs) (with the water table at about 5 m bgs). Hydraulic conductivity was very heterogeneous, ranging from 0.0011 to 0.11 cm/s. 15 injection/extraction wells, screened over a depth of 15 m, were installed at 1 m intervals perpendicular to and near the leading edge of the plume. After a one-time injection of CT-degrading microorganisms, 10 milligrams per liter (mg/L) phosphate, 100 mg/L acetate, and sodium hydroxide (added to obtain a pH of 8.2) were injected over a 5-h period weekly. The 15 wells were operated for a combined flow rate of 150 L/min each over the 5-h injection period. Injection and extraction wells were adjacent to each other, and every week the pumping assignment was switched (injection wells were used as extraction wells and vice versa). After 5 h of pumping, the flow field was reversed, by using injection wells as extraction wells, and extraction wells as injection wells, for 1-h. This pumping schedule was designed to help ensure uniform delivery of chemicals throughout the biocurtain.

Both a two-dimensional semi-analytical screening model (Tenney et al., 2004), and a three-dimensional numerical flow and transport model (Hyndman et al., 2000) were used to design the system and interpret results. The system was successful in achieving high CT removal efficiencies (>98%) over a 4-year period (Dybas et al., 2002). The intermittent pumping used in this system efficiently delivered substrates (and microorganisms) to the biocurtain, without encountering any biofouling problems.

6.5.2 Groundwater Circulation Well Application: Port Hueneme Naval Exchange Site, California

The vast majority of GCW applications have employed in-well vapor stripping (see Chapter 8) rather than chemical mixing. However, there have also been a number of cases where the wells have primarily been used to achieve chemical mixing. The latter case will be discussed here.

Approximately 11,000 gal of gasoline leaked into a shallow (depth to groundwater between 1 and 3.7 m below ground) perched aquifer at a Naval Exchange site at Port Hueneme Naval Air Station, California. The horizontal hydraulic conductivity at the site was measured at 3.84×10^{-2} cm/s, and the vertical conductivity estimated at 3.84×10^{-3} cm/s (ESTCP, 1999).

A GCW system was installed with four treatment wells; a single GCW with a 400 mm inner diameter well casing installed near the spill source, and three smaller GCWs (200 mm casings) installed down gradient. The three down gradient GCWs were installed with the primary purpose of mixing air into the fuel-contaminated groundwater, in order to stimulate *in situ* aerobic biodegradation and form a "biocurtain" to contain migration of the hydrocarbon plume (ESTCP, 1999).

The three biocurtain GCWs had upper and lower well screens installed between 2.2 and 5 m bgs and 7 and 8.7 m bgs, respectively. Based on modeling, the region of influence of each GCW was estimated, and the wells were installed 40 m apart to allow overlapping circulation cells. Eight monitoring wells, screened to sample at depths corresponding to the lower and upper screens of the GCWs, were placed around the biocurtain wells. Convergent and divergent dye tracer tests were conducted to verify groundwater circulation (ESTCP, 1999).

After 6 months of biocurtain operation, it was seen that benzene, toluene, ethylbenzene, and total xylenes (BTEX) concentrations measured in the shallow monitoring wells were reduced from 772 mg/L to less than 0.002 mg/L. BTEX concentrations measured in the deep monitoring wells were reduced from 118 mg/L to less than 0.001 mg/L. Based on these results, it was concluded that the GCWs were effectively containing the plume, and that containment of the contaminants was due, at least in part, to biodegradation that was stimulated by the addition of oxygen in the biocurtain GCWs (ESTCP, 1999).

6.5.3 Tandem Recirculating Well Application (Trichloroethene Bioremediation at Edwards AFB, California)

A field evaluation of a TRW system to treat a trichloroethene-contaminated groundwater plume was conducted at Edwards AFB (McCarty et al., 1998). An overall description of this study is provided in Section 2.4.6 of Chapter 2, and results of a detailed numerical model that followed data gathering from the study is given in Section 4.6.2 of Chapter 4. A pair of dual-screened wells was installed to treat TCE-contaminated groundwater in an 8 m thick upper aquifer that overlaid a 5 m thick lower aquifer. Both aquifers were contaminated with TCE at concentrations of 500–1,200 micrograms per liter (μg/L). The wells, which were placed approximately perpendicular to groundwater flow and located 10 m apart, were pumped at 25–80 L/min. Oxygen was continuously added (as both a gas, and using hydrogen peroxide as an oxygen source) into groundwater flowing through the wells. Toluene was pulsed into the groundwater, in order to stimulate indigenous microorganisms to aerobically cometabolize TCE, using toluene as the primary substrate. The toluene pulsing strategy was intended to minimize biofouling.

System design was accomplished using a two-dimensional flow code coupled with a one-dimensional transport code (McCarty et al., 1998). The transport code simulated advection, as well as the chemical and biological processes (rate-limited sorption and aerobic cometabolism) that affected the contaminant. As fully described in Section 4.6.2 of Chapter 4, the study results were then interpreted using a three-dimensional numerical model that incorporated all relevant flow and transport processes (Gandhi et al., 2002a, b).

Tracer tests were conducted to confirm that water recirculated between the two wells. With each pass of groundwater through the bioactive zone that was established in the upper aquifer around the upflow well injection screen, $87 \pm 8\%$ TCE removal was achieved. $83 \pm 16\%$ removals were achieved in the bioactive zone in the lower aquifer around the downflow well injection screen. These values represent the single-pass treatment efficiency (η) defined in Section 6.3.5. Because of recirculation between the two wells, much higher overall removals ($\eta_{overall}$, which compares TCE concentrations upgradient and downgradient of the TRW system) of 97–98% were attained (McCarty et al., 1998), a result one would expect through application of Equation 6.4.

Although pulsed addition of substrate and use of hydrogen peroxide as an oxygen source were intended to minimize biofouling around the injection well screens, head increases at the injection screens indicated these strategies were not fully successful, and periodic well redevelopment was required (McCarty et al., 1998).

6.6 CONCLUSIONS

Recirculation systems have been demonstrated to be an effective method for mixing amendments with contaminants in groundwater, in order to achieve *in situ* contaminant destruction. There are a number of important advantages to these systems, but there are also challenges that a remediation project manager must keep in mind when deciding whether to implement recirculation. System designers have to consider various factors when designing these systems. Fortunately, models are available to use as tools, so that the complex interactions between the various physical, chemical, and biological processes that affect system effectiveness can be considered in the design.

ACKNOWLEDGMENTS

The review comments of an anonymous reviewer were appreciated. In particular, Professor Jeffrey Cunningham is thanked for his extensive and detailed review of the manuscript. His comments contributed greatly to the final product. The views expressed in this chapter are those of the authors and do not reflect the official policy or position of the United States Air Force, Department of Defense, or the U.S. Government.

REFERENCES

Agarwal N, Semmens MJ, Novak PJ, Hozalski RM. 2005. Zone of influence of a gas permeable membrane system for delivery of gases to groundwater. Water Resour Res 41:W05017, doi:10.1029/2004WR003594.

Boman GK, Molz FJ, Boone KD. 1997. Borehole flowmeter application in fluvial sediments: Methodology, results, and assessment. Ground Water 35:443–450.

Butler JJ, Healey JM, McCall GW, Garnett EJ, Loheide II SP. 2002. Hydraulic tests with direct-push equipment. Ground Water 40:25–36.

Cho JS, Wilson JT, Beck Jr FP. 2000. Measuring vertical profiles of hydraulic conductivity with in situ direct-push methods. J Environ Eng 126:775–777.

Christ JA, Goltz MN, Huang J. 1999. Development and application of an analytical model to aid design and implementation of in situ remediation technologies. J Contam Hydrol 37:295–317.

Cirpka OA, Kitandis PK. 2001. Travel-time based model of bioremediation using circulation wells. Ground Water 39:422–432.

Cunningham JA, Reinhard M. 2002. Injection-extraction treatment well pairs: An alternative to permeable reactive barriers. Ground Water 40:599–607.

Cunningham JA, Hoelen TP, Hopkins GD, Lebrón CA, Reinhard M. 2004. Hydraulics of recirculating well pairs for ground water remediation. Ground Water 42:880–889.

Dacosta JA, Bennett RR. 1960. The Pattern of Flow in the Vicinity of a Recharging and Discharging Pair of Wells in an Aquifer Having Areal Parallel Flow. International Association of Scientific Hydrology, IUGG General Assembly of Helsinki, Publication No. 52, pp 524–536.

Davie MG, Cheng H, Hopkins GD, LeBron CA, Reinhard M. 2008. Implementing heterogeneous catalytic dechlorination technology for remediating TCE-contaminated groundwater. Environ Sci Technol 42:8908–8915.

Domenico PA, Schwartz FW. 1998. Physical and Chemical Hydrogeology, 2nd Edition. John Wiley and Sons, New York, NY, USA. 506 p.

Dybas MJ, Hyndman DW, Heine R, Tiedje J, Linning K, Wiggert D, Voice T, Zhao X, Dybas L, Criddle CS. 2002. Development, operation, and long-term performance of a full-scale biocurtain utilizing bioaugmentation. Environ Sci Technol 36:3635–3644.

Einarson MD, Mackay DM. 2001. Predicting impacts of groundwater contamination. Environ Sci Technol 35:66A–73A.

Elmore AC, DeAngelis L. 2004. Modeling a groundwater circulation well alternative. Ground Water Monit Remediat 24:66–73.

ESTCP (Environmental Security Technology Certification Program). 1997. Draft Technical Protocol for Implementing the Groundwater Circulating Well Technology for Site Remediation. ESTCP Project ER-9602. ESTCP, Arlington, VA, USA.

ESTCP. 2004. Rapid and Complete Treatment of Trichloroethene via Bioaugmentation in an Active Biobarrier. Appendix E.9. Principles and Practices of Enhanced Anaerobic Bioremediation of Chlorinated Solvents. Principles and Practices Manual: ESTCP Project ER-0125. ESTCP, Arlington, VA, USA.

ESTCP. 2005a. A Review of Biofouling Controls for Enhanced *In Situ* Bioremediation of Groundwater. Technical Report: ESTCP Project ER-0429. ESTCP, Arlington, VA, USA.

ESTCP. 2005b. Bioaugmentation for Remediation of Chlorinated Solvents: Technology Development, Status, and Research Needs. ESTCP White Paper. ESTCP, Arlington, VA, USA.

ESTCP. 1999. Groundwater Recirculating Well Technology Assessment. ER-9602 Final Report. Prepared by the Naval Research Laboratory, Washington DC for ESTCP, Arlington, VA, USA. http://www.serdp.org/content/advancedsearch/. Accessed August 30, 2011.

Fetter CW. 1999. Contaminant Hydrogeology, 2nd Edition. Prentice Hall, Upper Saddle River, NJ, USA.

Fienen MN, Luo J, Kitanidis PK. 2005. Semi-analytical anisotropic capture zone delineation. J Hydrol 312:39–50.

Gandhi RK, Hopkins GD, Goltz MN, Gorelick SM, McCarty PL. 2002a. Full-scale demonstration of in situ cometabolic biodegradation of trichloroethylene in groundwater, 1. Dynamics of a recirculating well system. Water Resour Res 38:10–1.

Gandhi RK, Hopkins GD, Goltz MN, Gorelick SM, McCarty PL. 2002b. Full-scale demonstration of in situ cometabolic biodegradation of trichloroethylene in groundwater, 2. Comprehensive analysis of field data using reactive transport modeling. Water Resour Res 38:11–1.

Garrett CA, Huang J, Goltz MN, Lamont GB. 1999. Parallel real-valued genetic algorithms for bioremediation optimization of TCE-contaminated groundwater. Proceedings, 1999 Congress on Evolutionary Computation, Washington, DC, USA, July 6–9, Vol. 3, pp 2183–2189.

Gentry TJ, Rensing C, Pepper IL. 2004. New approaches for bioaugmentation as a remediation technology. Crit Rev Environ Sci Technol 34:447–494.

Goltz MN, Mandalas GC, Hopkins GD, McCarty PL. 1998. Technology transfer of an innovative remediation technology from the laboratory to the field: A case study of in situ aerobic cometabolic bioremediation. Environ Eng Pol 1:117–124.

Goltz MN, Bouwer EJ, Huang J. 2001. Transport issues and bioremediation modeling for the *in situ* aerobic co-metabolism of chlorinated solvents. Biodegradation 12:127–140.

Goltz MN, Gandhi RK, Gorelick SM, Hopkins GD, Smith L, Timmins BH, McCarty PL. 2005. Field evaluation of in situ source reduction of trichloroethylene (TCE) in groundwater using bio-enhanced in-well vapor stripping. Environ Sci Technol 39:8963–8970.

Grindstaff M. 1998. Bioremediation of Chlorinated Solvent Contaminated Groundwater. Prepared for USEPA Technology Innovation Office, Washington DC, USA. 32 p. http://www.clu-in.org/download/studentpapers/meganfin.pdf. Accessed August 17, 2011.

Hatzinger PB. 2002. *In Situ* Bioremediation of Perchlorate. Final Report: Project CU-1163. Strategic Environmental Research and Development Program (SERDP), Arlington, VA, USA.

Hatzinger PB, Schaefer CE, Cox EE. 2009. Active Bioremediation. In Stroo HF, Ward, CH, eds, *In Situ* Bioremediation of Perchlorate in Groundwater. Springer Science + Business media, LLC, New York, NY, USA, pp 91–133.

Hazen TC, Looney BB, Enzien M, Dougherty JM, Wear J, Fliermans CB, Eddy CA. 1996. *In Situ* Bioremediation via Horizontal Wells. In Hickey RF, Smith G, eds, Biotechnology in Industrial Waste Treatment and Bioremediation, pp 79–86.

Herrling B, Stamm J. 1991. Numerical results of calculated 3D vertical circulation flows around wells with two screen sections for in situ and on-site remediation. Proceedings, IX International Conference on Computational Methods in Water Resources, Denver, CO, USA, June 9–12.

Hoelen TP, Cunningham JA, Hopkins GD, Lebron CA, Reinhard M. 2006. Bioremediation of cis-DCE at a sulfidogenic site by amendment with propionate. Ground Water Monit Remediat 26:82–91.

Huang J, Goltz MN. 2005. A three-dimensional analytical model to simulate groundwater flow during operation of recirculating wells. J Hydrol 314:67–77.

Huang J, Goltz MN, Hatzinger PB, Chosa PG, Diebold J, Farhan YH, Parr JC, Neville S. 2004. Modeling studies to support design of a field evaluation of in situ bioremediation of perchlorate-contaminated groundwater. In Proceedings, WEFTEC 2004, New Orleans, LA, USA, October 2–6.

Hyndman DW, Dybas MJ, Forney L, Heine R, Mayotte T, Phanikumar MS, Tatara G, Tiedje J, Voice T, Wallace R, Wiggert D, Zhao X, Criddle CS. 2000. Hydraulic characterization and design of a full-scale biocurtain. Ground Water 38:462–474.

Interstate Technology and Regulatory Council (ITRC). 2002. A Systematic Approach to *In Situ* Bioremediation in Groundwater Including Decision Trees on *In Situ* Bioremediation for Nitrates, Carbon Tetrachloride, and Perchlorate. August. http://www.itrcweb.org/Documents/ISB-8.pdf. Accessed August 15, 2011.

Jenal-Wanner U, McCarty PL. 1997. Development and evaluation of semicontinuous slurry microcosms to simulate in situ biodegradation of trichloroethylene in contaminated aquifers. Environ Sci Technol 31:2915–2922.

Johnson RL, Simon MA. 2007. Evaluation of groundwater flow patterns around a dual-screened groundwater circulation well. J Contam Hydrol 93:188–202.

Kabala ZJ. 1993. The dipole flow test: A new single-borehole test for aquifer characterization. Water Resour Res 29:99–107.

Knarr MR, Goltz MN, Lamont GB, Huang J. 2003. *In situ* bioremediation of perchlorate-contaminated groundwater using a multi-objective parallel evolutionary algorithm. Proceedings, 2003 Congress on Evolutionary Computation, Canberra, Australia, December 8–12.

Lang, MM, Roberts PV, Semprini L. 1997. Model simulations in support of field scale design and operation for cometabolic degradation. Ground Water 35:565–573.

Luo J, Kitanidis PK. 2004. Fluid residence times within a recirculation zone created by an extraction-injection well pair. J Hydrol 295:149–162.

Luo J, Weber FA, Cirpka OA, Wu WM, Nyman JL, Carley J, Jardine PM, Criddle CS, Kitanidis PK. 2007a. Modeling in situ uranium (VI) bioreduction by sulfate-reducing bacteria. J Contam Hydrol 92:129–148.

Luo J, Wu WM, Carley J, Ruan C, Gu B, Jardine PM, Criddle CS, Kitanidis PK. 2007b. Hydraulic performance analysis of a multiple injection-extraction well system. J Hydrol 336:294–302.

Mandalas GC, Christ JA, Goltz MN. 1998. Screening software for an innovative in situ bioremediation technology. Bioremediation J 2:7–15.

McCarty PL, Semprini L. 1994. Ground-water treatment for chlorinated solvents. In Norris RD et al., eds., Handbook of Bioremediation, Lewis Publishers, Boca Raton, FL, USA. 257 p.

McCarty PL, Goltz MN, Hopkins GD, Dolan ME, Allan JP, Kawakami BT, Carrothers TJ. 1998. Full-scale evaluation of in situ cometabolic degradation of trichloroethylene in groundwater through toluene injection. Environ Sci Technol 32:88–100.

Minsker BS, Shoemaker CA. 1998. Dynamic optimal control of in situ bioremediation of groundwater. ASCE J Water Resour Plan Manag 124:149–161.

Oya S, Valocchi AJ. 1998. Analytical approximation of biodegradation rate for in situ bioremediation of groundwater under ideal radial flow conditions. J Contam Hydrol 31:275–293.

Parr JC, Goltz MN, Huang J, Hatzinger PB, Farhan YH. 2003. Modeling in situ biodegradation of perchlorate-contaminated groundwater. In Wickramanayake GB et al., eds, Case Studies in the Remediation of Chlorinated and Recalcitrant Compounds. Battelle Press, Columbus, OH, USA.

Philip R, Walter G. 1992. Prediction of flow and hydraulic head fields for vertical circulation wells. Ground Water 30:765–773.

Pickens JF, Grisak GE. 1981. Scale-dependent dispersion in a stratified granular aquifer. Water Resour Res 17:1191–1211.

Pinto MJ, Gvirtzman H, Gorelick SM. 1997. Laboratory-scale analysis of aquifer remediation by in-well vapor stripping 2. Modeling results. J Contam Hydrol 29:41–58.

Renner R. 2004. Perchlorate contamination mystery in Massachusetts. 2004. Environ Sci Technol 38: 481A.

Scaturo DM, Widdowson MA. 1997. Experimental evaluation of a drive-point ground-water sampler for hydraulic conductivity measurement. Ground Water 35:713–720.

Shan C. 1999. An analytical solution for the capture zone of two arbitrarily located wells. J Hydrol 222:123–128.

Soga K, Page JWE, Illangasekare TH. 2004. A review of NAPL source zone remediation efficiency and the mass flux approach. J Hazard Mater 110:13–27.

Stallard WM, Wu KC, Shi N, Corapcioglu MY. 1996. Two-dimensional hydraulics of recirculating ground-water remediation wells in unconfined aquifers. J Environ Eng 122:692–699.

Stoppel CM, Goltz MN. 2003. Modeling Pd-catalyzed destruction of chlorinated ethenes in groundwater. J Environ Eng 129:147–154.

Suthersan SS. 1997. *In Situ* Bioremediation. In Suthersan SS, ed, Remediation Engineering: Design Concepts. Chapter 5. CRC Press LLC, Boca Raton, FL, USA.

Tenney CM, Lastoskie CM, Dybas MJ. 2004. A reactor model for pulsed pumping groundwater remediation. Water Res 38:3869–3880.

Travis BJ, Rosenberg ND. 1997. Modeling in situ bioremediation of TCE at Savannah River: Effects of product toxicity and microbial interactions on TCE degradation. Environ Sci Technol 31:3093–3102.

USEPA (United States Environmental Protection Agency). 1995. How to Evaluate Alternative Cleanup Technologies for Underground Storage Tank Sites: A Guide for Corrective Action Plan Reviewers. EPA 510-B-95-007. USEPA Solid Waste and Emergency Response, Washington, DC, USA.

USEPA. 1998. Field Applications of *In Situ* Remediation Technologies: Ground-Water Circulation Wells. EPA 542-R-98-009.

USEPA. 2000. Engineered Approaches to *In Situ* Bioremediation of Chlorinated Solvents: Fundamentals and Field Applications. EPA 542-R-00-008.

USEPA. 2001. Cost Analyses for Selected Groundwater Cleanup Projects: Pump and Treat Systems and Permeable Reactive Barriers. EPA 542-R-00-013.

USEPA. 2005a. Perchlorate Treatment Technology Update. Federal Facilities Forum Issue Paper. EPA 542-R-05-015.

USEPA. 2005b. Mass Flux Evaluation Finds SEAR Continues to Reduce Contaminant Plume. Technol News Trends 17:4–5.

Wilson JL. 1984. Double-cell hydraulic containment of pollutant plumes. Proceedings, Fourth National Symposium and Exposition on Aquifer Restoration and Ground Water Monitoring. National Water Well Association, Dublin, OH, USA, pp 65–70.

Wu WM, Carley J, Fienen M, Mehlhorn T, Lowe K, Nyman J, Luo J, Gentile ME, Rajan R, Wagner D, Hickey RF, Gu B, Watson D, Cirpka OA, Kitanidis PK, Jardine PM, Criddle CS. 2006. Pilot-scale in situ bioremediation of uranium in a highly contaminated aquifer. 1. Conditioning of a treatment zone. Environ Sci Technol 40:3978–3985.

Wu M, Smits KM, Goltz MN, Christ JA. 2008. A screening model for injection-extraction treatment well recirculation system design. Ground Water Monitor Remediat 28:63–71.

Zhan H. 1999. Analytical and numerical modeling of a double well capture zone. Math Geol 31:175–193.

Zlotnik VA, Ledder G. 1996. Theory of dipole flow in uniform anisotropic aquifers. Water Resour Res 32:1119–1128.

CHAPTER 7

PERMEABLE BARRIER WALLS

Robert W. Gillham[1] and John Vogan[2]

[1]University of Waterloo, Waterloo, ON, Canada
[2]Arcadis, Waterloo, ON, Canada

7.1 INTRODUCTION

A permeable reactive barrier (PRB) has been defined as "an *in situ* permeable treatment zone designed to intercept and remediate a contaminant plume" (ITRC, 2005). The PRB concept is illustrated schematically in Figure 7.1, which shows a contaminant plume moving, under natural hydraulic gradients, through a permeable "wall" of reactive material and exiting on the downgradient side with the contaminants removed. Remediation within the PRB can proceed through removal, as in the case of sorption or precipitation reactions, or by degradation as in the case of a range of biological or abiotic reactions for treatment of organic contaminants.

The earliest literature references to the PRB concept that we are aware of include McMurtry and Elton (1985) and Thomson and Shelton (1988). These authors considered applications at relatively shallow depths, and reactive materials that promote contaminant removal by sorption or by enhanced biodegradation. The concept gained considerable momentum in the early 1990s when it was recognized that granular metallic iron was effective in degrading a wide range of halogenated organic compounds, including industrial solvents, and that the reactions could proceed under ambient groundwater conditions (Gillham and O'Hannesin, 1992, 1994; Gillham and Burris, 1992; Matheson and Tratnyek, 1994). Since the early 1990s, the PRB concept has become a major topic of environmental research, resulting in a voluminous body of literature including both primary scientific papers and numerous review articles. Review papers of particular relevance include Blowes et al. (1998), Gaveskar et al. (1998), Scherer et al. (2000), Tratnyek et al. (2003) and Gillham et al. (2010).

Though stimulated initially by the potential for using granular iron to degrade industrial solvents such as trichloroethene (TCE), the concept has now been extended to a much broader range of contaminants, using various reactive materials. The *in situ* use of iron to degrade chlorinated hydrocarbons has now been applied at over 200 contaminated sites. Though there is no reliable estimate of the number of PRB sites at which other reactive materials have been used, it is quite likely similar.

The primary advantage of the PRB concept is associated with its passive nature and thus low operating and maintenance costs over long periods of time. Other advantages include no aboveground structures, and thus reuse of the land; water and energy conservation; and ease of performance monitoring. Furthermore, in most applications, space in the subsurface is created which is then filled with a porous and immobile solid-phase reactive material. With mixing achieved by passive migration of groundwater through the PRB, and thereby coming into contact with the reactive material, the technology is largely immune to the limitations of mixing and mass transfer associated with the injection of liquids containing dissolved reactants and biostimulants. Though the removal/degradation processes generally require mass transfer between the aqueous phase and the reactive surfaces, these occur at the pore scale and are,

P.K. Kitanidis and P.L. McCarty (eds.), *Delivery and Mixing in the Subsurface: Processes and Design Principles for In Situ Remediation*, doi: 10.1007/978-1-4614-2239-6_7, © Springer Science+Business Media New York 2012

in most cases, rapid relative to the chemical reaction rate and groundwater velocity and are therefore generally not of practical consequence.

While the PRB concept offers several advantages, depending upon the reactive material, the capital cost of construction of a conventional trenched system is relatively high, often comparable to, or greater than, that of a pump-and-treat system and thus the economic advantage rests primarily with the low operating and maintenance costs. Furthermore, the rate of cleanup is generally slow, with the natural groundwater velocity as a primary determining factor. Thus, while a PRB can be particularly effective in preventing off-site migration or in protecting sensitive receptors such as wells, streams and wetlands, complete remediation of a large contaminant plume could require several decades following treatment or removal of the source zone.

Primary considerations in applying the PRB technology include identification of a reactive material that is effective in removing/degrading the contaminants of concern and the design and engineering necessary for installation at a particular site. These topics will be considered in the following sections of this chapter as well as issues related to long-term performance.

7.2 REACTIVE MATERIALS

The PRB concept has stimulated considerable research and innovation concerning the identification and testing of various reactive materials for treatment of a wide range of groundwater contaminants. Table 7.1, reproduced from ITRC (2005), summarizes the major

Table 7.1. Examples of Reactive Materials Used in PRBs (from ITRC, 2005)

Treatment material categories	Example materials	Constituents treated (examples, not comprehensive)
Metal-enhanced reductive dechlorination for organic compounds	Zero-valent metals (Fe)	Chlorinated ethenes, ethanes, methanes, and propanes; chlorinated pesticides, freons, nitrobenzene
Metal-enhanced reduction of metal contaminants	Zero-valent metals (Fe), basic oxygen furnace slag, ferric oxides	Cr, U, As, Tc, Pb, Cd, Mo, U, Hg, P, Se, Ni
Sorption and ion-exchange	Zero-valent iron, granular activated carbon, apatite (and related materials), bone char, zeolites, peat, humate	Chlorinated solvents (some), BTEX, Sr-90, Tc-99, U, Mo
pH control	Limestone, zero-valent iron	Cr, Mo, U, acidic water
In situ redox manipulation	Sodium dithionite, calcium polysulfide	Cr, chlorinated ethenes
Enhancements for bioremediation (including carbon, oxygen, and hydrogen sources)	(Includes solid, liquid, and gaseous sources): oxygen-release compounds, hydrogen-release compounds, carbohydrates, lactate, zero-valent iron, compost, peat, sawdust, acetate, and humate	Chlorinated ethenes and ethanes, nitrate, sulphate, perchlorate, Cr, MTBE, polyaromatic hydrocarbons

Note: *As* arsenic, *BTEX* benzene, toluene, ethyl benzene, and total xylenes, *Cd* cadmium, *Cr* chromium, *Fe* iron, *Hg* mercury, *Mo* molybdenum, *MTBE* methyl tertiary-butyl ether, *Ni* nickel, *P* phosphorus, *Pb* lead, *Se* selenium, *Sr*-90 strontium-90, *Tc* technetium, *Tc*-99 technetium-90, *U* uranium classes of treatment materials and the contaminants for which they are effective. The same ITRC report includes comprehensive tables that list specific applications. The following discussion provides a summary of the most commonly used reactive materials.

7.2.1 Granular Metallic Iron

Zero-valent metals are in a highly reduced state and therefore reactions with a wide range of reducible compounds such as chlorinated solvents and certain electroactive metal ions are thermodynamically favorable. Though several metals, including aluminum, zinc and tin have this potential, attention has been focused largely on iron because of its effectiveness, availability and relatively low cost.

Granular iron has been shown to be effective in degrading a wide range of halogenated aliphatic compounds, including those most frequently encountered at hazardous waste sites, TCE and perchloroethene (PCE). Recent reviews of this topic include Tratnyek et al. (2003) and Gillham et al. (2010). When granular iron is added to an aqueous solution containing a halogenated hydrocarbon, two redox reactions proceed: oxidation of iron accompanied by reduction of water (Equation 7.1), and oxidation of iron accompanied by reduction of the halogenated organic (Equation 7.2).

$$\begin{array}{c} Fe^{o} \rightarrow Fe^{2+} + 2e^{-} \\ 2H_2O + 2e^{-} \rightarrow H_2 + 2\,OH^{-} \\ \hline Fe^{o} + 2H_2O \rightarrow Fe^{2+} + 2OH^{-} + H^{2} \end{array} \qquad \text{(Eq. 7.1)}$$

$$\begin{array}{c} Fe^{o} \rightarrow Fe^{2+} + 2e^{-} \\ R\,Cl + 2e^{-} + H^{+} \rightarrow Fe^{2+} + RH + Cl^{-} \\ \hline R\,Cl + Fe^{o} + H^{+} \rightarrow Fe^{2+} + RH + Cl^{-} \end{array} \qquad \text{(Eq. 7.2)}$$

Since the iron is directly in contact with water, Reaction 1 proceeds continuously and throughout the PRB. Reaction 2, on the other hand, proceeds most rapidly where contaminant concentrations are highest, and thus mass removal of contaminants declines with distance from the influent face of the PRB. Though Equation 7.1 does not involve the chlorinated organic compound, it nevertheless has important consequences. For stoichiometric calculations, it is necessary to account for the consumption of metallic iron through oxidation by water. Reardon (1995) indicated that rates of consumption could be on the order of 0.1–0.6 millimoles per kilogram of iron per day (mmol/kg Fe/day). Also, Equation 7.1, as well as Equation 7.2, contributes dissolved iron (Fe^{2+}) to the solution phase. While this suggests that water exiting a PRB could have high concentrations of dissolved iron, this is generally not the case and experience has shown that the iron concentration in the effluent is frequently lower than that of the influent water. Two identified sinks for Fe^{2+} include $Fe(OH)_2$, which is unstable and is quickly converted to magnetite (Fe_3O_4) and siderite ($FeCO_3$). A further consequence of Equation 7.1 is high pH values, often in the range of 9–10, as a result of OH^{-} (hydroxyl ion) production. This has important consequences concerning the inorganic chemistry of the water, and will be addressed further in Section 7.4, dealing with long-term performance. It should also be noted that while the pH is high within the PRB and in the water exiting the PRB, because of the natural buffering capacity of geologic materials, background pH values are generally reached within 1–2 m down gradient of the PRB.

Equation 7.2 shows the reductive dechlorination of a generalized chlorinated hydrocarbon (RCl). Though the process is generally agreed to be reductive dehalogenation, through extensive and continuing research, various mechanisms and pathways have been proposed (see Tratnyek et al., 2003, for example), depending upon the particular contaminant of interest.

It is generally agreed that the reactions occur on the surface of the iron, and while exceptions have been reported, in most situations degradation follows pseudo-first-order kinetics with respect to the contaminant concentration, with the rate reflected in the first order rate constant (k_{obs}, hour [h^{-1}]).

That is,

$$\ln\left(\frac{c}{c_o}\right) = k_{obs}\, t \qquad \text{(Eq. 7.3)}$$

where C is concentration (milligrams per liter [mg/L]), C_o is the initial concentration (mg/L) and t is contact time (h). The half-life, $t_{1/2}$ (h), the time required to remove one half of the initial concentration can be derived from Equation 7.3 as:

$$t_{1/2} = \ln 2/k_{obs} \qquad \text{(Eq. 7.4)}$$

Gillham et al. (2010) summarized the results of numerous laboratory column tests in which commercial granular iron materials were used with waters obtained from contaminated sites. The rate constants for the common chlorinated solvents ranged from $3.4 \pm 1.6\ h^{-1}$ for carbon tetrachloride (CT) to $0.62 \pm 0.9\ h^{-1}$ for vinyl chloride (VC). Corresponding half-lives are 0.26 ± 0.17 h and 3.2 ± 2.9 h for CT and VC respectively. The high degree of variability about the mean values is believed to reflect variability in the iron materials, as well as variability in the organic and inorganic matrix of the various waters tested. As a further example, for TCE, the most commonly identified contaminant at hazardous waste sites, an average rate constant of $0.71 \pm 0.38\ h^{-1}$ (half-life of $1.3 \pm 0.9\ h^{-1}$) was reported. Using the half-life for TCE as an example, a reduction in concentration from 1,000 micrograms per liter (μg/L) to 1 μg/L would require about ten half-lives, corresponding to a residence time of about 13 h.

As reported in O'Hannesin and Gillham (1998), the first field trial of a granular iron PRB was initiated in 1991 at Canadian Forces Base Borden. The PRB was 1.6 meters (m) in flow-through thickness and consisted of a mixture of 22% granular iron and 78% sand (by volume). The influent groundwater contained 268 mg/L TCE and 58 mg/L PCE. Approximately 90% of the TCE and PCE were removed within the wall, and though about 1% of the initial contaminants appeared as dichloroethene isomers, primarily *cis*-dichloroethene (*cis*-DCE), these also degraded within the PRB. Of particular importance, the performance was reasonably consistent with expectations based on laboratory tests, and there was no apparent decline in performance over the 5-year period of the study.

While granular iron is effective in degrading a wide range of halogenated aliphatic compounds, with dehalogenated hydrocarbon(s) as the final product, there are compounds that are not degraded by iron and in some cases the final products of degradation are as hazardous as the parent compound. For example, 1,2-dichloroethane (1,2-DCA) is not susceptible to dechlorination by iron and CT is sequentially dechlorinated through chloroform to dichloromethane (DCM), but DCM remains as a final product, normally at about 40 mol% of the original CT. Similarly, iron will degrade a wide range of nitrogen-containing compounds, but degradation is frequently incomplete, with toxic residuals. For example, nitrobenzene degrades rapidly to analine, but analine, which is also toxic, is persistent (Agrawal and Tratnyek, 1996). Similarly, compounds used in explosives such as 2,4,6-trinitrotoluene (TNT) and dinitrotoluene (DNT) degrade to triamino- and diaminotoluene respectively, both of which are toxic and persistent (Bandstra et al., 2005; Oh et al., 2002). As a further example, chlorinated aromatics such as polychlorinated biphenyls (PCBs) and dioxins do not degrade at perceptible rates in the presence of conventional granular iron materials, though substantial rates have been reported for nano-sized iron particles and bimetallic materials (Lowry and Johnson, 2004; Kim et al., 2008).

If a granular iron PRB is being considered for degradation of a particular organic contaminant it is essential that information be available or be acquired concerning possible rates and products of degradation. While much of this information is available for the most common groundwater contaminants, there are many halogenated hydrocarbons which have not yet been investigated. It should also be noted that should rates of degradation be favorable, but the products are not, a sequential PRB system could be considered. In the case of DNT for example, granular iron is highly effective in reducing DNT to diaminotoluene (DAT) and the DAT could be subsequently mineralized through oxidative biodegradation.

Granular iron has also been shown to be highly effective in removing a wide range of metals from solution through processes of reduction, precipitation and sorption. Several electroactive metals such as As(V), Cr(VI), Se(VI), Tc(VIII) and U(VI) occur in oxic groundwaters as oxyanions, but under strongly reducing conditions form sparingly soluble precipitates. Of these, Cr is the most frequently encountered inorganic contaminant at hazardous waste sites, and has received the greatest attention. For example, in tests using Fe^0, FeS_2 and $FeCO_3$ as potential reductants, Blowes and Ptacek (1992) reported Fe^0 to give the highest rates of removal. Hexavalent chromium (Cr(VI)) was reduced to Cr(III) in the presence of iron and precipitated as mixed Fe-Cr oxyhydroxides.

In 1996, a full-scale PRB was installed for Cr(VI) removal at the US Coast Guard Facility in Elizabeth City, North Carolina. The design, installation and performance are well documented in Blowes et al. (1999), Blowes and Mayer (1999) and Wilken et al. (2003). The PRB, 46 m long, 7.3 m deep and 0.6 m thick was installed across the path of the contaminant plume. Cr(VI) concentrations were observed to decline from influent values of up to 8 to <0.01 mg/L within the first few centimeters of the PRB. As is commonly the case at metal plating facilities, the Cr (VI) plume was accompanied by a chlorinated solvent plume that contained up to 19 mg/L TCE and lesser amounts of *cis*-DCE and VC. With the exception of a small area where underflow was suspected, the chlorinated solvents were treated to maximum contaminant levels (MCLs). It is noteworthy that the iron PRB proved to be effective in treating both the organic and inorganic contaminants.

In laboratory tests, rapid rates of removal in the presence of granular iron have also been reported for other metals including As(V) (McRae, 1999; Lackovic et al., 2000; Su and Puls, 2001), Se(V) (Amrhein et al., 1998; McRae, 1999), Tc(VII) (Bostick et al., 1990; Del Cul et al., 1993; Cantrell et al., 1995) and U(VI) (Cantrell et al., 1995; Gu et al., 1998; Morrison et al, 2001).

With the exception of chromium, much of the available information concerning the use of iron for removing metals from groundwater is based on laboratory tests and small-scale field trials. An exception is the well-documented study of metals removal in an iron PRB installed in Monticallo Canyon, Utah (Morrison et al., 2002; Morrison, 2003). Uranium mill tailings had previously been stored at the site, and though the tailings had been removed, the groundwater continued to contain high concentrations of several metals. Morrison et al. (2002) indicated the following removals across the PRB: uranium, 396 mg/L to <0.24 mg/L; arsenic, 10.3 mg/L to <0.2 mg/L; selenium, 18.2–0.1 mg/L; molybdenum, 62.8–17.5 mg/L; vanadium, 395–1.2 mg/L; manganese, 308–177 mg/L and nitrate, 60.7 to <0.065 mg/L. Following 2.7 years of operation, removal of metals remained excellent, and as reported in Morrison (2003), there was no apparent decrease in the hydraulic conductivity of the PRB as a consequence of the accumulation of secondary minerals.

It should be noted that while iron is frequently listed as being effective in nitrate removal, laboratory tests (Cheng et al., 1997 and Huang et al., 1998, for example) have demonstrated abiotic reduction of nitrate with ammonia as the final product and thus the environmental benefit is questionable. Furthermore, as a consequence of iron oxide formation during nitrate reduction, there is a rapid reduction in the reactivity of the iron (Schlicker et al., 2000; Lu et al., 2005). Thus the evidence suggests that granular iron is not likely to be effective for nitrate removal over significant periods of time.

7.2.2 Organic Carbon Amendments

Solid-phase organic carbon that is capable of supporting biological activity has been used in PRBs for removal of several groundwater contaminants. The organic carbon acts as a bio-stimulant, with the subsequent biological activity resulting in the direct or indirect removal of the contaminants. The organic carbon can act as a primary substrate, providing energy for biological growth, with subsequent metabolism or co-metabolism of organic contaminants; can stimulate growth of denitrifying or sulfate reducing bacterial populations or can induce geochemical environments (reducing conditions, for example) that enhance other contaminant removal processes. A useful review of this topic is included in Scherer et al. (2000).

For example, Robertson and Cherry (1995) used sawdust in a PRB configuration to promote nitrate removal (denitrification) in septic systems and Craig (2004) showed several wood chip-based materials to be effective in promoting the degradation of perchlorate in contaminated groundwater. In addition, a solid-phase organic material modified through the addition of a small percentage of microscale iron has proven to be effective in degrading chlorinated solvents, including some that are degraded slowly, or not at all, by granular iron, such as polynuclear aromatic hydrocarbons (PAHs), pentachlorophenol and DCM (USEPA, 1996; Mueller et al., 2004). The addition of iron is believed to accelerate the onset of strongly reducing conditions and also to neutralize the organic acids released through decomposition of the carbon source, thus maintaining favorable conditions for biological growth.

Acidic drainage from mine tailings (acid mine drainage [AMD]) is a major environmental problem in many parts of the world. Oxidation of pyrite and other reduced sulphur minerals produces highly acidic conditions in many mine tailings and waste rock disposal areas. Characteristically, seepage from these areas has very low pH, high sulfate and high dissolved iron concentrations, and because of the low pH, frequently contains unacceptable concentrations of various metals. As discussed in Blowes et al. (1995, 1998, 2000), the addition of organic carbon can promote the reduction of sulfate to sulfide, with subsequent precipitation of metal sulfides. Waybrant et al. (1995, 1998), in laboratory tests, showed that the addition of organic carbon to stimulate reduction of an anion-forming species, was effective in promoting the indirect removal of several metals including silver (Ag), Cd, cobalt (Co), copper (Cu), Fe, Ni, Pb and zinc (Zn).

Two full-scale applications of the technology are discussed in Blowes et al. (2000). In the first, a barrier consisting of municipal compost, leaf compost and wood chips (as organic substrate) and pea gravel (to maintain permeability) was constructed across the path of an AMD plume at the Nickel Rim mine site near Sudbury, Ontario (Benner et al., 1997, 1999). Sulfate concentrations decreased from 2400–3800 to 110–1900 mg/L across the barrier accompanied by decreases in iron concentration of from 740–1000 to $<$1–91 mg/L and a decrease in Ni concentration from 30 mg/L to $<$0.2 mg/L. Alkalinity, expressed as calcium carbonate ($CaCO_3$), increased from 60–22 to 850–2700 mg/L across the barrier, and it was shown that the water was transformed from acid generating (upon exposure to oxygen) to acid consuming as it passed through the barrier. The second application (McGregor et al., 1999) involved the construction of a compost-based barrier at an industrial site in Vancouver, Canada. Concentrations of Cu declined from 300 mg/L to <5 μg/L across the barrier and concentrations of the other metal constituents (Cd, Ni, Pb and Zn) showed similar declines with concentrations in the effluent generally below the detection limits.

Permeable mulch biowalls have been installed at several Department of Defense (DoD) facilities to promote the biodegradation of chlorinated solvents, perchlorate and explosives (AFCEE, 2008).

7.2.3 Oxygen Addition

The remediation schemes discussed to this point generally rely on stimulating highly reducing conditions within the PRB. However, some contaminants degrade much more readily under oxidizing conditions. Notable among these are the soluble constituents of petroleum products including benzene, toluene, ethylbenzene and total xylenes (BTEX), and additives such as methyl tertiary-butyl ether (MTBE) and ethanol. Biodegradation of these compounds generally consumes, very rapidly, any oxygen that is present naturally in the groundwater. While biodegradation can proceed under anaerobic conditions, rates of removal are much slower than under aerobic conditions. Thus, to accelerate mass removal, the addition of oxygen (O_2) can be advantageous. Two approaches include the addition of solid phases that release O_2 upon contact with water and the direct addition of gaseous O_2. Solid phases for this purpose are based on magnesium-peroxide or calcium-peroxide formulations and include products such as ORC™ by Regenesis Inc. or ECH-O™ by Adventus Americas Inc. These materials can be placed in trenches, but more commonly are mixed in a slurry and injected, or placed in permeable cylindrical containers that can be lowered into, and retrieved from, the screened sections of remediation wells. The rate of oxygen release from these materials declines over time and thus periodic replacement may be required (every few months).

A common method of direct O_2 addition is through diffusion of pure oxygen through polyethylene tubing installed within the flow path of the contaminant plume. The polyethylene tubing is generally wrapped around a cylindrical frame that can be lowered into remediation wells and subsequently recovered. For both the solid-phase oxygen release materials and diffusive addition of O_2, the PRB frequently consists of a series of closely spaced boreholes, rather than a "wall" of reactive material. Mixing of the contaminant and oxygen then depends on lateral dispersion of these constituents between the respective plumes as discussed in more detail in Chapter 3. Good mixing here depends on close spacing between boreholes. Notably, remediation in this case does not occur within the PRB, but at some distance downgradient. It should also be noted that the effectiveness can be compromised if there are high concentrations of oxidizable organic matter or minerals naturally present in the remediation zone.

Useful case studies that demonstrate the removal of both BTEX and MTBE are given in Wilson et al. (2002) (Vandenberg Air Force Base, California) and Johnson et al. (2003) (Port Hueneme Naval Air Station, California).

7.2.4 Sorptive Materials

A variety of materials have been proposed for use in a PRB configuration for removal of contaminants by sorptive processes. Because sorptive materials have a finite capacity, a major consideration in design is the selection of a material that has a very high capacity for the contaminant(s) of concern. For organic contaminants, granular activated carbon (GAC) is perhaps the most obvious choice. This is particularly the case for contaminants that have high organic partitioning coefficients such as PAHs.

Zeolites, naturally occurring minerals with a particularly high negative surface charge density, have been proposed for the removal of cationic contaminants. Two applications we are aware of include removal of Sr-90 from a contaminant plume at the Chalk River Nuclear Laboratory of Atomic Energy Canada (Lee et al., 1998) and a pilot PRB at the West Valley facility in New York State (Warner et al., 2004).

7.2.5 Other Materials

The PRB concept has stimulated an active search for materials that can degrade or sequester particular contaminants and indeed, there remains ample opportunity for innovation in this area. Some of these materials will be presented here very briefly, and in some cases, additional information is provided in ITRC (2005).

Plating a small amount of a more noble metal such as palladium (Muftikian et al., 1995) or nickel (Odziemkowski et al., 1998) onto the surface of iron particles can increase the degradation rate of many organic contaminants by a factor of 10 or more relative to non-catalytic iron. However, these materials have been shown to lose their catalytic effect over relatively short periods of time (Muftikian et al., 1996). Thus, because of the need for frequent replacement or reactivation, these materials have not been applied in PRB configurations.

The use of metallic iron in the particle size range of a few tens of nanometers was introduced by Wang and Zhang (1997) and has received considerable attention over the past 10 years. Because of the small particle size, these materials have a very high specific surface area (on the order of 35 square meters per gram [m^2/g]) and are therefore highly reactive. As a consequence of the high reactivity, large amounts of contaminant can be degraded in short periods of time and as noted in Section 7.2.1, there is evidence that nano-scale iron can degrade contaminants, such as the chlorinated aromatics, that are not degradable by conventional iron. On the other hand, following Equation 7.1, large amounts of the iron can also be oxidized by water in relatively short periods of time, raising questions concerning the persistence of the iron nanoparticles in the subsurface. It has also been proposed that because of its small size, nano-scale iron can migrate substantial distances from the point of injection and thus remediate areas substantially down gradient from the point of injection. To date, there is limited field evidence that shows this to be the case and Tratnyek and Johnson (2006) suggest that under normal subsurface conditions, migration distances of more than 1–2 m are unlikely. Because of the high reactivity but limited persistence relative to granular iron, the use of nano-scale iron may be most appropriate in situations where high concentrations of contaminants are required to be removed over relatively short periods of time. A useful review of the applicability of nano-iron for groundwater remediation is provided in Tratnyek and Johnson (2006).

The injection of various fluids to alter the subsurface redox conditions such that certain contaminants will be removed has been proposed, and in some cases, tested. For example, Rouse et al. (2001) investigated the injection of calcium polysulfide for removal of reducible metals and a large-scale application of the technology for removal of Co(VI) has been completed (Zawislanski et al., 2002). Similarly, sodium dithionite has been injected for treatment of a chromium plume at the Hanford Site in Richland Washington (Fruchter et al., 2000).

Basic oxygen furnace (BOF) slag is a granular waste product from steel production, rich in iron and calcium oxyhydroxides. It has been shown to be highly effective in phosphate removal from waste water and from groundwater plumes associated with septic systems (Baker et al., 1997, 1998; Smyth et al., 2002). BOF slag has also been used effectively for removal of arsenic from a large contaminant plume near Chicago (Wilkens et al., 2003).

7.3 DESIGN CONSIDERATIONS

The primary objective in the design of a PRB is to ensure sufficient residence time within the reactive material such that the contaminant concentrations are reduced to acceptable values. A further criterion, and the subject of Section 7.4, is that acceptable performance be maintained over an appropriate period of time. Through its effect on both quantity and cost of reactive material and also on the choice of construction method, the required thickness of the PRB is a primary outcome of the design process.

For contaminants that are removed through chemical or biological processes, the required thickness can be calculated by dividing the required residence time by the groundwater velocity, with the residence time determined from the influent and objective concentrations and the rate of reaction. Though simple in concept, determination of the key parameters, reaction rates and groundwater velocity, can present difficult technical challenges. For contaminants removed through sorptive processes, contaminant loading (concentration and groundwater flux) and sorptive capacity of the reactive material are primary design parameters.

For more detailed discussion of PRB design, the reader is referred to Battelle (1997), ITRC (2005) and Gillham et al. (2010). The following discussion briefly concerns the reaction rates and hydrogeologic considerations.

7.3.1 Reaction Rates

Design consideration for granular iron PRBs for treating halogenated organic contaminants has received the greatest attention and is perhaps the most advanced. As noted previously, the weight of evidence supports first-order kinetics for the degradation of these compounds and thus the first-order rate constant (or half-life) provides a quantitative basis for calculating required residence times. In practice, laboratory column tests are normally conducted using the anticipated reactive material (granular iron or iron-sand mixtures) and groundwater collected from the particular site. The results of these tests indicate the rate of degradation of the parent compound, the proportion of the parent that passes through intermediate degradation products and rates of degradation of the intermediates.

Mathematical expressions can be fit to the data, giving first-order rate constants for each dechlorination step and the fraction of the parent compound that participates in each step. Having established the rate constants and conversion factors, the model can then be used to calculate the time required for all compounds to reach concentrations that are lower than the objective concentration (the residence time). Because less chlorinated compounds (VC for example) generally have lower rate constants than the more chlorinated hydrocarbon (TCE for example), it is not uncommon for the residence time to be determined by the degradation rate of intermediate products rather than by the parent compounds. In most cases the column tests are conducted at room temperature, requiring that correction factors be applied such that the results are applicable at the groundwater temperatures at the site. The correction factor follows the Arrhenius equation, resulting in lower rate constants at the lower field temperatures. That is,

$$\ln\left(\frac{k_{T2}}{k_{T1}}\right) = \left(\frac{E_a}{R}\right)\left(\frac{1}{T_2}\right) - \left(\frac{1}{T_1}\right) \qquad \text{(Eq. 7.5)}$$

where k_{T2} and k_{T1} are the rate constants at two temperatures (T_1 and T_2 in Kelvin [K]), E_a (kilojoules per mole [kJ/mol]) is the activation energy and R is the universal gas constant. Thus knowing the rate constant at one temperature, and the activation energy, the rate constant at any other temperature can be calculated. The value of E_a depends upon the organic compound, the character of the iron material and other conditions. Based on a survey of literature values, Tratnyek et al. (2003) suggest E_a values to range from 15 to 55 kJ/mol.

Though first-order kinetic models provide a reasonably robust means of quantifying required residence times, there are nevertheless uncertainties. In particular, water collected in the field and transported to the laboratory cannot be fully preserved (degassing or oxygen invasion, for example), the surface area of iron to volume ratio in the laboratory column will differ to some degree from that of the PRB when installed and some variation in the quality and reactivity of the iron is unavoidable. Furthermore, unless the column tests are run for exceptional periods of time, the results will generally not give an indication of the manner in which the reactivity of the iron is likely to change over time. A more detailed discussion, including examples is provided in Gillham et al. (2010).

Metallic oxyanions are reduced rapidly in the presence of granular iron. For chromate reduction, Gould (1982) and Mayer et al. (2001) used the kinetic expression:

$$\frac{d[Cr(VI)]}{dt} = -k_{SA-cr-Fe^o}\ S[Cr(VI)]^{0.5}\ [H^+]^{0.5} \qquad \text{(Eq. 7.6)}$$

where $k_{SA-Cr-Fe^o}$ is the rate constant for Cr removal normalized to iron surface area (L H_2O m^2 iron s^{-1}), S is the reactive surface area concentration of Fe^0 (m^2 iron L^{-1} bulk) and [Cr(VI)] and [H^+] are molar concentrations of Cr(VI) and [H^+] (mol L^{-1} H_2O). The reaction rate, dCr(VI)/dt, has units of (mol L^{-1} bulk s^{-1}). Thus the reaction is clearly not first order and the rate depends on the concentration of Cr(VI) and also on pH. Gould (1982) reported a rate constant of 9.08×10^{-3} L H_2Om^{-2} iron s^{-1} while Jeen et al. (2008) reported a somewhat lower value (4.41×10^{-4} L H_2O m^{-2} iron s^{-1}). While the reported values differ by more than an order of magnitude, both indicate very rapid removal of Cr(VI) and thus, for practical purposes, the magnitude of the rate constant is of little relevance. Indeed, in column experiments, Jeen et al. (2008) showed Cr(VI) to be removed within the first 2.5 centimeters (cm) of the column and in the PRB installation at Elizabeth City, North Carolina, Puls et al. (1998) reported Cr(VI) removal to occur "within the first few centimeters of the wall". It follows that the initial reaction rates would seldom be a significant consideration concerning residence time or PRB thickness. Declining reactivity over time, as discussed in Section 7.4, may nevertheless require attention in PRB design.

Reaction rates for most other types of PRBs are less well understood and difficult to quantify. Those involving the stimulation of biological activity through the addition of organic carbon (nitrate removal, and remediation of acid mine drainage, for example) depend on numerous factors including temperature, inorganic composition of the groundwater being treated, composition of the indigenous microbial population, and the degradability of the organic carbon that is used. However, because of the low cost of the materials commonly used (wood chips or leaf mulch, for example) there is little cost penalty associated with overdesign.

For yet other materials, it is recognized that regular replacement will be required (solid-phase oxygen-release materials, for example) and the required frequency of replacement is normally determined through experience and performance monitoring rather than from rates of reaction.

7.3.2 Hydrogeologic Considerations

For many reactive materials, the required thickness of the PRB is directly proportional to the groundwater velocity. Certainly this is the case for granular iron when used to treat chlorinated organic contaminants. Thus errors and uncertainty in velocity determinations can be a significant cause for over design or under design.

Direct measurements of groundwater velocity can be made using various types of velocity probes, with borehole-dilution perhaps the most common. However, for reasons that remain unclear, these methods are generally not considered standard practice and are not widely used. Though tracer tests should provide the most accurate measure of velocity, they can be time-consuming and costly, and in complex hydrogeologic environments, the results are frequently difficult to interpret.

Current practice generally relies on estimates of velocity obtained using the Darcy equation:

$$q = Ki \quad \text{(Eq. 7.7)}$$

where q is the groundwater flux [L/T], K is the hydraulic conductivity of the geologic material [L/T] and i is the hydraulic gradient [-]. Velocity in the aquifer is obtained by dividing the groundwater flux by the porosity, and recognizing that the porosity of the PRB may differ from that of the aquifer, the velocity in the PRB, V_{PRB} is given by

$$V_{PRB} = V_{aq}\frac{n_{aq}}{n_{PRB}} \quad \text{(Eq. 7.8)}$$

where n_{aq} and n_{PRB} are the porosities of the aquifer and PRB respectively. Of the various parameters in Equations 7.7 and 7.8, considerable uncertainty is normally associated with both K and i. In the close vicinity of a proposed PRB, determination of i generally requires that very small differences in water elevation be measured over relatively short distances. As discussed in Devlin and McElwee (2007), this can lead to significant errors. Of greater concern is the accuracy and variability in K. Determination of K from pump tests is common practice, but results in an integrated value over a large volume of the aquifer and is therefore generally not useful for design of PRBs. Single-well response tests are preferable, though these should be performed at several locations along the proposed line of the PRB, as well as at several depths. Detailed stratigraphic analysis of core samples combined with single-well response tests is perhaps the best method for determining the variability in K and thus the expected spatial variability in velocity.

Even when great care is taken in the hydrogeologic characterization of a site, thin zones of highly contrasting hydraulic conductivity may be present and may contribute to unexpected flow conditions. In addition, flow velocity (magnitude and direction) may vary seasonally or may be influenced by changes in rates of pumping of neighboring wells. Further discussion of hydrogeologic factors for consideration in PRB design is provided in ITRC (2005).

7.4 LONG-TERM PERFORMANCE

The capital cost of a granular iron PRB is relatively high, and depending upon site conditions and remediation objectives, can be comparable to or greater than a pump-and-treat system. Thus the potential economic advantage relies on low operating and maintenance costs over substantial periods of time. While these savings can be substantial, selection of the PRB remedy requires confidence that adequate performance will be maintained well into the future. For very shallow installations, and particularly in situations where the reactive materials are inexpensive (wood chips, for example), the importance of long-term performance is somewhat diminished.

Calculating cost on the basis of net present value indicates that replacement of a PRB every 10–15 years adds very little to the long-term total cost. Thus, for the purpose of this discussion, it will be assumed that in order to be economically viable, a PRB should operate effectively and without appreciable maintenance costs for periods of at least 10–15 years.

In general, the factors that may compromise long-term performance include consumption of the reactive material, declining reactivity and physical changes such as reduced porosity and hydraulic conductivity. With the record of performance of several of the early PRBs (granular iron in particular) now approaching or exceeding 10 years, there is a growing, and generally favorable, body of evidence concerning long-term performance. Furthermore, there is a growing body of literature, based on laboratory and theoretical studies that addresses the issue of long-term performance. Nevertheless, as a consequence of the numerous interacting processes that can occur, and the changes in these processes that can occur over time, the ability to make reliable site-specific and quantitative predictions of performance, well into the future, remains questionable.

7.4.1 Granular Iron

As the material most frequently used in PRBs, the long-term performance of granular iron has received the greatest attention. Discussion of this topic, particularly with respect to the degradation of chlorinated organic contaminants is included in Gillham et al. (2010). Based on corrosion rates of iron it can readily be shown that the commercial iron materials most commonly used in PRB construction will persist in the subsurface for periods of up to 100 years. Though corrosion rates can increase in the presence of strong oxidants such as nitrate, these effects will be countered by gradual passivation and thus, within the 10–15 year criterion for long-term performance, persistence of the metallic iron is not an issue. The primary consideration is the formation of secondary minerals and the effects that these may have on permeability and reactivity.

As shown in Equations 7.1 and 7.2, oxidation of iron by chlorinated organics, as well as by water, produces Fe^{2+}. Through the reaction with water, Fe^{2+} will be produced throughout the PRB, while additional Fe^{2+} will be produced in the region of most rapid degradation of the chlorinated organic. Further, from Equation 7.1, reduction of water contributes OH^-, resulting in pH values frequently in the range of 9–10. At the elevated pH, Fe^{2+} precipitates as $Fe(OH)_2$, which transforms to magnetite (Fe_3O_4). Magnetite is conducting with respect to electrons and therefore has a minor effect on reactivity of the iron. In addition, since metallic iron is consumed in the production of magnetite, the net loss in porosity is also small. Of greater importance is the effect that the elevated pH has on the chemical characteristics of the influent water. In particular, most groundwater contains significant alkalinity, particularly in the form of bicarbonate. Upon exposure to the high pH of an iron PRB, the bicarbonate-carbonate equilibrium shifts towards carbonate, resulting in the precipitation of various carbonate minerals. These include $CaCO_3$ generally as aragonite and iron carbonate minerals, possibly including siderite ($FeCO_3$) and iron hydroxy carbonate. Core samples collected from PRBs generally show carbonate minerals to be the most abundant secondary mineral phases (Vogan et al., 1999 and McMahon et al., 1999, for example) though, depending upon the composition of the influent groundwater, various other precipitates such as iron sulfide and green rust have been observed (Blowes et al., 1999; Wilkin et al., 2003).

Because of the rapid shift in equilibrium from bicarbonate to carbonate, one might anticipate a relatively rapid accumulation of carbonate minerals at the influent surface of a PRB, accompanied by a relatively rapid decline in hydraulic conductivity. However, in laboratory column tests, Zhang and Gillham (2005) showed that the accumulation of carbonate minerals gradually passivates the iron. With passivation, the increase in pH and thus the region of precipitate formation moves further into the column and further, that passivation occurs at precipitate concentrations that are not likely to cause a serious decrease in hydraulic conductivity. This is highly favorable in that it suggests that PRBs are not likely to become impermeable barriers. On the other hand, it shows that as a consequence of passivation, the effective thickness of a PRB is likely to decline over time. Using what were considered to be typical groundwater conditions, Zhang and Gillham (2005) suggested that the effective thickness would decrease by about 1 cm per year. This rate of passivation could be readily accommodated in the design of a PRB, but further indicates the need to consider passivation in the design process.

Early attempts to develop mathematical models for predicting geochemical changes and precipitate formation within PRBs were based on geochemical equilibrium models (Gaveskar et al., 1998; Morrison et al., 2001), while more recent models are more complex, including reactive transport processes (Mayer et al., 2001; Yabusaki et al., 2001; Li et al., 2006). In all cases however, the reactivity of the iron is assumed to be constant over time; clearly this is not the case.

Based on the results of column tests, Jeen et al. (2006, 2007) extended the model of Mayer et al. (2001) to include a decline in iron reactivity over time. Comparing with the results of the column tests, the model reproduced the changing reactivity towards TCE, as a consequence of carbonate precipitation, reasonably well. The model is highly complex however and its ability to predict performance under normal field conditions has yet to be demonstrated.

Competing oxidants, such as dissolved oxygen and nitrate, can also cause passivation with respect to the degradation of chlorinated hydrocarbons. In laboratory column tests (Mackenzie et al., 1999, for example), significant declines in hydraulic conductivity have been attributed to the formation of iron oxide precipitates near the influent end. Possibly because of the generally low concentrations of oxygen in groundwater at contaminated sites, significant accumulations of Fe(III) oxides have not been observed in core samples collected from field PRBs (O'Hannesin and Gillham, 1998; Vogan et al., 1998) and thus dissolved oxygen does not appear to be a passivating agent of practical concern.

On the other hand, NO_3^- can be present in groundwater at concentrations of several tens of mg/L NO_3^-–N. Though nitrate reduction does not form nitrogen-containing precipitates, it can nevertheless have a strong passivating effect. As observed by Ritter et al. (2003) and Lu et al. (2005), nitrate in the influent water causes an upward shift in the iron corrosion potential (less reducing conditions) such that hematite and maghemite, rather than magnetite, become the stable iron oxide phases. Unlike magnetite, the Fe (III) oxyhydroxides are not conducting with respect to electrons and are therefore strongly passivating with respect to both further reduction of nitrate and degradation of chlorinated organic compounds. Though the evidence is clear that increasing nitrate concentrations will result in higher rates of passivation, it is not known if there is a threshold, below which nitrate does not pose a significant problem. Where high nitrate concentrations are encountered, it may be necessary to use sequential treatment for removal of nitrate in advance of the iron PRB.

Other oxidants such as chromate and permanganate could affect performance with respect to chlorinated organic degradation in a manner similar to that of nitrate. These could have additional effects however in that reduced chromium and manganese form additional solid phases.

When using granular iron to treat metallic anions, additional passivating processes need to be considered; in particular, the precipitates that form through reduction of the metal. For example, Blowes et al. (1997) indicated the reduction of Cr(VI) to result in formation of a Fe-Cr oxyhydroxide phase having a structure similar to goethite (α FeOOH) with both Cr and Fe present in the plus three oxidation state. As demonstrated by Blowes et al. (1997), the oxyhydroxide coatings can have a passivating effect towards further reduction of Cr(VI). More recently Gui et al. (2009) showed that with the introduction of both Cr(VI) and dissolved calcium carbonate to columns of granular iron, the passivating effect of the Cr-Fe oxyhydroxide precipitates was substantially greater than that of $CaCO_3$.

While it is clear that both $CaCO_3$ and precipitated Cr minerals can passivate granular iron (with respect to Cr(VI) reduction), the empirical evidence suggests the rates of passivation to be slow. In particular, for the Elizabeth City installation referred to previously, with an influent Cr (VI) concentration of almost 8 mg/L and a groundwater velocity of 0.1 m/day, after 8 years of operation, almost complete removal of Cr(VI) continues to occur within the first 0.1 m of the PRB (Wilkin et al., 2005).

Because of the complexity of biological processes, there is no clear theoretical basis for anticipating or predicting the long-term performance of PRBs in which organic carbon is used to stimulate biological treatment. As discussed in Blowes et al. (2000), at the Nickel Rim installation, where a mixture of municipal compost, leaf compost and wood chips was installed for treatment of acid mine drainage, iron removal (through sulfide precipitation) had declined by 45%, 3 years after installation. Consumption of the more labile components of the organic carbon, as well as preferential flow as a result of non-uniform placement were suggested as the primary causes. On the other hand, Robertson et al. (2008) reported effective performance of a sand-sawdust PRB for nitrate removal that was in continuous operation for 15 years.

Monitoring of mulch biowalls at DoD facilities indicates consistent performance over the first few years of operation. A recent Air Force Center for Engineering and the Environment (AFCEE) protocol document suggests that these systems may need to be replenished with fluid carbon substrates on a periodic basis, every 3–5 years, rather than incur the larger expense of reconstructing the mulch PRB (AFCEE, 2008).

7.5 METHODS OF INSTALLATION

7.5.1 PRB Configuration

An early consideration in the design and implementation of a PRB is the configuration of the installation. The two most common configurations are a continuous “wall”, where the wall of reactive material is constructed across the entire path of the contaminant plume (as indicated in Figure 7.1), and funnel and gate systems. A funnel and gate configuration uses impermeable materials (slurry walls or sheet piling) to direct the contaminant plume through one or more reactive sections (gates). Theoretically, the amount of reactive material required is the same for both configurations. The primary advantage of a funnel and gate is that the reactive material is isolated to relatively small areas and therefore can be replaced more easily if required. There may also be site-specific factors that favor the funnel and gate configuration. There are several disadvantages however. As the hydraulic head builds up behind the funnel sections, the contaminant plume will be directed to greater depths, possibly causing contaminants to flow under the treatment system. This can be avoided by constructing the funnel and gate system to considerably greater depths than that of the initial plume. A more favorable solution exists where an aquitard underlies the contaminant plume. In these situations the funnel and gate can be keyed into the aquitard, thus preventing underflow. Hydraulic considerations in the construction of a funnel and gate treatment system are discussed in Starr and Cherry (1994). A further limitation concerns precipitate formation within the reactive material. Where natural constituents of the groundwater precipitate within a PRB (calcium carbonate, for example), the rate of precipitate formation is proportional to the flux of the precipitating solutes and the flux of solutes is proportional to the flux of water. Through the funneling effect of funnel and gate systems, the flux of water through the face of the treatment zone is much greater than the natural flux of groundwater in the absence of the funnel and gate system. This greater flux will lead to greater amounts of precipitate resulting in more rapid rates of passivation and declines

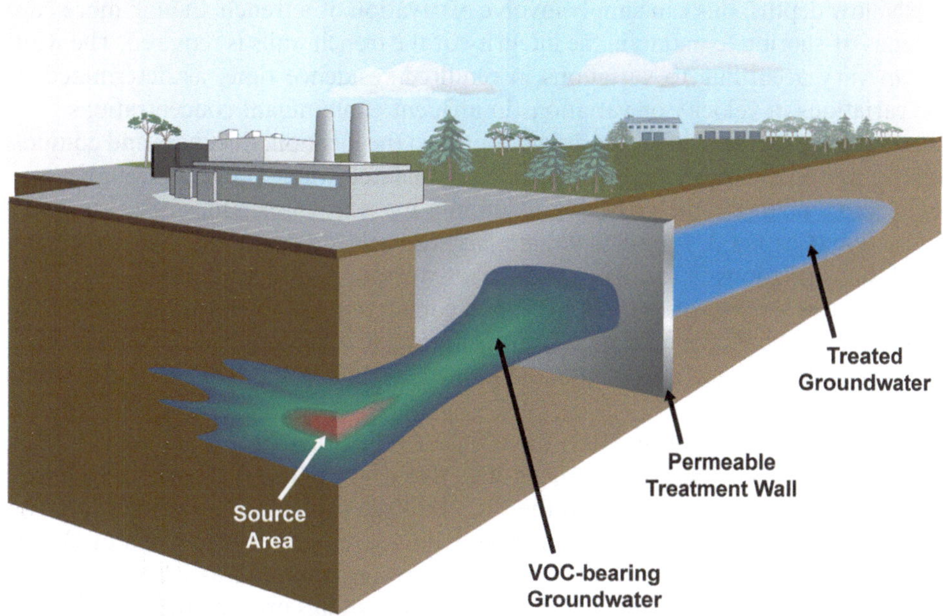

Figure 7.1. Conceptual drawing of a permeable reactive barrier (adapted from ETI).

in hydraulic conductivity. Because of greater difficulties in controlling/predicting the hydraulics, as well as the accelerated precipitate formation in funnel and gate systems, by far the majority of PRBs have been installed as continuous walls.

Closely related to the funnel and gate design are *in situ* reaction vessels, which normally include funnels or collection trenches to divert the contaminated water, through differences in hydraulic head, to a buried vessel containing the treatment material. The applicability of this approach is generally determined by site-specific conditions. Particular examples are given in Primrose et al. (2004) and Phifer et al. (1999).

Yet a fourth approach to implementation considers horizontal layers of reactive material. Provided the hydraulic conductivity of the reactive material is substantially greater than that of the geologic material, the flow lines will converge on the reactive zones. By this means, and depending upon the hydrogeologic conditions, contaminated water from significant distances, both above and below the high permeability layer, can be drawn through the treatment material. For very shallow plumes, this concept could be implemented by excavation methods, or for deep plumes one might consider injection of reactive materials as several discrete and horizontal high-permeability zones. This concept and particular design considerations are introduced in Robertson et al. (2005).

7.5.2 Construction Methods

In seeking greater cost effectiveness in implementing the PRB technology and in attempting to extend the applicability to greater depths, various methods of installation have developed and evolved over the past 15 years. These are discussed in some detail in Battelle (1997) and ITRC (2005), and in somewhat less detail in Gillham et al. (2010). The purpose here is not to repeat those discussions, but to comment on the most common methods of installation from the particular perspective of mixing and mass transfer.

The most obvious and direct method of installation is to first create an open trench across the path of the contaminant plume, then fill the trench with the reactive material. In competent soils at shallow depths, this can simply involve excavation of a trench, though more commonly, some means of shoring to maintain the integrity of the trench walls is required. The width of the trench can vary according to variations in required residence time, as determined by stratigraphic variations in velocity or variations in influent contaminant concentrations.

This method provides for the least uncertainty in the location, thickness and composition of the reactive material. Furthermore, with the contaminated water flowing through a continuous wall of reactive material, the possible mixing and mass transfer issues often associated with liquid injection (Chapter 3, for example) are largely avoided. Variations in velocity through the PRB, and thus variations in residence time, does warrant consideration however. These variations can occur as a result of varying hydraulic conductivity in the adjacent aquifer, as well as non-uniformity in the PRB materials. Major stratigraphic variations should be determined in the initial site investigation and accommodated in the design; however smaller-scale variations are certain to persist.

Interestingly, granular iron PRBs have a self-regulating process. Initially, because the hydraulic conductivity of the PRB is greater than that of the native materials, velocity variations in the aquifer tend to be dampened in the PRB through diverging and converging flow lines. Perhaps more importantly, precipitate accumulations are greater in regions of the PRB that receive the greatest amount of water (high velocity regions) and as a result, the hydraulic conductivity in these zones will decline over time. Through this process, the velocity through the PRB will tend towards a uniform value. This process is demonstrated in Li et al. (2005) where, through numerical simulation, the precipitation of calcium carbonate was shown to result in a more uniform velocity field.

Continuous PRBs have also been constructed by continuous "dig and fill" procedures. Two examples are continuous trenchers, and trenching using a biopolymer to stabilize the trench walls. Continuous trenchers excavate a trench using a cutting chain and the trench is immediately filled with the reactive material from a trench box or "boot" that is attached immediately behind the cutting chain. In biopolymer trenching, as the trench is dug by an excavator, it is filled with a biopolymer slurry to stabilize the walls. As the trench advances, the reactive material is added through the slurry. Continuous trenchers are generally limited to depths of about 10 m, while slurry trenching has been used to depths greater than 20 m. Though cost becomes a factor, depending upon the excavation method that is used, slurry trenching is applicable to considerably greater depths.

In both the continuous trenches and biopolymer methods, there is no opportunity for visual confirmation of placement of the reactive material. In both cases, and depending upon the geologic materials, variations in PRB thickness should be anticipated. Flowing sands can be particularly problematic when using continuous trenches and while major collapses of the walls can be readily identified in biopolymer trenching, minor collapses could go unnoticed. The result could be small areas of the PRB where reactive materials are absent. Where there is significant bypass of the reactive zone, subsequent repair may be required. It should be noted however, that of the several PRBs that have been installed using both continuous trenchers and biopolymer trenching, bypass has not been identified as a significant problem and no repairs have been required.

The biopolymer, typically guar gum, is highly biodegradable and thus residual polymer can contribute to high levels of biological activity following installation. While in some cases this may be advantageous in removing contaminants it can also result in reduced permeability. For example, Johnson et al. (2008) suggested that reduced permeability upgradient of an iron PRB may have been a consequence of biological activity stimulated by the guar used during construction. To minimize the possibility of these effects, efforts should be made to flush as much of the polymer material as possible from the PRB during construction.

In the continuing attempts to reduce cost, particularly for deep installations, various injection methods have been used for installing continuous PRBs. Examples include vertical hydraulic fracturing (Hocking et al., 2004; ITRC, 2005), jetting, direct push injection and pneumatic fracturing and injection (McCall et al., 2004). The extent to which the reactive material penetrates the geologic formation from the point of injection depends upon the geologic material and particularly on variations and heterogeneity in the geologic materials. Thus ensuring continuity of the treatment zone can be difficult and though various hydraulic and geophysical testing methods have been used to evaluate continuity (Hocking et al., 2004 for example) verifying placement remains a challenge.

Oxygen addition (Section 7.2.3) is commonly used to treat groundwater contaminated by petroleum products and is therefore generally implemented at relatively shallow depths. Using large-diameter screened wells, there is little uncertainty concerning placement of the oxygen source; however, treatment downgradient from the source wells requires convergence of the oxygen plumes that are emanating from the source wells. Thus, development of a uniform treatment zone, particularly in heterogeneous geologic materials, is subject to many of the mixing and mass-transfer limitations discussed in Chapter 3.

In Section 7.2.5, the use of injection fluids to modify the *in situ* redox conditions for removal of metals was introduced. Particular examples included calcium polysulfide and sodium dithionite, both of which would generate reduced zones within which metals (Co(VI) and Cr(VI)) in these particular cases), would precipitate. Because the technology involves the injection of fluids, the mixing and mass transfer processes of Chapter 3 are particularly relevant. Thus in heterogeneous geologic materials, which are clearly the norm rather than the exception, creating a uniformly reduced zone is challenging and in some cases may not be possible. While a uniformly reduced zone may not be required in order to meet the objectives of the project, the potential limitations implied by Chapter 3 need to be recognized during the design stage.

7.6 SUMMARY COMMENTS

When introduced in the early 1990s, the use of granular iron in a PRB configuration for *in situ* treatment of organic solvents met with considerable skepticism. This resulted in part from the apparent simplicity of the concept, but in turn spawned major research efforts directed at both the fundamental and applied aspects of the technology. Ultimately these efforts, combined with early success at industrial sites, led to acceptance of the technology as standard practice for groundwater remediation. Furthermore, the concept was quickly adopted for other types of contaminants and reactive materials.

Though simple in concept, both research and experience have shown that one must nevertheless be judicious in application of the PRB technology. Issues common to all types of PRBs include the transferability and scale-up of laboratory reactivity tests to field conditions, adequate characterization of the hydrogeologic and geochemical conditions and a thorough understanding of the factors that are likely to influence the long-term performance of the PRB. An even more fundamental consideration is to insure that the project goals are consistent with the capabilities of the PRB concept. In particular, if rapid cleanup throughout a large contaminant plume is required, then it is unlikely that a PRB will provide the required results.

As research relevant to the PRB concept continues, progress in several areas can be anticipated. These could include:

- Application to a wider range of groundwater contaminants,
- Development of alternate reactive materials,

- Greater understanding of potential uses of nano-scale iron,
- Improved and more cost-effective methods of installation, particularly for fractured rock, and
- Improved understanding and predictive capability with respect to long-term performance.

REFERENCES

AFCEE (Air Force Center for Environmental Excellence [renamed the Air Force Center for Engineering and the Environment]). 2008. Technical Protocol for Enhanced Anaerobic Bioremediation Using Permeable Mulch Biowalls and Bioreactors. Prepared by Parsons Infrastructure & Technology Group, Inc., Denver, CO, USA. http://www.afcee.af.mil/resources/technologytransfer/programsandinitiatives/enhancedinsituanaerobicbioremediation/resources/index.asp . Accessed August 6, 2011.

Agrawal A, Tratnyek PG. 1996. Reduction of nitro aromatic compounds by zero-valent iron metal. Environ Sci Technol 30:153–160.

Amrhein C, Hunt ML, Roberson JA, Yarmoff SR, Qui SR, Lai H-F. 1998. The use of XANES, STM, and XPS to identify the precipitation products formed during the reaction of U, Cr, and Se with zero-valent iron. Proceedings, V.M. Goldschmidt Conference, Toulouse, France, pp 51–52.

Baker MJ, Blowes DW, Ptacek CJ. 1997. Phosphorous adsorption and precipitation in a permeable reactive wall: Applications for wastewater disposal systems. Land Contam Reclam 5:189–193.

Baker MJ, Blowes DW, Ptacek CJ. 1998. Laboratory development of permeable reactive mixtures for the removal of phosphorous from onsite wastewater disposal systems. Environ Sci Technol 32:2308–2316.

Bandstra JZ, Miehr R, Johnson RL, Tratnyek PG. 2005. Reduction of 2,4,6-trinitrotoluene by iron metal: Kinetic controls on product distributions in batch experiments. Environ Sci Technol 39:230–238.

Battelle. 1997. Design Guidance for Application of Permeable Barriers to Remediate Dissolved Chlorinated Solvents. CU-107 Report. Prepared for the Strategic Environmental Research and Development Program (SERDP), Arlington, VA, USA. http://www.serdp.org/content/advanced search/. Accessed August 6, 2011.

Benner SG, Blowes DW, Ptacek CJ. 1997. A full-scale porous reactive wall for prevention of acid mine drainage. Ground Water Monit Remediat 17:99–107.

Benner SG, Herbert RB Jr, Blowes DW, Ptacek CJ, Gould D. 1999. Geochemistry and microbiology of a permeable reactive barrier for acid mine drainage. Environ Sci Technol 33:2793–2799.

Blowes DW, Mayer KU. 1999. An In-Situ Permeable Reactive Barrier for the Treatment of Hexavalent Chromium and Trichloroethylene in Groundwater: Vol 3, Multicomponent Reactive Transport Modeling. EPA 600-R-99-095c. USEPA Office of Research and Development, Washington, DC, USA. http://www.epa.gov/nrmrl/pubs/600R99095/prb model_v3.pdf. Accessed August 6, 2011.

Blowes DW, Ptacek CJ. 1992. Geochemical remediation of groundwater by permeable reactive walls: Removal of chromate by reaction with iron-bearing solids. Proceedings, Subsurface Restoration Conference, 3rd International Conference on Ground Water Quality Research, June 21–24, Dallas, TX, USA, pp 214–216.

Blowes DW, Ptacek CJ, Bain JG, Waybrant KR, Robertson WD. 1995. Treatment of mine drainage water using in situ permeable reactive walls. Proceedings, Sudbury '95, Mining and the Environment, CANMET, Ottawa, Ontario, Vol 3, pp 979–987.

Blowes DW, Ptacek CJ, Jambor JL. 1997. In-situ remediation of chromate contaminated groundwater using permeable reactive walls. Environ Sci Technol 31:3348–3357.

Blowes DW, Ptacek CJ, Benner SG, Waybrant KR, Bain JG. 1998. Porous reactive walls for the prevention of acid mine drainage: A review. Min Proc Ext Met Rev 19:25–37.

Blowes DW, Puls RW, Gillham RW, Ptacek CJ, Bennett TA, Bain JG, Hanton-Fong CJ, Paul CJ. 1999. An In Situ Permeable Reactive Barrier for the Treatment of Hexavalent Chromium and Trichloroethylene in Groundwater: Vol 2, Performance Monitoring. EPA 600-R-99-095b. U.S. Environmental Protection Agency (USEPA) Office of Research and Development, Washington, DC, USA. http://www.epa.gov/nrmrl/pubs/600R99095/prb performance_ v2.pdf. Accessed August 6, 2011.

Blowes DW, Ptacek CJ, Benner SG, McRae CWT, Bennett TA, Puls RW. 2000. Treatment of inorganic contaminants using permeable reactive barriers. J Contam Hydrol 45:123–137.

Bostick WD, Shoemaker JL, Osborne PE, Evans-Brown B. 1990. Treatment and disposal options for a heavy metals waste containing soluble technetium-99. In Tedder DW, Pohland FG, eds, Am Chem Soc Symp Ser 422:345–367.

Cantrell KJ, Kaplan DI, Wietsma TW. 1995. Zero-valent iron for the in situ remediation of selected metals in groundwater. J Hazard Mater 42:201–212.

Cheng F, Muftikian R, Fernando Q, Korte N. 1997. Reduction of nitrate to ammonia by zero-valent iron. Chemosphere 35:2689–2695.

Craig M. 2004. Biological PRB used for perchlorate degradation in ground water. USEPA Technology News and Trends. EPA 542-N-04-001. Issue 10 (February):2–3.

Del Cul GD, Bostick WD, Trotter DR, Osborne PE. 1993. Technetium-99 removal from process solutions and contaminated groundwater. Sep Sci Technol 28:551–564.

Devlin JF, McElwee CM. 2007. Effects of measurement error on horizontal hydraulic gradient estimates in an alluvial aquifer. Ground Water 45:62–73.

Fruchter JS, Cole CR, Williams MD, Vermeul VR, Amonette JE, Szecsody JE, Istok JD, Humphrey MD. 2000. Creation of a subsurface permeable treatment zone for aqueous chromate contamination using in situ redox manipulation. Ground Water Monit Remediat 20:66–77.

Gaveskar AR, Gupta N, Sass BM, Janosy RJ, O'Sullivan D. 1998. Permeable Barriers for Groundwater Remediation: Design, Construction and Monitoring. Battelle Press, Columbus, OH, USA. 176 p.

Gillham RW, Burris DR. 1992. In situ treatment walls: Chemical dehalogenation, denitrification, and bioaugmentation. Proceedings, Subsurface Restoration Conference Dallas, TX, USA, June 21–24, pp 66–68.

Gillham RW, O'Hannesin SF. 1992. Metal-catalysed abiotic degradation of halogenated organic compounds. IAH Conference – Modern Trends in Hydrogeology, Hamilton, Ontario, Canada, pp 94–103.

Gillham RW, O'Hannesin SF. 1994. Enhanced degradation of halogenated aliphatics by zero-valent iron. Ground Water 32:958–967.

Gillham RW, Vogan J, Gui L, Duchene M, Son J. 2010. Iron barrier walls for chlorinated solvent remediation. In Stroo HF, Ward CH, eds, In Situ Remediation of Chlorinated Solvent Plumes. Springer Science + Business Media, New York, NY, USA, pp 537–572.

Gould JP. 1982. The kinetics of hexavalent chromium reduction by metallic iron. Water Res 16:871–877.

Gu B, Liang L, Dickey MJ, Yin X, Dai S. 1998. Reductive precipitation of uranium VI by zero-valent iron. Environ Sci Technol 32:3366–3373.

Gui L, Yang Y, Jeen SW, Gillham RW, Blowes DW. 2009. Reduction of chromate by granular iron in the presence of dissolved $CaCO_3$. Appl Geochem 24:677–686.

Hocking G, Wells SL, Givens CN. 2004. A deep iron permeable reactive barrier constructed below 100-foot depth. Proceedings, 4th International Conference on Remediation of Chlorinated and Recalcitrant Compounds, Monterey, CA, USA, May 24–27, 3A.09/1-3A.09/7.

Huang C, Wang H, Chiu P. 1998. Nitrate reduction by metallic iron. Water Res 32:2257–2264.

ITRC (Interstate Technology & Regulatory Council). 2005. Permeable Reactive Barriers: Lessons Learned/New Directions. PRB-4. ITRC, Washington DC, USA. http://www.itrcweb.org/Documents/ PRB-4.pdf. Accessed August 6, 2011.

Jeen SW, Gillham RW, Blowes DW. 2006. Effects of carbonate precipitates on long-term performance of granular iron for reductive dechlorination of TCE. Environ Sci Technol 40:6432–6437.

Jeen SW, Mayer KU, Gillham RW, Blowes DW. 2007. Reactive transport modeling of trichloroethene treatment with declining reactivity of iron. Environ Sci Technol 41:1432–1438.

Jeen SW, Blowes DW, Gillham RW. 2008. Performance evaluation of granular iron for removing hexavalent chromium under different geochemical conditions. J Contam Hydrol 95:76–91.

Johnson PC, Bruce CL, Miller KD. 2003. In Situ Bioremediation of MTBE in Groundwater. CU-0013 Cost and Performance Report. Prepared for the Environmental Security Technology Certification Program (ESTCP), Arlington, VA, USA. http://www.serdp.org/ content/advancedsearch/. Accessed August 6, 2011.

Johnson RL, Thoms RB, O'Brien Johnson R, Nurmi JT, Tratnyek PG. 2008. Mineral precipitation and flow reduction up-gradient from a zero-valent iron permeable reactive barrier. Ground Water Monit Remediat 28:56–64.

Kim, J-H, Tratnyek PG, Chang Y-S. 2008. Rapid dechlorination of polychlorinated dibenzo-p-dioxins (PCDDs) by bimetallic and nanosized zerovalent iron. Environ Sci Technol 42:4106–4112.

Lackovic JA, Nikolaidis NP, Dobbs GM. 2000. Inorganic arsenic removal by zerovalent iron. Environ Eng Sci 17:29–39.

Lee DR, Smyth DJA, Shikaze SG, Jowett RJ, Hartwig DS, Milloy C. 1998. Wall-and-curtain for passive collection/treatment of contaminated plumes. Proceedings, First International Conference on Remediation of Chlorinated and Recalcitrant Compounds, Monterey, CA, USA. Battelle Press, Columbus, OH, USA.

Li L, Benson CH, Lawson EM. 2005. Impact of mineral fouling on hydraulic behavior of permeable reactive barriers. Ground Water 43:582–596.

Li L, Benson CH, Lawson EM. 2006. Modeling porosity reductions caused by mineral fouling in continuous-wall permeable reactive barriers. J Contam Hydrol 83:89–121.

Lowry GV, Johnson KM. 2004. Congener-specific dechlorination of dissolved PCBs by microscale and nanoscale zerovalent iron in a water/methanol solution. Environ Sci Technol 38:5208–5216.

Lu Q, Gui L, Gillham RW. 2005. Effects of nitrate on trichloroethylene degradation by granular iron. Earth Sci Front 12:176–183.

Mackenzie PD, Horney DP, Sivavec TM. 1999. Mineral precipitation and porosity losses in granular iron columns. J Hazard Mater 68:1–17.

Matheson LJ, Tratnyek PG. 1994. Processes affecting remediation of contaminated groundwater by dehalogenation with iron. Environ Sci Technol 28:2045–2052.

Mayer KU, Blowes DW, Frind EO. 2001. Reactive transport modeling of an in situ reactive barrier for the treatment of hexavalent chromium and trichloroethylene in groundwater. Water Resour Res 37:3091–3103.

McCall J, Swanson G, Cheng HC, Brooks P, Chan DB, Chen S, Menack J. 2004. Demonstration of ZVI injection for in-situ remediation of chlorinated solvents at Hunter Shipyard. Proceedings, Fourth International Conference on Remediation of Chlorinated and Recalcitrant Compounds, Monterey, CA, USA. Battelle Press, Columbus, OH, USA.

McGregor RG, Blowes DW, Ludwig R, Pringle E, Pomery M. 1999. Remediation of a heavy metal plume using a reactive wall. Proceedings, In Situ and On-Site Bioremediation Conference, San Diego, CA, USA, April 19–22.

McMahon PB, Dennehy KF, Sandstrom MW. 1999. Hydraulic and geochemical performance of a permeable reactive barrier containing zero-valent iron, Denver Federal Center. Ground Water 37:396–404.

McMurtry D, Elton RO. 1985. New approach to in-situ treatment of contaminated groundwaters. Environ Prog 4:168–170.

McRae CWT. 1999. Evaluation of Reactive Materials for In Situ Treatment of Arsenic III, Arsenic V and Selenium VI using Permeable Reactive Barriers: Laboratory Study. MSc Thesis, University of Waterloo, Waterloo, Ontario, Canada.

Morrison SJ. 2003. Performance evaluation of a permeable reactive barrier using reaction products as tracers. Environ Sci Technol 37:2302–2309.

Morrison SJ, Metzler DR, Carpenter CE. 2001. Uranium precipitation in a permeable reactive barrier by progressive irreversible dissolution of zero-valent iron. Environ Sci Technol 35:385–390.

Morrison SJ, Metzler DR, Dwyer BP. 2002. Removal of As, Mn, Mo, Se, U, V and Zn from groundwater by zero-valent iron in a passive treatment cell: Reaction progress modeling. J Contam Hydrol 56:99–116.

Mueller J, MacFabe S, Vogan J, Duchene M, Hill D, Bolanos-Shaw K, Seech A. 2004. Reactive dechlorination of solvents in groundwater using controlled release of carbon with micro-scale ZVI. Proceedings, Fourth International Conference on Remediation of Chlorinated and Recalcitrant Compounds, Monterey, CA, USA. Battelle Press, Columbus, OH, USA.

Muftikian R, Fernando Q, Korte NE. 1995. A method for the rapid dechlorination of low molecular weight chlorinated hydrocarbons in water. Water Res 29:434–2439.

Muftikian R, Nebesny K, Fernando Q, Korte N. 1996. X-ray photoelectron spectra of the palladium-iron bimetallic surface used for the rapid dechlorination of chlorinated organic environmental contaminants. Environ Sci Technol 30:3593–3596.

Odziemkowski MS, Gillham RW, Focht R. 1998. Electroless hydrogenation of trichloroethylene by Fe-Ni(P) galvanic couples. Electrochem Soc Proc 90:91–103.

Oh SY, Cha DK, Chiu PC. 2002. Graphite-mediated reduction of 2,4-dinitrotoluene with elemental iron. Environ Sci Technol 36:2178–2184.

O'Hannesin SF, Gillham RW. 1998. Long-term performance of an in situ "iron wall" for remediation of VOCs. Ground Water 36:164–170.

Phifer MA, Sappington FC, Nichols RL, Ellis N, Cardoso-Neto JE. 1999. GeoSiphon/Geoflow Treatment Systems. Waste Management 99 Symposia, Tuscon, AZ, USA, February 28-March 4.

Primrose A, Butler L, Wiemelt K, Castaneda N, Gavaskar AR, Chen ASC. 2004. Successful field implementation of permeable reactive barriers at Rocky Flats. Proceedings, Fourth International Conference on Remediation of Chlorinated and Recalcitrant Compounds, Monterey, CA, USA, May.

Puls RW, Blowes DW, Gillham RW. 1998. Emplacement verification and long-term performance monitoring of a permeable reactive barrier at the USCG Support Center, Elizabeth City, North Carolina. Groundwater Quality: Remediation and Protection. In Herbert M, Kovar K, eds, IAHS Publication, Vol 250, pp 459–466.

Reardon EJ. 1995. Anaerobic corrosion of granular iron: Measurement and interpretation of hydrogen evolution rates. Environ Sci Technol 29:2936–2945.

Ritter K, Odziemkowski MS, Simpraga R, Gillham RW, Irish DE. 2003. An in situ study of the effect of nitrate on the reduction of trichloroethylene by granular iron. J Contam Hydrol 6:121–136.

Robertson WD, Cherry JA. 1995. In situ denitrification of septic system nitrate using reactive porous media barriers: Field trials. Ground Water 33:99–111.

Robertson WD, Yeung N, vanDriel PW, Lombardo PS. 2005. High-permeability layers for remediation of ground water; Go wide, not deep. Ground Water 43:574–581.

Robertson WD, Vogan JL, Lombardo PS. 2008. Nitrate removal rates in a 15-year-old permeable reactive barrier treating septic system nitrate. Ground Water Monit Remediat 28:1–8.

Rouse JV, Davies IN, Hutton J, DeSantis A. 2001. In-situ hexavalent chromium reduction and geochemical fixation in varied geohydrological regimes. First International Conference on Oxidation and Reduction Technologies for In-Situ Treatment of Soil and Groundwater. Niagara Falls, Ontario, Canada.

Scherer MM, Richter S, Valentine RL, Alvarez PJJ. 2000. Chemistry and microbiology of permeable reactive barriers for in situ groundwater cleanup. Crit Rev Environ Sci Technol 30:363–411.

Schlicker O, Ebert M, Fruth M, Weidner M, Wust W, Dahmke A. 2000. Degradation of TCE with iron: The role of competing chromate and nitrate reduction. Ground Water 38:403–409.

Smyth DJA, Blowes DW, Ptacek CJ, Baker MJ, Ford G, Foss S, Bernstene E. 2002. Removal of phosphate and waterborne pathogens from wastewater effluent using permeable reactive materials. In Ground and Water: Theory to Practice. Proceedings, 55th Canadian Geotechnical and 3rd Joint IAH-CNC and CGS Groundwater Specialty Conferences, Niagara Falls, Ontario, Canada, October 20–23.

Star RC, Cherry JA. 1994. In situ remediation of contaminated groundwater: the funnel-and-gate system. Ground Water 32:465–76.

Su C, Puls RW. 2001. Arsenate and arsenite removal by zero-valent iron: Kinetics, redox transformation, and implications for in situ groundwater remediation. Environ Sci Technol 35:1487–1492.

Thomson BM, Shelton SP. 1988. Permeable barriers: A new alternative for treatment of contaminated ground waters. Proceedings, Focus Conference of Southwestern Ground Water Issues, Albuquerque, NM, USA, March 23–25.

Tratnyek PG, Johnson RL. 2006. Nanotechnologies for environmental cleanup. Nano Today 1:44–48.

Tratnyek PG, Scherer MM, Johnson TL, Matheson LJ. 2003. Permeable reactive barriers of iron and other zero-valent metals. Environ Sci Pollut Control Ser 26 (Chemical Degradation Methods for Wastes and Pollutants):371–421.

USEPA (U.S. Environmental Protection Agency). 1996. GRACE Bioremediation Technologies, Daramend^{TM} Bioremediation Technology. Superfund Innovative Technology Evaluation (SITE) Program Innovation Technology Evaluation Report. EPA/540/R-95/536.

Vogan JL, Butler BJ, Odziemkowski MS, Friday G, Gillham RW. 1998. Inorganic and biological evaluation of cores from permeable iron reactive barriers. Proceedings, First International Conference on Remediation of Chlorinated and Recalcitrant Compounds, Monterey, CA, May 18–21. Battelle Press, Columbus, OH, USA, Paper C1–6.

Vogan JL, Focht RM, Clark DK, Graham SL. 1999. Performance evaluation of a permeable reactive barrier for remediation of dissolved chlorinated solvents in groundwater. J Hazard Mater 68:97–108.

Wang C-B, Zhang W-X. 1997. Synthesizing nanoscale iron particles for rapid and complete dechlorination of TCE and PCBs. Environ Sci Technol 31:2154–2156.

Warner SD, Mok C, Bennett P, Frappa R, Steiner R, Bohan C, Rabideau A, Steiner A. 2004. Performance assessment of a zeolitic permeable treatment wall designed to remove strontium-90 from groundwater. Proceedings, Fourth International Conference on Remediation of Chlorinated and Recalcitrant Compounds, Monterey, CA, USA. Battelle Press, Columbus, OH, USA.

Waybrant KR, Blowes DW, Ptacek CJ. 1995. Selection of reactive mixtures for the prevention of acid mine drainage using in situ porous reactive walls. Proceedings, Sudbury '95, Mining and the Environment, CANMET, Ottawa, Ontario, Canada, Vol 3, pp 945–953.

Waybrant KR, Blowes DW, Ptacek CJ. 1998. Selection of reactive mixtures for use in permeable reactive walls for treatment of mine drainage. Environ Sci Technol 32:1972–1979.

Wilkens J, Shoemaker SH, Bazela WB, Egler AP, Sinha R, Bain JG. 2003. Arsenic removal from groundwater using a PRB of BOF slag at the DuPont East Chicago (In) site. The Research Technology Demonstration Forum Permeable Reactive Barriers Action Team Meeting, Niagara Falls, NY, USA, October 16.

Wilkin RT, Puls RW, Sewell GW. 2003. Long-term performance of permeable reactive barriers using zerovalent iron: Geochemical and microbiological effects. Ground Water 41:493–503.

Wilkin RT, Su C, Ford RG, Paul CJ. 2005. Chromium-removal processes during groundwater remediation by a zerovalent iron permeable reactive barrier. Environ Sci Technol 39:4599–4605.

Wilson RD, Mackay DM, Scow KM. 2002. In situ MTBE degradation supported by diffusive oxygen release. Environ Sci Technol 36:190–199.

Yabusaki S, Cantrell K, Sass B, Steefel C. 2001. Multicomponent reactive transport in an in situ zero-valent iron cell. Environ Sci Technol 35:1493–1503.

Zawislanski PT, Beatty JJ, Carson WL. 2002. In situ treatment of low pH and metals in groundwater using calcium polysulfide. Second International Conference on Oxidation and Reduction Technologies for In-Situ Treatment of Soil and Groundwater. Toronto, Ontario, Canada, November 17–21.

Zhang Y, Gilham RW. 2005. Effects of gas generation and precipitates on performance of Fe^0 PRBs. Ground Water 43:113–121.

CHAPTER 8

IN SITU SPARGING FOR DELIVERY OF GASES IN THE SUBSURFACE

Richard L. Johnson[1] and Paul C. Johnson[2]

[1]Division of Environmental and Biomolecular Systems, Oregon Health & Science University, Beaverton, OR, USA
[2]School of Sustainable Engineering and the Built Environment, Arizona State University, Tempe, AZ, USA

8.1 INTRODUCTION

In situ sparging (ISS)[1] has been used extensively as a tool for restoration of aquifers contaminated with a range of organic chemicals. The primary mechanisms of contaminant removal are volatilization and enhanced *in situ* degradation. Volatilization has been discussed extensively (Johnson, 1998; Johnson et al. 1998, 1999; Rutherford and Johnson, 1996) and will only be discussed briefly here. In the context of enhanced degradation, the most common sparging approach is to deliver oxygen (as air) to the subsurface to facilitate aerobic biodegradation. This approach is, of course, best suited for compounds that can be readily biodegraded under aerobic conditions (e.g., petroleum hydrocarbons) and is less well suited for other important classes of contaminants (e.g., chlorinated hydrocarbons, energetics, etc.).

In order to expand the range of applicability of ISS to chlorinated hydrocarbons and other contaminants, a range of gases in addition to air can be delivered to the subsurface to stimulate both biotic and abiotic reactions. These include hydrogen, methane, propane, butane, ammonia, pure oxygen and ozone. Sometimes one or more of these will be blended with air. There are a number of issues relating to the delivery of these gases that will control their effectiveness, including: (1) their solubilities in water; (2) the hazards associated with the gas (e.g., the concentration in air at which they become explosive); and (3) the lifetime of their reactivity in the subsurface. Injection of gases other than pure air and issues associated with them will be discussed within this chapter.

When delivering gases other than air to the subsurface, there are a number of operational issues that may differ from conventional ISS with air. These may include: (1) pulsed injections on widely-separated time intervals; (2) injection of limited gas volumes during each injection cycle; and (3) elimination of any off-gas vapor treatment system. Limited-volume and pulsed conditions may be desirable to increase the efficiency of gas utilization and/or mitigate the risk posed by excessive injection of those gases. As is discussed below, the physics of gas flow in otherwise water-saturated porous media results in trapped gas within the aquifer pore spaces. Because of solubility limits, dissolution of trapped gas into flowing groundwater may occur over periods of weeks or more following injection. This reservoir of trapped gas can then act as a steady source of reactants and minimizes the need for frequent injections. Using

[1] The term "*in situ* air sparging" is frequently represented by the acronym IAS. However, since this chapter will discuss the injection of a range of gases, we have used the more general acronym ISS.

P.K. Kitanidis and P.L. McCarty (eds.), *Delivery and Mixing in the Subsurface: Processes and Design Principles for In Situ Remediation*, doi: 10.1007/978-1-4614-2239-6_8, © Springer Science+Business Media New York 2012

short-duration, infrequent injection conditions, fugitive emissions from the groundwater zone can typically be managed to avoid the need for off-gas collection systems.

Much of the discussion in this chapter will focus on the mechanics of gas delivery to the subsurface and subsequent mass transfer from the gas phase into the groundwater. While ISS of reactive gases can potentially facilitate abiotic reactions, the emphasis in this chapter will be on the enhancement of biotic processes.

8.2 BRIEF OVERVIEW OF THE PHYSICS OF *IN SITU* SPARGING

There is a significant amount of information regarding the detailed physics of sparging (Battelle, 2001; Clayton, 1998, 1999; ESTCP, 2005; Johnson et al., 1993, 2001a; McCray and Falta, 1997; Thomson and Johnson, 2000; Tomlinson et al., 2003; van Dijke and van der Zee, 1998) which will only be reviewed briefly here. The physics of gas flow in saturated media can be summarized as follows:

1. Buoyancy and the pressure applied at the injection well are the driving forces for gas movement in saturated porous media.
2. In most unconsolidated granular media (i.e., sands, silts, gravels, etc.), gases move outward and upward from the injection point in continuous pathways (i.e., channels) rather than as bubbles (even if they are injected as bubbles through special well diffusers).
3. The horizontal structure of saturated media (e.g., layering) often has a controlling effect on the lateral distribution of the gases and may result in the formation of laterally-extensive "pools" of gas beneath lower-permeability layers.
4. At the same time, the presence of high-permeability structures (including man-made ones) can provide conduits for rapid preferential gas movement. This is of particular significance when hazardous gases are being injected.
5. When gas injection is terminated (e.g., in pulsed sparging), continuous air channels collapse, leaving trapped residual air within the water-saturated medium.
6. Dissolution of gases into groundwater occurs at the interface between the groundwater and either gas channels or residual trapped gas.

Depending upon the structure of the aquifer and the gas injection rate, the time required to reach a quasi-steady-state air distribution may be minutes to days. As is discussed below, there are a number of diagnostic tools available to determine both the distribution of gas in the subsurface and the time-frame required to reach steady-state. For the injection of gases other than pure air, both steady-state distribution and time-frame are important because optimum system efficiency occurs when the best possible air distribution is achieved with a minimum of total gas flow.

To a limited extent, both the gas distribution and time to steady-state can be impacted by the rate at which gas is injected into the subsurface. For example, if gas is injected at a very slow rate, movement of the gas may occur in a relatively small number of channels in the vicinity of the injection well and, as a result, mass transfer to the groundwater will be limited. In contrast, rapid injection of gas will generally result in more channels, and may also facilitate vertical movement through horizontal layers that would otherwise inhibit vertical movement. At the same time, it is important to note that excessive injection pressures can lead to fracturing or fluidization of the medium, which may have undesirable consequences.

Gas flow in the subsurface can be strongly influenced by wells, and conversely the behavior of wells can be strongly influenced by gas flow. This occurs, in part, because wells can represent significant vertical conduits for the gas, particularly if it is moving laterally beneath lower-permeability strata. If gas intercepts a well (or the "filter pack" around a well) it will move upward, potentially directly to ground surface. In the process, it may strip volatile contaminants from water within the well, resulting in a false indication of very rapid "success" of the sparging system.

8.3 APPLICATIONS OF GAS DELIVERY SYSTEMS

8.3.1 Air Biosparging

As mentioned above, delivery of oxygen through the injection of air with a blower or compressor (Figure 8.1a) to the subsurface represents the most common application of sparging as a delivery system. Air contains about 20% oxygen by volume, and the solubility of oxygen in water contacting air is approximately 10 milligrams (mg)-O_2/liter (L)-H_2O.

8.3.2 Oxygen Biosparging

Sparging with pure oxygen (instead of air) has been used successfully to minimize fugitive gas emissions while maximizing delivery of oxygen to the subsurface (e.g., Johnson et al., 2003). As discussed in the examples section below, maintaining oxygenated conditions within naturally-anoxic groundwater is particularly important if aerobic biodegradation is desired either by aerobic microorganisms naturally present or that have been introduced into the treatment zone. Pure oxygen from cylinders can be used or it can be produced from air on-site using commercially-available molecular sieve-based separation systems. Because oxygen production rates from the separation systems are usually low (<200 cubic feet per hour [ft^3/h]), oxygen storage tanks associated with each well can be filled and pressurized over time and then discharged periodically in a higher-flow pulsed sparging mode (Figure 8.1b).

8.3.3 Cometabolic Biosparging

A number of different gases have been injected into the subsurface to meet the metabolic needs of subsurface microorganisms. Carbon and energy sources (e.g., propane, butane, hydrogen have all been added). In addition, the co-injection of some aerobically biodegradable gases (e.g., propane) with air can result in the production of enzymes that can degrade a number of otherwise recalcitrant organic compounds (e.g., chlorinated solvents). This approach has been examined by a number of researchers at both the laboratory and field scales (ESTCP, 2001, 2003, 2005; SERDP, 2003). From a sparging perspective, almost any gas-phase amendment can be delivered by mixing ambient air (or inert gas) with the amendment gas delivered under pressure to the flow system (Figure 8.1c). In practice, as discussed below, there may be a number of operational issues that need to be addressed to accomplish this safely.

8.3.4 Gas Injection of Chemical Oxidants

Ozone injection has been proposed by several vendors using a range of possible oxidant delivery schemes. The lifetime of ozone in the subsurface is probably less than 1 h (Yu et al., 2005), thus transport of ozone in groundwater will probably not be significant and as a result, any reactions with ozone will need to occur in the immediate vicinity of the air channels.

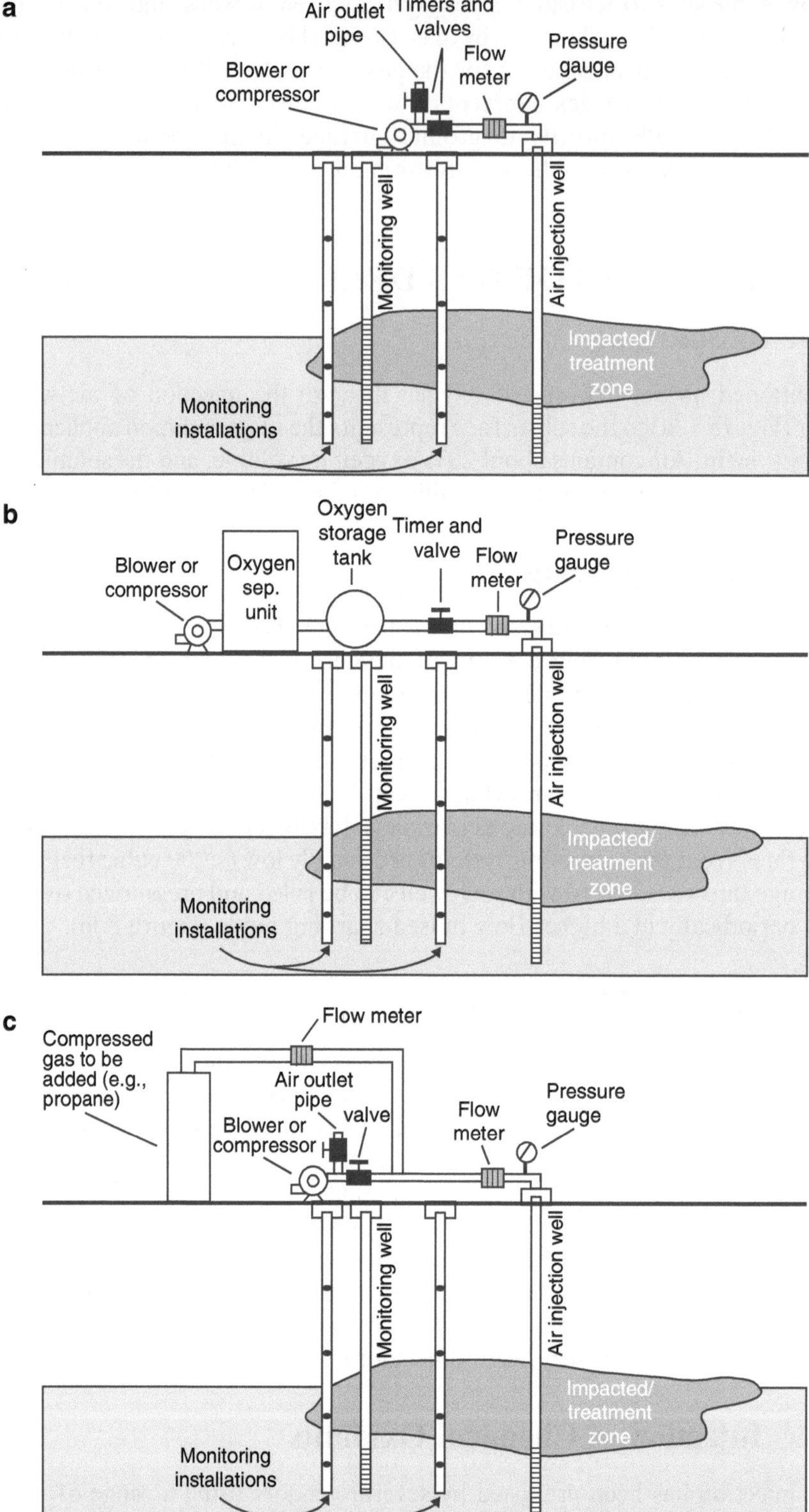

Figure 8.1. Schematic drawings of *in situ*; (a) air sparging, (b) oxygen sparging, and (c) cometabolic sparging systems.

8.4 DESIGN PRINCIPLES

8.4.1 Conceptual Design

Both the conceptual and practical aspects of ISS system design are discussed in detail elsewhere (Johnson et al., 2010; U.S. Army, 2008), and the reader is again directed to those sources for a complete discussion. In the context of this chapter we will focus on aspects of design that are of particular relevance to sparging as a delivery system, and in particular to the delivery of gases other than ambient air to enhance subsurface biological reactions.

8.4.1.1 Physical Characteristics

In the course of developing the design for a sparging system, it is important to develop a robust conceptual model for the behavior of gases in the target treatment zone. Given the strong dependence of gas distribution on the aquifer's physical structure, it is important to have an accurate conceptual picture of that structure. This includes the stratigraphy of the site, as well as the permeabilities of those strata. For lower-permeability media, it is important to know the extent to which preferential pathways (including man-made ones) will allow gases to move upwards through the strata.

Figure 8.2 represents a conceptual example of gas distribution in a source zone contaminated with a dense nonaqueous phase liquid (DNAPL). The DNAPL tends to move downward in small fingers or channels due to its density, surface tension, and viscosity, and to accumulate on top of lower-permeability layers. In contrast, injected gases tend to move upward through water-saturated media as channels or fingers (again due to density, surface tension and

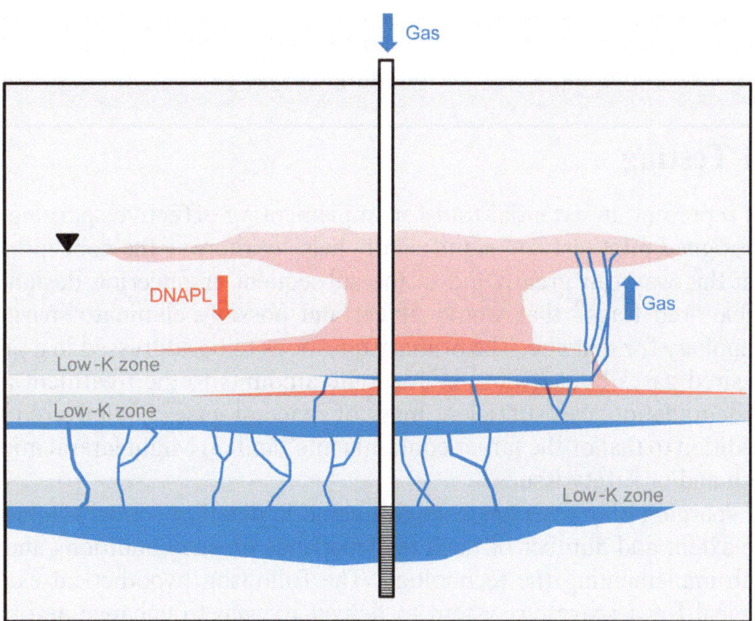

Figure 8.2. Schematic representation of downward movement of a dense nonaqueous phase liquid (DNAPL) and upward movement of gas in a horizontally-stratified aquifer.

viscosity) and to accumulate in pockets beneath lower-permeability strata. In both the DNAPL and gas cases, the extent to which the fluids will move into and through the strata depends on the permeability and capillary property differences between adjacent layers. If flow pathways through the strata exist, or if sufficient pressure is exerted by the fluid, it may move through the layer, re-form channels on the other side of the layer and move until the next layer is encountered. The gas accumulating beneath strata can grow to fill perhaps 40% of the pore space. At such levels, water flow through the gas pool zone will be substantially reduced. Within vertical channels, which in most cases will have dimensions of millimeters to a few grain diameters, the degree of gas saturation will be high, reducing water flow within the channel and as a result, mass transfer occurs primarily to the water as it flows around the channels. For the case depicted in Figure 8.2, the fraction of the medium occupied by gas during sparging will probably be a few percent or less of the pore space. When gas injection stops (i.e., between pulses), gas concentration in both the pools and channel water will decrease, although water flow through both of those zones will probably continue to be lower than through the water-saturated media.

8.4.1.2 Biological Characteristics

The rate at which microbial degradation occurs can have an important impact on the conceptual design of a sparging delivery system. For example, if the intent of the sparging system is to deliver oxygen to the groundwater, and if the degradation half-lives of the contaminants being degraded are on the order of weeks, then the system needs to be designed to meet the oxygen demand required for oxidizing such materials that may be contained on aquifer solids as well as in the groundwater over those timeframes. The oxygen supply must not only meet the oxygen demand occurring within the sparging zone itself, but also downgradient of that zone in order to achieve the extent of degradation desired (e.g. a transport time of perhaps five degradation half-lives). Additionally, oxygen demand by non-targeted materials within the groundwater and aquifer material will likely have to be met as well. Such additional competitive demand for all reactive gases must be considered in the design of the sparging system.

8.4.2 Pilot Testing

Pilot tests represent an essential tool for implementing effective sparging systems. Typically, a well-designed pilot test can significantly help to improve the conceptual model of the system, and in this way can greatly aid in the subsequent engineering design. It can help to identify possible "red flags" that would affect and possibly eliminate sparging as a viable treatment technology for that site. The primary questions to be addressed in a pilot test include: (1) Can the desired gases be delivered in adequate amounts to the treatment zone? (2) In the context of biodegradation, can sufficient mass of reactant gases be delivered to meet system demands in addition to that of the target contaminants? and (3) Might lateral migration of gases pose any health and/or safety issues?

Design of sparging pilot tests has been discussed in detail previously (Johnson et al., 1997, 2001b, c). The extent and number of such tests depends on site conditions and potential risks associated with implementing the technology. The following hypothetical example might be considered typical for a sparging system to deliver oxygen to enhance aerobic degradation. Suppose a site to be treated contains a dissolved groundwater plume of contaminants at 5 meters (m) below the water table, and that the treatment objective is to eliminate plume migration off-site without the need for off-gas control, then the following pilot task activities might be undertaken:

1. Collection of soil core. In many settings where sparging is to be implemented, it is desirable to collect continuous core samples as part of the pilot well installation process. These cores can be used both qualitatively in the field to help determine the depth interval for gas injection and quantitatively in the laboratory to determine depth-specific values for horizontal and vertical permeability.
2. Installation of sparging pilot well. Based on core analysis and site conceptual model, the pilot sparging well can be installed (e.g., in the same hole used for core collection). In most cases, the design of the pilot well should be similar to the anticipated design for the treatment wells. Based on published design guidelines (e.g., U.S. Army, 2008), this will probably involve 1-inch (in) to 2-in polyvinyl chloride (PVC) pipe with a relatively short (e.g., 1–2 ft long) screened interval. Depending upon the manner of installation, the sparge well may or may not have a "filter pack" associated with it. From the perspective of gas distribution, the filter pack probably will not have a major impact. However, it should be recognized that gas leaving the well screen will probably move upward within the filter pack and may accumulate at the top of the filter pack before entering the aquifer. In this context, it is important to ensure that the seal at the top of the filter pack is of good quality and at the desired depth.
3. Installation of monitoring wells. Depending upon specific pilot tests to be conducted, one or more monitoring wells should be installed. For the tests discussed below the configuration shown in Figure 8.3 is recommended. If significant layering is present at the site, it is recommended that discrete monitoring wells be completed at whatever depths are important for treatment. The monitoring wells can be used for both gas distribution tests (discussed below) and to determine whether mass transfer of gases will be adequate to meet system needs.

   If lateral migration of gases has potential health and safety impacts, it is recommended that a temporary vapor recovery well be installed and that an air recovery test be conducted. Pressure monitoring wells represent an important component of the pilot test in that they can provide both real-time short-term information on air-flow and information useful for full system design.
4. Pressure response test. Typically, guidelines for flow rates in the range of 5–20 standard cubic feet per minute (SCFM) are appropriate for pressure response testing (Johnson et al., 2001e). As discussed below, gas injection equipment should be selected to exceed the hydrostatic pressure expected based on depth of the sparging well screen. In addition, it is common for injection pressures to exceed hydrostatic pressure for some period following start-up, so this should be taken into account. If injection pressures approach the soil over-burden pressure, then pneumatic fracturing of the subsurface is possible (U.S. Army, 2008). It is worth noting here that for safety reasons pilot tests can, and probably should, be conducted initially with air and non-reactive tracers, rather than the reactive gases that may ultimately be injected.
5. During steady flow injection in homogeneous porous media, groundwater pressures in the aquifer typically increase for a period of a few minutes before they return to near-hydrostatic conditions (Figure 8.4). In layered systems, particularly if the layers are continuous, the groundwater pressure can remain above hydrostatic pressure for periods of hours or even days. In both cases, the period of elevated pressure indicates the interval over which the total volume of gas in the subsurface is increasing. Thus, if elevated pressures persist for long periods, it is likely that a significant volume of gas is trapped below layers within the system. If, as is the case for many sites, the layers

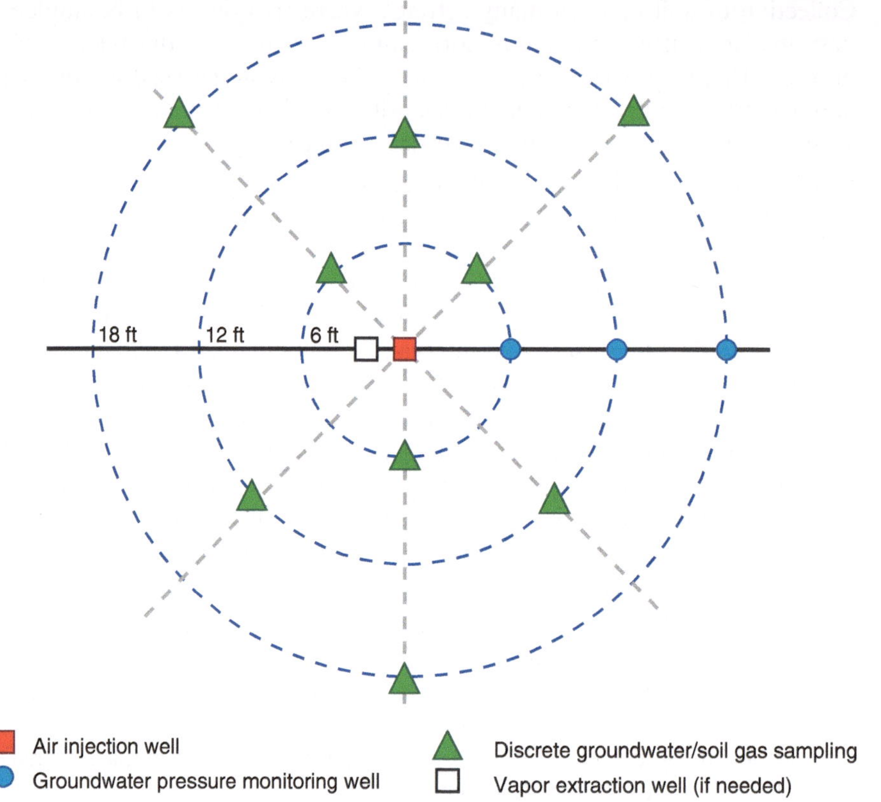

Figure 8.3. Schematic plan view layout of sparging, monitoring, vapor extraction and pressure monitoring wells (Johnson et al., 2010).

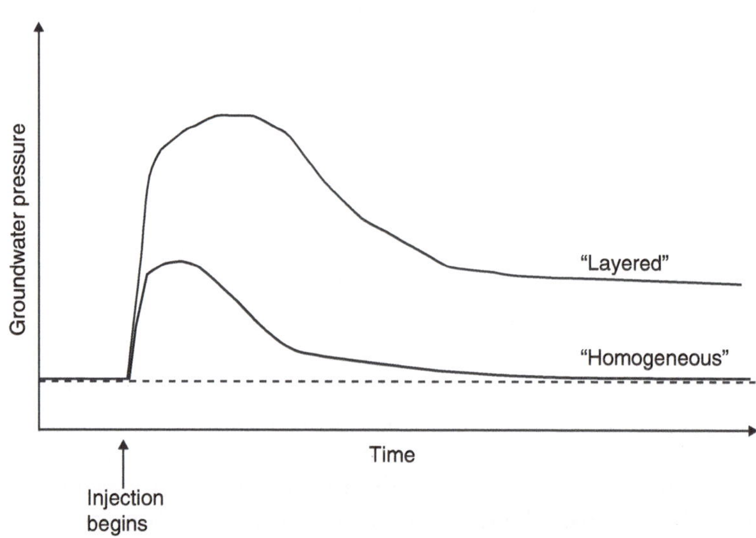

Figure 8.4. Conceptual plot of pressure versus time for sparging in homogeneous and layered aquifers.

are not completely horizontal, then lateral migration of gas under those layers can extend for hundreds of feet in the "up–dip" direction. In this context it is important to note that groundwater flow has little if any impact on movement of the air. From the perspective of engineering design, the time interval required for pressure to return to hydrostatic conditions approximates the time required to achieve quasi-steady-state gas flow, thus, this time interval is useful in determining the duration of gas injection "pulses".

6. Sparge gas recovery test. Lateral migration of gases is a distinct possibility in any sparging system and may be of particular concern when injecting reactive gases. To evaluate the extent to which lateral migration occurs, a relatively short air recovery test can be conducted. As described by Johnson et al. (2001d), this may involve the addition of helium to the injected gas, and subsequent measurement of its recovery by a soil vapor extraction (SVE) system. If the helium injection flow rate, the gas extraction rate and the helium concentration in the extracted gas are known, then a mass balance on the helium can be performed once the helium in the extracted gas reaches a steady concentration. This can occur in as little as an hour, but it is also possible that it will take much longer if significant accumulation of gas occurs in the subsurface.
7. Sparge gas distribution test. In many cases, the distribution of sparged gases in the subsurface can be estimated by measuring the dissolved gas concentration directly. However, in the case of reactive gases (e.g., oxygen), consumption of the gas can mask its distribution, particularly over the timeframes of typical pilot tests. In addition, it can be difficult to measure small changes in gas concentrations (again, especially for oxygen) if the gas is naturally present in the groundwater. As a result, it is often useful to conduct a gas distribution test in which a non-reactive tracer with low background concentrations is used. Historically the tracer of choice has often been sulfur hexafluoride (SF_6), (Bruce et al., 2001). The duration of these tests can vary from hours to days and involves the addition of small quantities of SF_6 to the sparge gas as it is injected into the subsurface. Water samples from a network of discrete monitoring points are then analyzed for the presence of the tracer. This test is primarily intended to be qualitative (i.e., detect the presence or absence of sparge gas) rather than quantitative. However, it is possible to estimate mass transfer information from the tests (Bruce et al., 2001).
8. Dissolved reactive gas measurements. As mentioned above, oxygen or other injected reactive gases can be used to examine sparge gas distribution. However, consumption of such gases by abiotic or biotic reactions may complicate the interpretation of occurrence data. On the other hand, the combination of a sparge gas distribution test and reactive gas measurements can help to understand the effectiveness with which the reactive gas is delivered to the treatment zone. For example, if a sparge gas distribution test following injection of air containing SF_6 shows the rapid appearance of SF_6 but no oxygen, even after extended operation, then it can be concluded that the oxygen demand by aquifer solids and groundwater may be so high that it will limit the potential for aerobic degradation of the target contaminants. This is particularly the case if contaminant reaction kinetics indicates days or weeks would be required to oxidize the target contaminants. In this case, it may be important to design the tests to examine the extent to which the reactive gas can be transported by groundwater advection to locations downgradient from the injection point.

8.4.3 System Design

The diagnostic pilot tests described above can provide a robust conceptual model for engineering design of a full-scale sparging delivery system. The focus of this section will be on those aspects of design that are most applicable to sparging as a delivery system.

1. Construction. If oxygen in air is the reactive gas to be delivered, then "standard" sparging system design and construction practices can be used (Johnson et al., 2010; U.S. Army, 2008). Since it will often be desirable to eliminate the need for an off-gas collection system, sparging delivery systems may be optimized by installing a larger number of more-closely-spaced sparge wells and to pulse the sparge gas. In this context, the use of direct-push well installation techniques can be very cost effective. If potentially-explosive or toxic gases are to be injected, then additional safety considerations must be taken into account (see item (4) below). The chemical compatibility of the injected gases with well construction materials should also be considered.
2. Operation. Since reactive gases may be both expensive and hazardous, it will often be desirable to minimize the injected volumes of such gases, while maximizing their effectiveness. In most cases this will involve pulsed injection, perhaps for only short periods of time (e.g., minutes). The duration of pulses can typically be determined based on the pressure and tracer data from the pilot test, coupled with longer-term monitoring of full system performance. The frequency of the pulses will depend, in large part, on the properties of the reactive gases and the desired delivery rate into the groundwater. For most gases of interest (oxygen, propane, butane, hydrogen), dissolution from gas mixtures trapped within the aquifer is likely to be the primary mechanism involved with mass transfer from the gas to the aqueous or aquifer solid phases. Such trapped gases may persist for many days, and as a result the intervals between pulses can be long. However, if mass transfer during inactive sparging periods is insufficient to accomplish remediation goals, then the pulsing frequency may need to be increased.
3. Calculations regarding residence time in the unsaturated zone. Another constraint on both duration and frequency of pulses is the acceptable residence time for injected gases in the unsaturated zone above the water table. In most regulatory contexts, in order to avoid the need for a vapor recovery system, calculations of residence time for vapors in the subsurface under proposed operating conditions need to be compared to the expected lifetimes of both contaminants and reactants in order to determine whether fugitive emissions will pose a health and/or safety problem. In addition with volatile contaminants, the unsaturated zone may be used as part of the remediation design to degrade volatile organics because oxygen is typically not a limiting reactant there.
4. Safety. Since many of the reactive gases of interest may have explosion and/or exposure limits, it is important to consider this in the design of sparging delivery systems. Table 8.1 lists the Lower-Explosion-Levels (LELs) and Recommended Exposure Limits (RELs) for a number of gases of potential interest for delivery using sparging (NIOSH, 2005). In addition, some combinations of injection gases (especially oxidants) and contaminants may result in increased risk of explosion. As a result, it may be important to install LEL meters and explosion-proof compressors/blowers.

8.4.4 System Operation and Maintenance

Most conventional air sparging systems can be operated with relatively little maintenance (e.g., periodic servicing of the air delivery blower, compressor, valves, and the vapor recovery system components). However, for most systems in which reactive gases are being injected

Table 8.1. Lower Explosion Levels (LELs) and Recommended Exposure Limits (RELs) for Selected Reactive Sparge Gases

Substance	NIOSH LEL (ppmV[a])	NIOSH REL (ppmV)
Acetylene	25,000	2,500
Ammonia	150,000	25
Butane	18,000	800
Ethanol	31,000	1,000
Hydrogen	40,000	–
Methane	50,000	–
Propane	21,000	1,000
Toluene	33,000	100

[a]*ppmV* parts per million by volume

additional attention may be required. This can be minimized if periods between pulses are on the order of days or weeks. During injection periods, system parameters (temperature, flow rate, reactive gas injection concentration, etc.) should be monitored. In addition, it may be necessary to actively monitor fugitive emissions to ensure that reactive gases do not create a safety hazard. (This can be minimized if injection concentrations are kept below the LEL. However, if the reactive gas is delivered from a pressurized tank, failure of the air injection system can still lead to release of explosive/hazardous concentrations of reactive gases).

The details of system operation will depend upon a large number of factors, including subsurface conditions, contaminant type and distribution, reaction rates, safety and other factors. For a number of these, actual site data will be required, and as a consequence well designed pilot tests are considered critical. In addition, as discussed next, ongoing performance monitoring of the system will be important to ensure that both safety and effectiveness are maintained during operation.

8.4.5 Performance Monitoring

Routine monitoring of groundwater will generally be an integral part of sparging system operation. In the case where reactive gases are injected, this should include measurement of dissolved concentrations of these gases as well as the contaminants of concern. Because of the potential for active injection to impact measurements in monitoring wells, it is generally recommended that measurements be made after the sparging system has been turned off for some period (e.g., just prior to initiation of a pulse if the period between pulses is on the order of days or after the system has been turn off for several days) and that discrete interval monitoring points be used when possible.

8.5 CASE STUDIES

8.5.1 Air Biosparging: Environmental Security Technology Certification Program (ESTCP) Multi-Site *In Situ* Air Sparging Project

The Multi-Site *In Situ* Air Sparging Project (ESTCP 2002a, b) documents performance of "conventional" sparging and biosparging applications at a number of sites. Some of the sites were on-going Department of Defense (DoD) remediation sites and some were sites where

Table 8.2. Diagnostic Tests Conducted at Selected DoD Sites as Part of the ESTCP Multi-site Air Sparging Project

	Hydraulic pressure response test	Air recovery test (helium)	Air distribution test (SF_6)	Reactive gas delivery test (oxygen)
Ammonia	X	X	X	X
Butane	X	X		
Ethanol	X	X	X	
Hydrogen	X			
Propane	X	X		

sparging/biosparging was proposed as a remedy. In most cases the objectives of sparging systems has been to remove contaminant mass by volatilization and to enhance aerobic biodegradation. (However, in several cases the intent was to accomplish both of these while minimizing fugitive emissions in order to avoid the need for an off-gas collection system).

A primary objective of this project was to develop tools to assess design and performance at sparging sites, and based on that to develop guidelines for implementation of sparging at large (e.g., DoD) sites. This resulted in the suite of diagnostic tools discussed above (Section 8.4.2).

In the context of sparging as a delivery system, it is most useful to look at the range of test results observed at these sites (Table 8.2) and to discuss briefly the implications of those tests.

8.5.1.1 Pressure Response Test Data

Data from five sites are shown in Figure 8.5. These tests provided an initial assessment of whether or not the presence of stratagraphic layering will affect air distribution at the site. In Figure 8.5a (Fairchild Air Force Base [AFB], Washington), pressure increases due to the injection of air are short-lived and for most monitoring points the pressure dropped temporarily below hydrostatic pressure after about 4 min. This is characteristic of a system without significant layering. In this case, air injection caused the initial pressure increase, but once the air began to break through the water table and into the unsaturated zone, the air "pocket," temporarily formed in the medium, collapsed. As water returned to fill the pocket space, the pressure dropped below hydrostatic pressure until water and air flow became stabile. In significant contrast to the first example, at Hill AFB, Utah (Figure 8.5b), subsurface pressure remained elevated for a period of days reflecting the sustained accumulation of air beneath the strata at the site. Figure 8.5c–e (Eielson AFB, Alaska; McClellan AFB, California; and Port Hueneme Naval Air Station [NAS], California) present intermediate cases where air accumulation persisted for hours.

8.5.1.2 Sparge Gas Recovery Test Data

As discussed above, sparge gas recovery tests can be used both to assess the performance of an SVE system for capturing sparge air, and as a measure of the lateral extent to which sparge air moves from the injection well. Figure 8.6a shows helium recovery for the air sparging/SVE system at Eielson AFB. In this case, injected helium was rapidly detected at the SVE system, and essentially all of the injected helium was recovered. These data, taken together with the pressure response data (Figure 8.5c) suggest that there is some layering (as evidenced

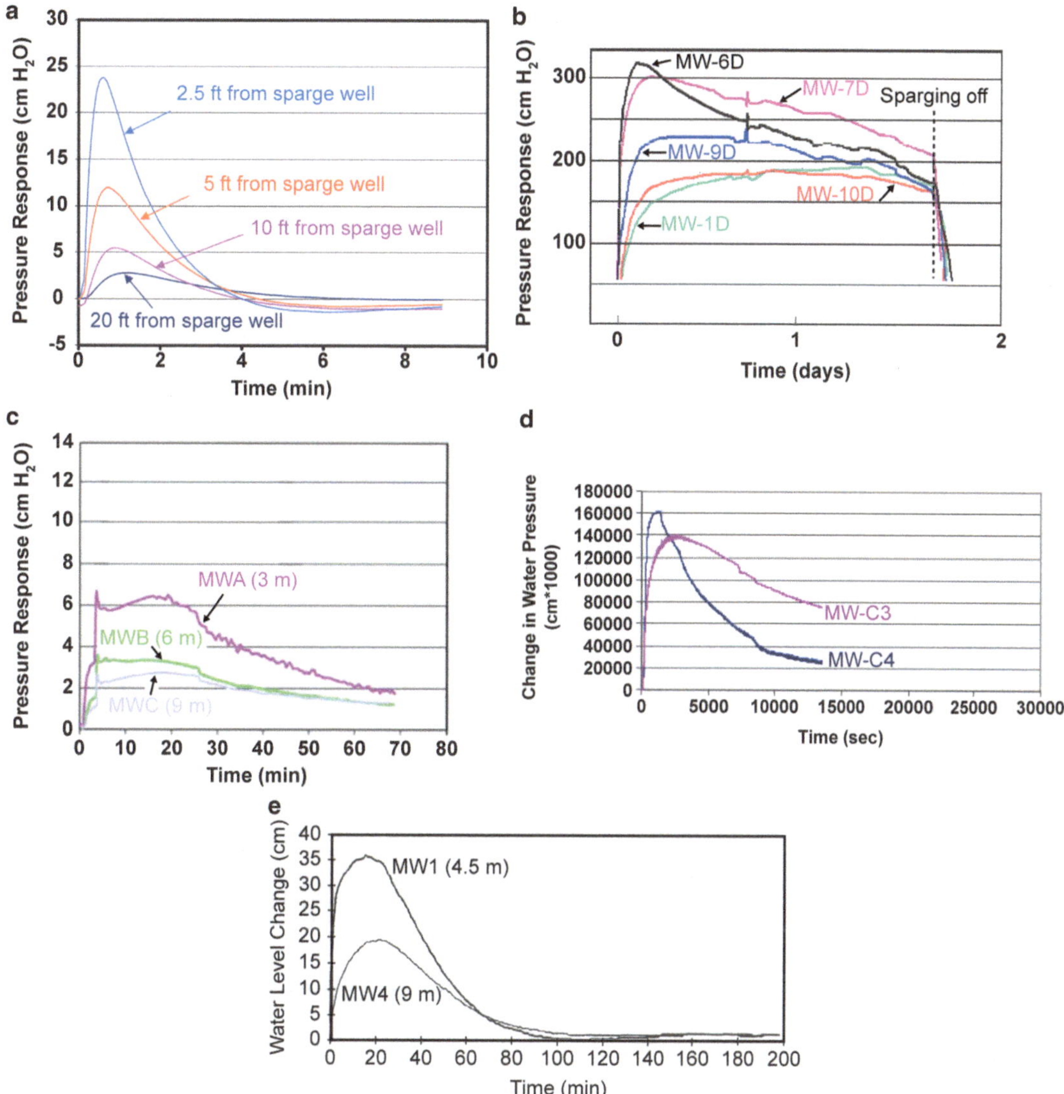

Figure 8.5. Pressure responses during sparging well startup at: (a) Fairchild AFB, Washington; (b) Hill AFB, Utah; (c) Eielson AFB, Alaska; (d) McClellan AFB, California; and (e) Port Hueneme NAS, California. Parts (a) and (d) from ESTCP, 2002a; parts (b), (c), and (e) from ESTCP, 2002b.

by the pressure data), but that the lateral extent of the layering is small compared to the lateral extent of the SVE system. Figure 8.6b, again from Eielson AFB, shows the aerial distribution of helium in the unsaturated zone during air sparging at 10 SCFM. Air recovery data from Hill AFB (Figure 8.6c) show that little of the injected air was recovered by the SVE system. In this case, most of the sparge air moved laterally beyond the reach of the SVE system and found a short circuit through overlying layers directly to ground surface via a monitoring well. This is not an uncommon result, and points to the importance of conducting such tests, particularly when there is the risk of causing adverse health effects. The air recovery test at Port Hueneme (Figure 8.6d, e) shows the effects of flow rate on helium recovery. In this case, an increased air injection rate caused air to break through a confining layer that trapped air when using a lower flow rate. It is important to note that these changes do not necessarily reflect an improved

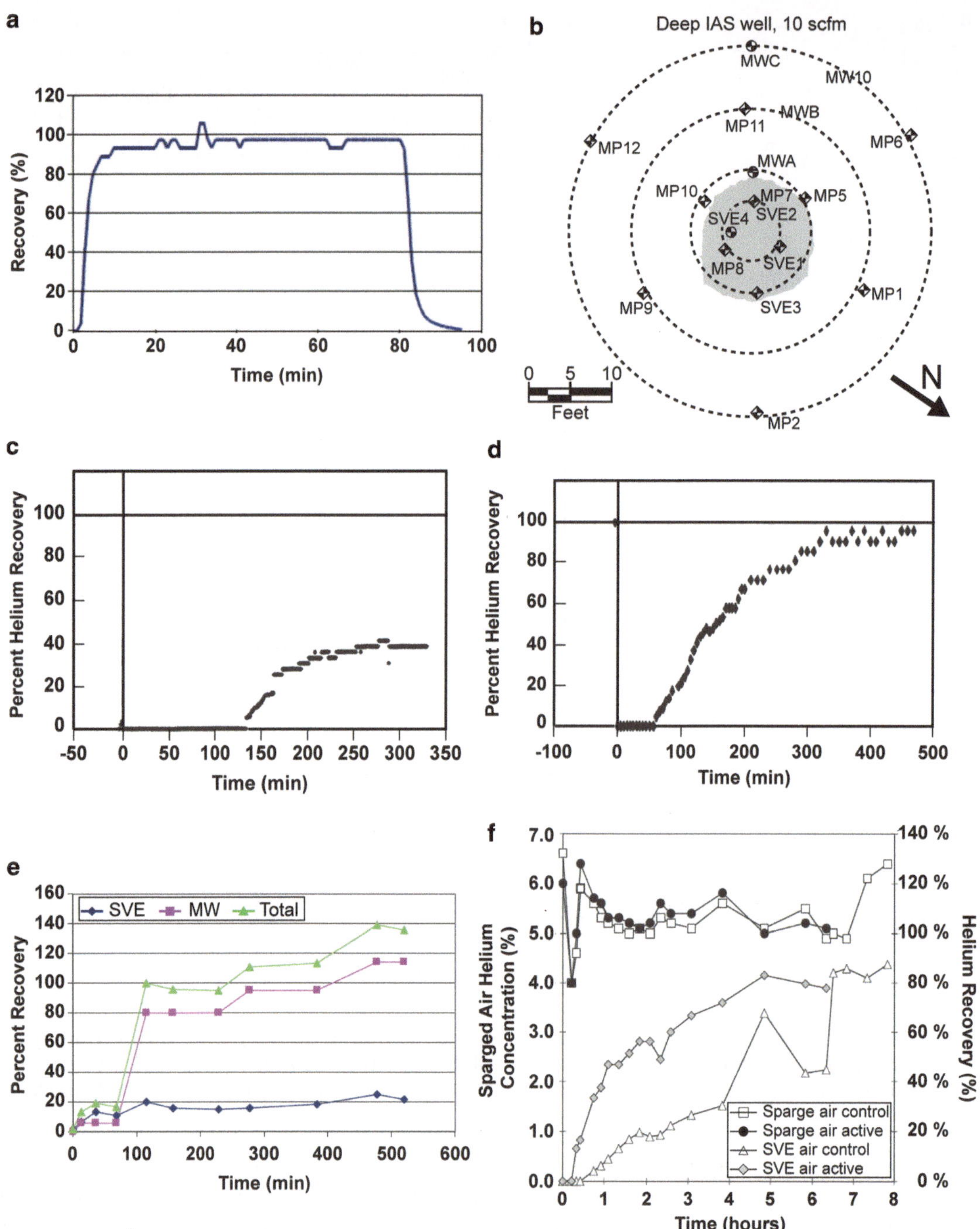

Figure 8.6. Helium recovery measurements at: (a) Eielson AFB; (b) aerial distribution of helium measured in the unsaturated zone at Eielson AFB; (c) Port Hueneme NAS at 5 scfm; (d) Port Hueneme NAS at 10 scfm; (e) Hill AFB; and (f) McClellan AFB. Parts (a), (b), and (e) from ESTCP, 2002a; parts (c) and (d) from ESTCP, 2002b; part (f) from ESTCP, 2001.

efficiency of oxygen transfer to groundwater. This assessment should be carried out as part of sparge gas distribution tests. The final air recovery test (McClellan AFB, Figure 8.6f) is typical of many tests. In this case, most of the injected air was ultimately recovered, however, the relatively long delay noted may, in part, be due to significant lateral spreading of the injected sparge air.

8.5.1.3 Gas Distribution Test Data

As discussed above, injected gas is rarely dispersed uniformly around injection wells (either laterally or vertically). This can be due to a variety of processes that are difficult to predict, even with careful field measurements. As a result, gas distribution tests are important for assessing actual ISS performance. The data in Figure 8.7b show the percent saturation of SF_6 measured in groundwater at discrete depths intervals around a sparge will at Port Hueneme. These data demonstrate where a reactive gas could be delivered under the conditions tested. (Figure 8.7a shows dissolved oxygen concentrations at the same locations). There is general agreement between the two, although significant differences do occur. In a number of cases, SF_6 is present but oxygen is not. (This is probably due to oxygen removal from the groundwater in this naturally-anoxic system). Figure 8.7b shows a condition not uncommon at sparging sites. In this case, essentially all of the sparge air appears to move in one general direction from the sparge well. This kind of behavior can occur, for example, if stratagraphic bedding at the site dips at an angle of more than a few degrees, in which case injected gas will move "up–dip" beneath confining layers.

The data presented in these examples point to the complexity of gas movement in the subsurface and the importance of characterizing movement from both performance and safety perspectives. As discussed below, the latter is particularly important when injecting reactive gases.

8.5.2 Oxygen Biosparging: Methyl Tertiary-Butyl Ether (MTBE) Biodegradation at Port Hueneme NAS, California

Johnson et al. (2003) conducted a full-scale study to assess the effectiveness of the injection of pure oxygen to facilitate aerobic degradation of MTBE while minimizing injection volumes in order to avoid the need for a fugitive vapor recovery system. The target treatment area was within a broad groundwater plume of MTBE, and the goal was to "cut-off" movement of the MTBE from the source zone into the emerging groundwater plume.

Unlike most sparging treatment systems, this application involved the addition of mixed- and single- microbial cultures that were capable of degrading MTBE (Salanitro et al., 2000). Since groundwater at the site was naturally anoxic, and the microbes were strict aerobes, it was essential that sparging provide a consistent dissolved oxygen concentration.

8.5.2.1 System Configuration

In part due to the need for constant aerobic conditions, sparging was conducted at two depths (18–20 ft below ground surface [bgs] and 14–15 ft bgs), with wells placed at close spacing across the plume (approximately 2 ft apart, Figure 8.8). This approach was practical because small-diameter sparge wells could be rapidly installed at the site using direct-push technology. Two additional biobarrier systems were subsequently installed with larger gas injection well spacings, with equal success.

As mentioned above, since it was important to minimize fugitive gas emissions at this site, gas injections were of short duration with relatively wide intervals in between injections (30 s of injection once every 6 h). Subsequent pilot-scale experiments conducted on site demonstrated that more widely-spaced injection intervals of a few weeks should also be successful. To minimize equipment requirements, a 20-gallon oxygen storage tank associated with approximately six wells was charged to approximately 45 pounds per square inch (psi) using an oxygen generator, and then its contents were discharged at once to two wells. This was

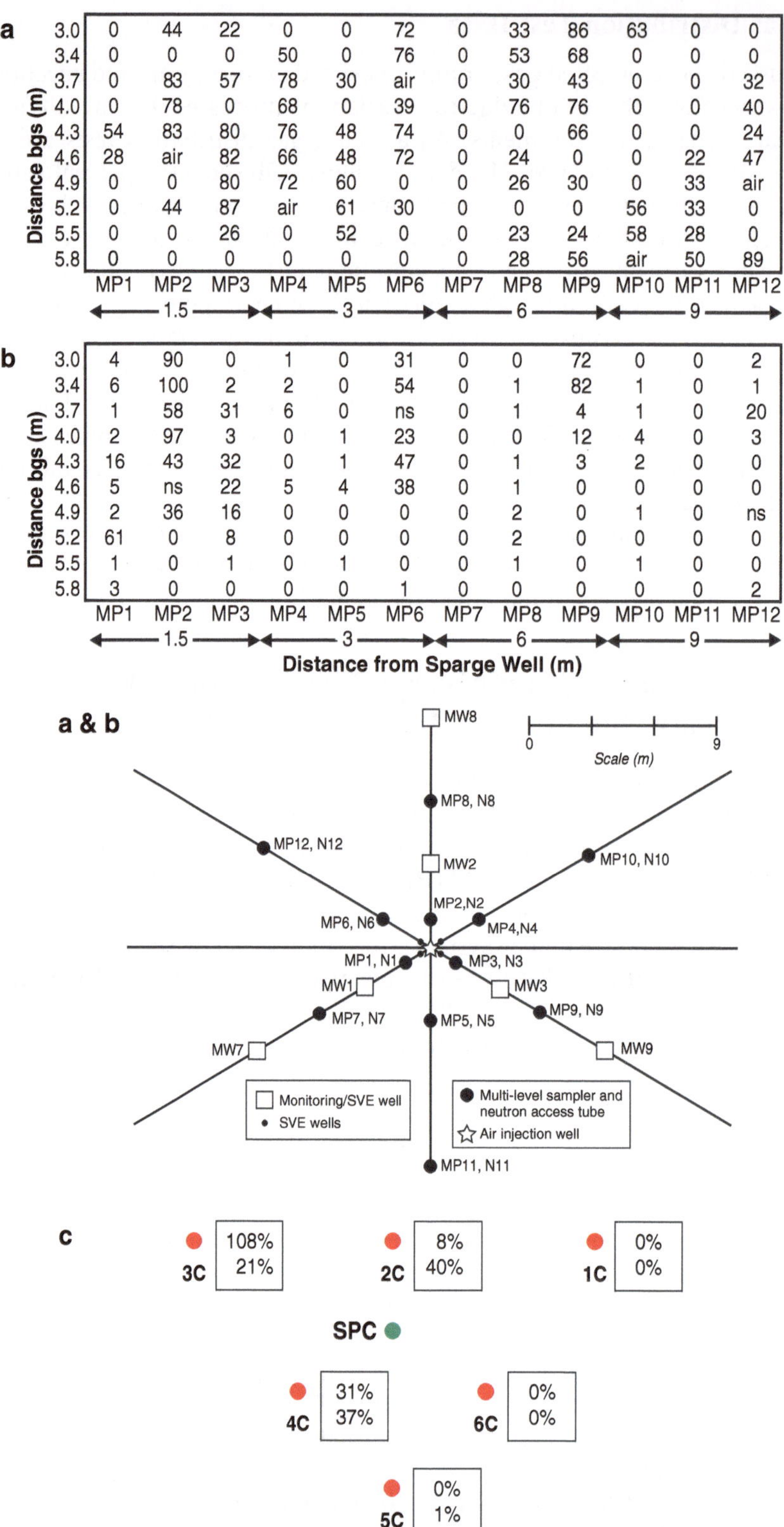

a

Distance bgs (m)	MP1	MP2	MP3	MP4	MP5	MP6	MP7	MP8	MP9	MP10	MP11	MP12
3.0	0	44	22	0	0	72	0	33	86	63	0	0
3.4	0	0	0	50	0	76	0	53	68	0	0	0
3.7	0	83	57	78	30	air	0	30	43	0	0	32
4.0	0	78	0	68	0	39	0	76	76	0	0	40
4.3	54	83	80	76	48	74	0	0	66	0	0	24
4.6	28	air	82	66	48	72	0	24	0	0	22	47
4.9	0	0	80	72	60	0	0	26	30	0	33	air
5.2	0	44	87	air	61	30	0	0	0	56	33	0
5.5	0	0	26	0	52	0	0	23	24	58	28	0
5.8	0	0	0	0	0	0	0	28	56	air	50	89
	←	1.5	→	←	3	→	←	6	→	←	9	→

b

Distance bgs (m)	MP1	MP2	MP3	MP4	MP5	MP6	MP7	MP8	MP9	MP10	MP11	MP12
3.0	4	90	0	1	0	31	0	0	72	0	0	2
3.4	6	100	2	2	0	54	0	1	82	1	0	1
3.7	1	58	31	6	0	ns	0	1	4	1	0	20
4.0	2	97	3	0	1	23	0	0	12	4	0	3
4.3	16	43	32	0	1	47	0	1	3	2	0	0
4.6	5	ns	22	5	4	38	0	1	0	0	0	0
4.9	2	36	16	0	0	0	0	2	0	1	0	ns
5.2	61	0	8	0	0	0	0	2	0	0	0	0
5.5	1	0	1	0	1	0	0	1	0	1	0	0
5.8	3	0	0	0	0	1	0	0	0	0	0	2
	←	1.5	→	←	3	→	←	6	→	←	9	→

Distance from Sparge Well (m)

Figure 8.7. (a) Dissolved oxygen saturations (%) measured during sparging at Port Hueneme NAS; (b) SF_6 saturations (%) measured at Port Hueneme NAS; and (c) SF_6 saturations (%) measured at two depths at McClellan AFB. The portion labeled "a & b" shows a plan view of the locations of the monitoring points in parts a and b. Parts (a) and (b) from ESTCP, 2002b; part (c) from ESTCP, 2002a.

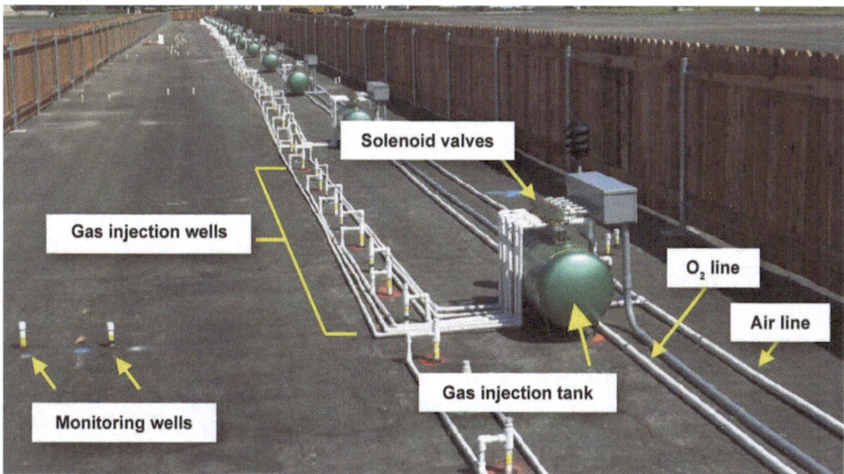

Figure 8.8. Photograph showing the oxygen sparging system installed at Port Hueneme NAS (Johnson et al., 2003).

accomplished on a schedule that was dictated by the rate that oxygen could be produced, but as the data discussed below indicate, the 6-h interval was adequate to keep the groundwater oxygenated.

8.5.2.2 Oxygen Concentration Data

From the point of view of sparging as a mixing system, perhaps the most important aspect of this project is its ability to deliver high dissolved oxygen concentrations over a contiguous area. As the plan-view data in Figure 8.9 indicate, this was successful over most of the length of the sparging "barrier."

8.5.2.3 Contaminant Removal Data

Data in Figure 8.10 indicate that essentially all of the MTBE was removed by the sparging system.

8.5.3 Cometabolic Biosparging: McClellan AFB, California

One of the best-documented demonstrations of cometabolic biosparging is the injection of propane at McClellan AFB near Sacramento, California (ESTCP, 2001). Propane stimulated cometabolism of chlorinated compounds represents a conceptually simple example of cometabolic biosparging. Because of the wide usage of propane as a fuel, it is relatively easy to design and permit these systems. In principle, cometabolic sparging requires simply the addition of a suitable carbon/energy source to stimulate the microbiological processes that lead to the removal of otherwise-recalcitrant compounds.

In practice, as with all sparging applications, this approach can be challenging because of the complexity of the subsurface (e.g., aquifer heterogeneity). It can also be challenging because more than one nutrient may be necessary to stimulate microbial growth (e.g., both a nitrogen and a carbon source). A third complexity is the frequently limited availability of oxygen in groundwater (due in part to its limited solubility). As a consequence, it can be

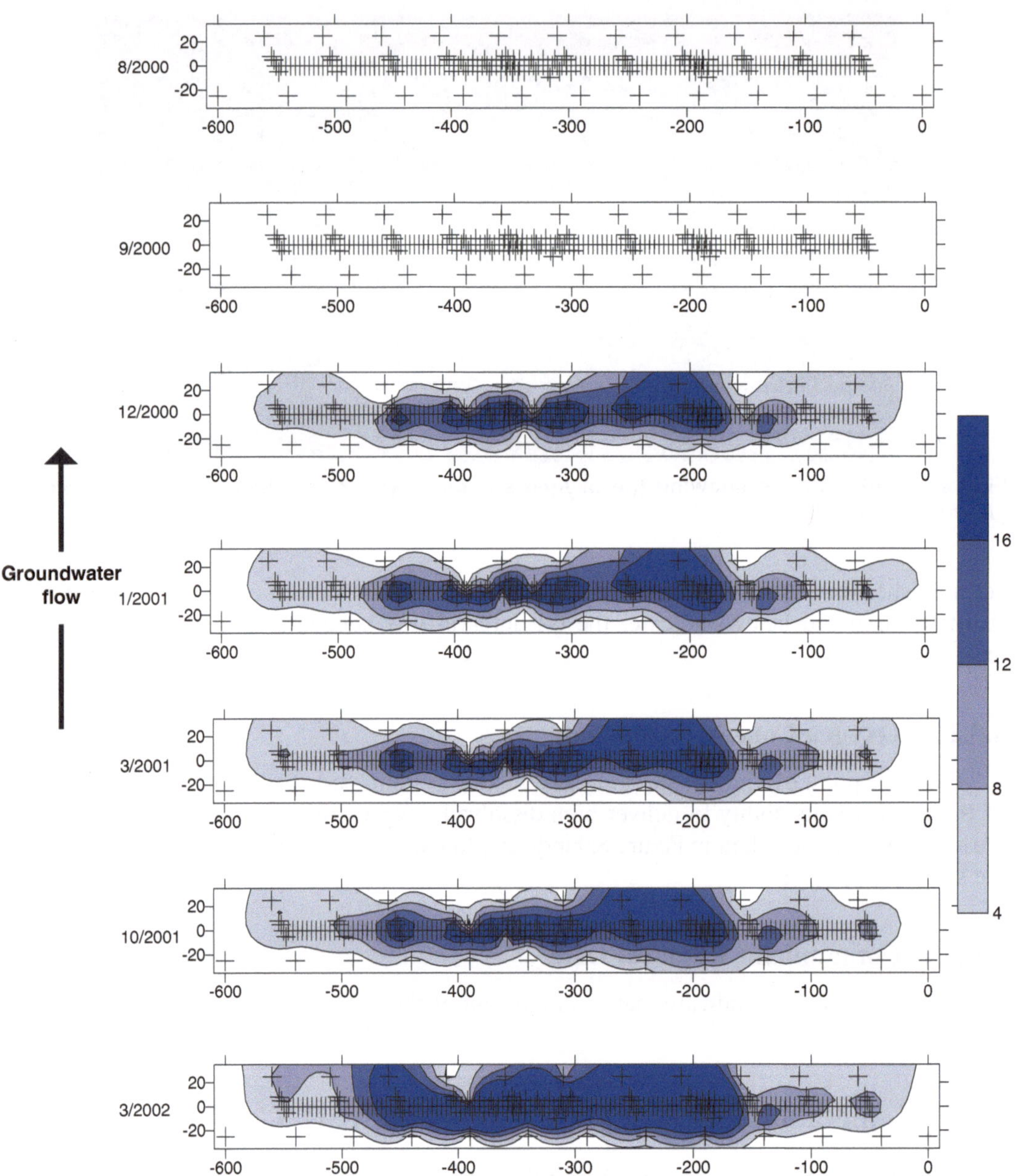

Figure 8.9. Plan-view plots of dissolved oxygen concentrations (mg/L) at several time snapshots at the Port Hueneme NAS site (Johnson et al., 2003). Lateral dimensions for monitoring well locations are in feet (ft).

important to design the sparging system in a manner that utilizes the unsaturated zone as a bioreactor.

8.5.3.1 System Configuration

Figure 8.11 shows the general configuration of the site. A central sparging well was surrounded by multi-level monitoring wells. In addition, at this site there was an SVE system. Because the unsaturated zone at this site was quite thick (>100 ft), it was anticipated that the

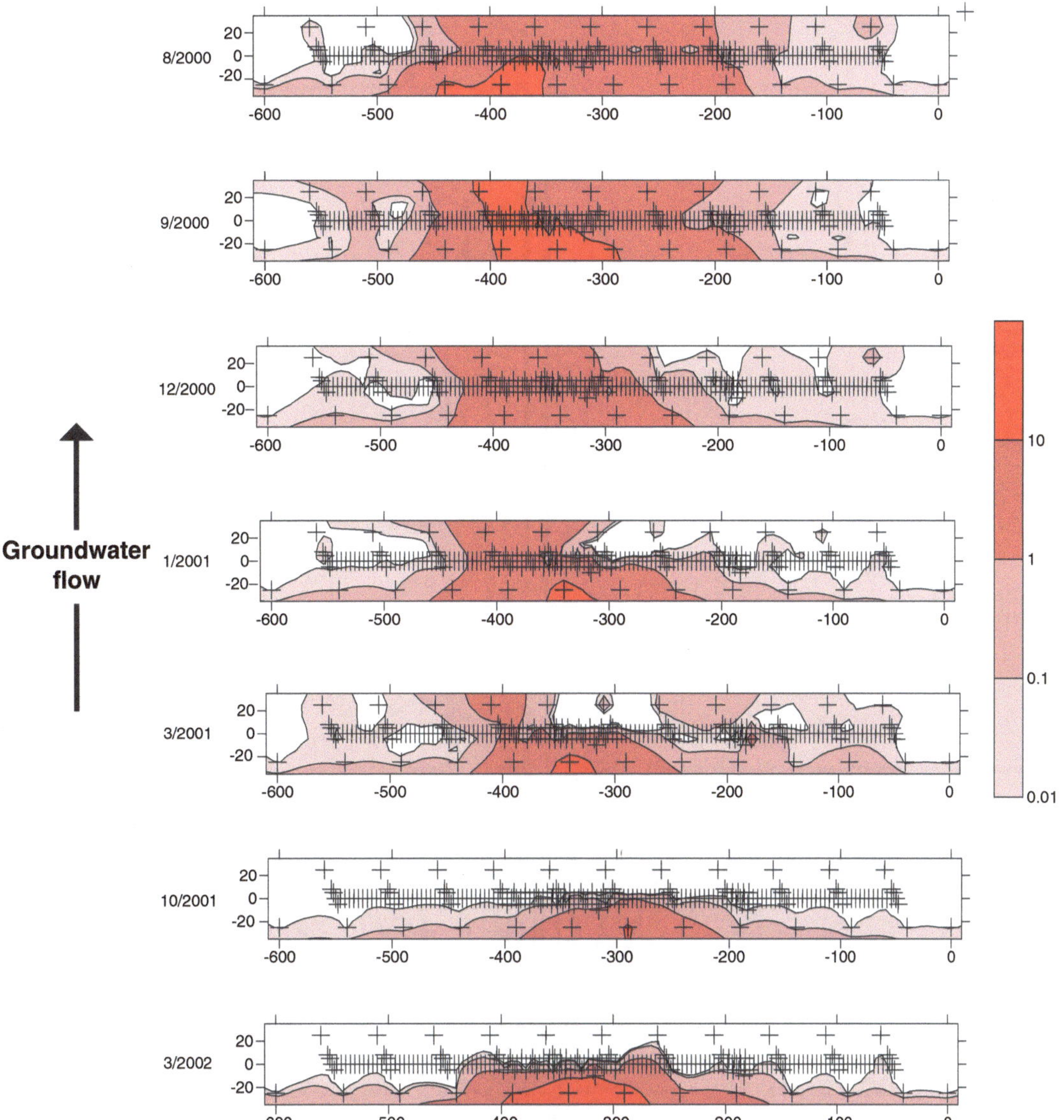

Figure 8.10. Plan view MTBE concentrations (mg/L) at several time snapshots at Port Hueneme NAS (Johnson et al., 2003). Lateral dimensions for monitoring well locations are in feet (ft).

unsaturated zone would serve as a bioreactor (in particular, because the oxygen concentration would not be limited in that portion of the system).

The site was divided into two portions. One was designated a control site in which sparging occurred without the addition of propane or other nutrients. The other portion of the site received propane and, for a limited period of time, ammonia.

8.5.3.2 Tracer Tests to Assess Cometabolic Performance

As with all sparging activities, prior to the initiation of cometabolic air sparging it is advisable to conduct tracer tests to assess both the behavior of injected gases and their impact

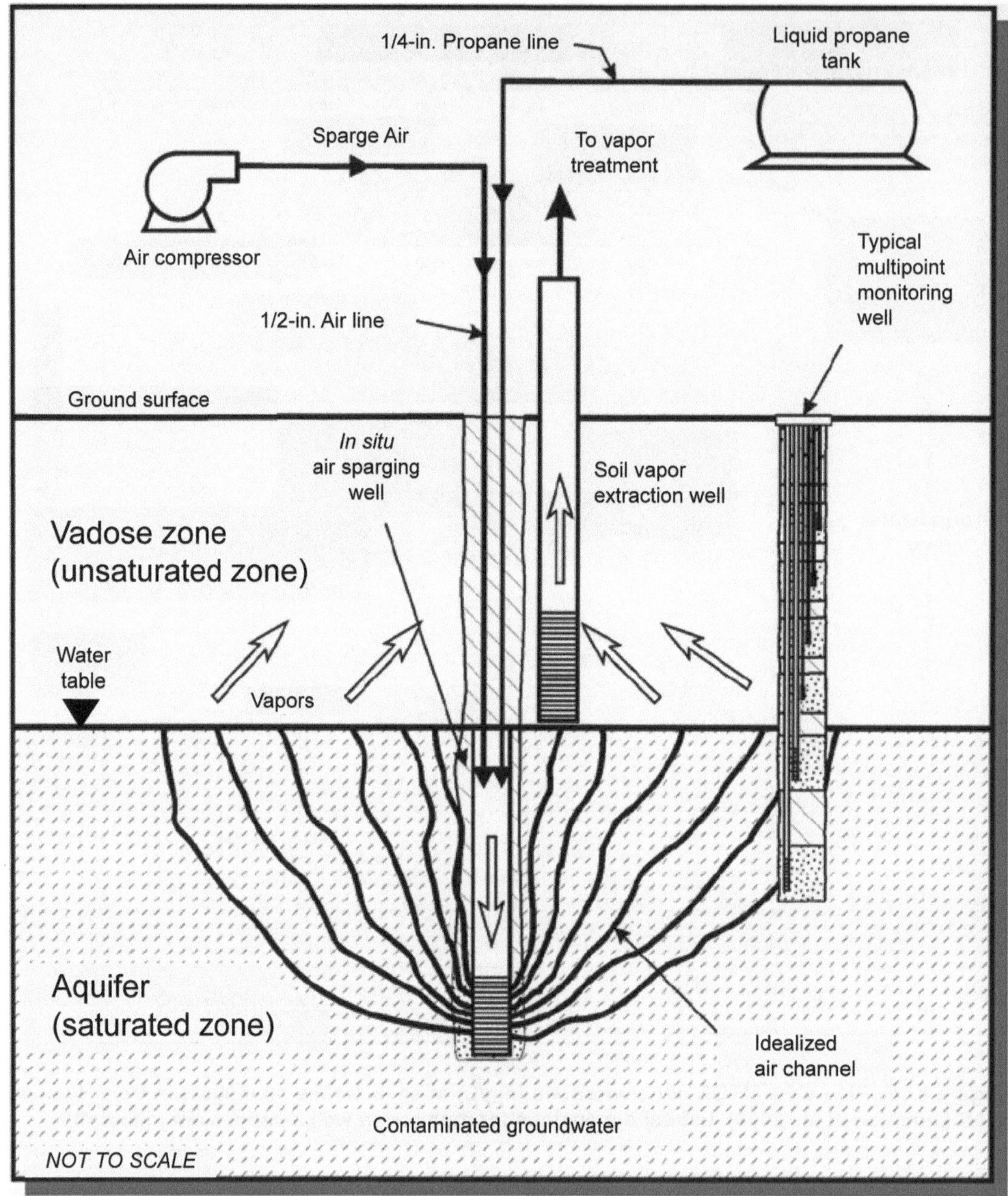

Figure 8.11. Schematic drawing of the cometabolic sparging system used at McClellan AFB (ESTCP, 2001).

on degradation of contaminants. As discussed above, both air recovery and air distribution tests were conducted at the McClellan AFB site. In addition, at this site a "push–pull" test for evaluating the aerobic cometabolism of chlorinated aliphatic hydrocarbons was conducted (ESTCP, 2005). The push–pull tests demonstrated that injection of propane could stimulate microbial growth and oxygenation of surrogate compounds (ethylene and propylene), however, it was concluded that "transformation of *cis*-DCE and trichloroethene (TCE) proved more difficult to assess".

8.5.3.3 Contaminant Removal Data

The difficulty in interpretation of cometabolic sparging at the site arose, in large part, to the challenges in implementing any sparging system. As discussed above Figure 8.7c, sparging was not effective at delivering gases to all of the monitoring wells, with apparent directionally-preferential flow. In some wells (e.g., MW-C4-113 in Figure 8.12a) significant reductions in both TCE and *cis*-1,2-dichloroethylene (*cis*-DCE) concentrations were observed. However, the TCE concentrations remained well above the cleanup objective. As might be expected from the air distribution tracer test, some of the wells in the active treatment area did not receive any injected propane (e.g., well MW-C1-113 in Figure 8.12b), and as a consequence, there was no significant reduction in contaminant concentrations. In contrast, at the control site at monitoring locations where air was known to be present (again based on the air distribution test, Figure 8.7c), significant concentration reductions were observed without the addition of cometabolites. These data suggest that volatilization of the contaminants probably played a significant role in concentration reductions. The extent to which this was also the case at the active site is not known.

8.5.3.4 Unsaturated Zone "Bioreactor"

The site has a relatively thick unsaturated zone, and as a consequence it was anticipated that chlorinated aliphatic hydrocarbon degradation could be effective there (i.e., both oxygen and propane could be maintained at high concentrations). However, with a few exceptions, the rates of consumption of contaminants in the unsaturated zone appeared to be quite slow. It was hypothesized that this could have been due to nitrogen limitations in the soil. To test this, ammonia was sparged into the groundwater along with the propane. However, stimulation with ammonia did not appear to significantly enhance degradation rates. In addition, it was concluded that "continuous ammonia addition was not possible, because ammonia addition led to a significant pH increase above pH 10 in some monitoring points." Nevertheless, the use of sparging to deliver reactants to the unsaturated zone to enhance biodegradation may be effective at some sites.

8.6 SUMMARY

As the last example demonstrates, it can be very difficult to implement ISS as a delivery system and it can be difficult to assess the relative effectiveness of enhanced degradation compared to volatilization. Nevertheless, delivery of reactants in addition to oxygen may represent an important tool for subsurface restoration at sites contaminated with recalcitrant compounds such as chlorinated solvents. However, to an even greater extent than for conventional air sparging, the addition of reactive gases makes it important to conduct pilot tests to improve the conceptual model of sparge gas flow at the site and to consider safety in the design of the air injection system.

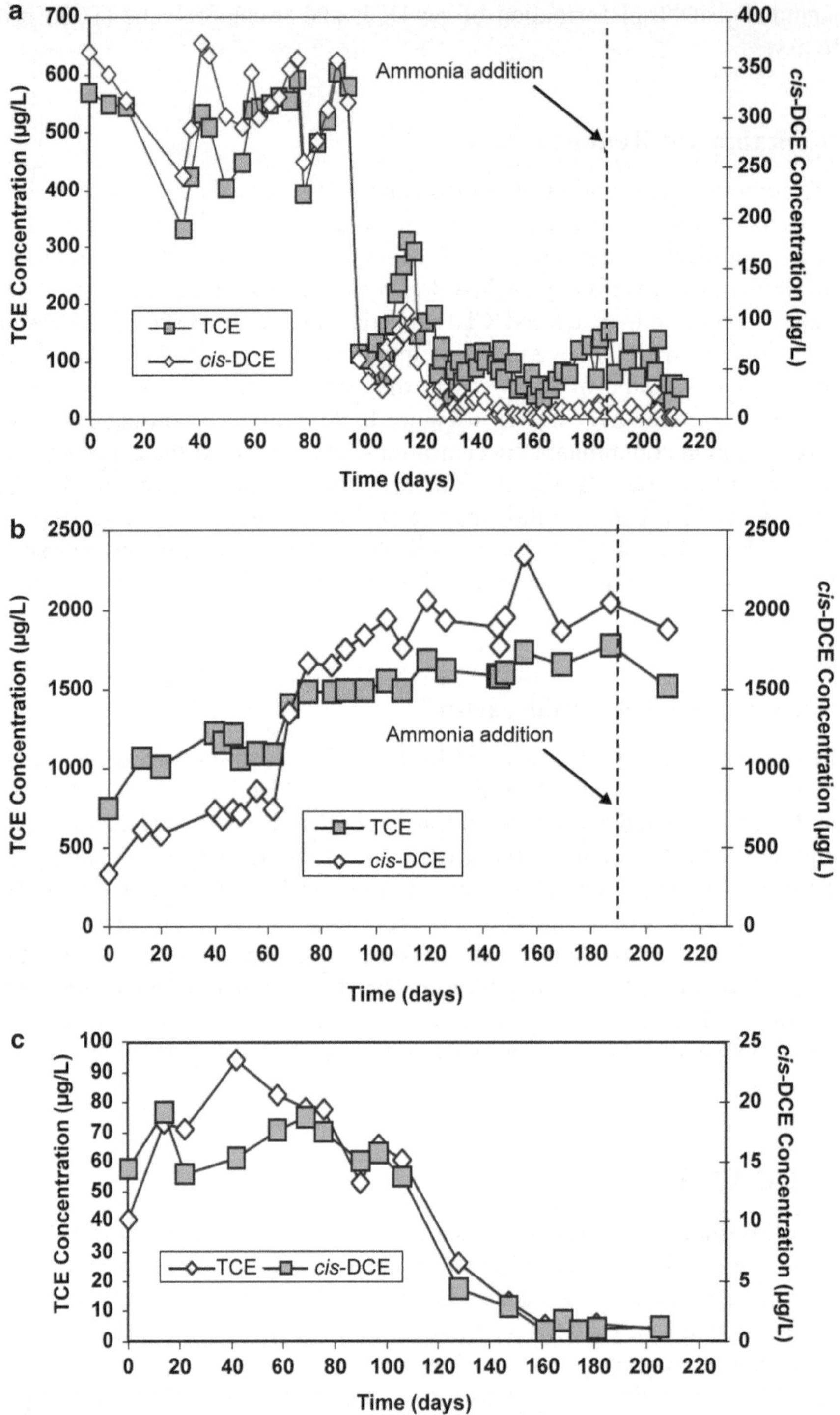

Figure 8.12. Dissolved contaminant concentrations as a function of time at McClellan AFB: (a) active zone well MW-C4-113, (b) active zone well MW-C1-113; and (c) control zone well MW-A1-113 (ESTCP, 2001).

REFERENCES

Battelle. 2001. Final Air Sparging Guidance Document. Naval Facilities Engineering Service Center (NFESC) Technical Report TR-2193-ENV.

Bruce CL, Amerson IL, Johnson RL, Johnson PC. 2001. Use of an SF6-based diagnostic tool for assessing air distributions and oxygen transfer rates during IAS operation. Bioremediation J 5:337–347.

Clayton WS. 1998. A field and laboratory investigation of air fingering during air sparging. Ground Water Monit Remediat 17:134–145.

Clayton WS. 1999. The effects of pore scale dead-end air fingers on relative permeabilities for air sparging in soils. Water Resour Res 35:2909–2919.

ESTCP (Environmental Security Technology Certification Program). 2001. Use of Cometabolic Air Sparging to Remediate Chloroethene-Contaminated Groundwater Aquifers: Project CU-9810 Final Report. Prepared by Battelle, Columbus, OH, USA. http://www. serdp.org/content/advancedsearch/. Accessed August 5, 2011.

ESTCP. 2002a. Multi-Site In Situ Air Sparging: Project CU-9808 Final Report. Prepared by Battelle, Columbus, OH, USA. http://www.serdp.org/content/advancedsearch/. Accessed August 5, 2011.

ESTCP. 2002b. Design Paradigm: Air Sparging – Technology Transfer and Multi-Site Evaluation. Prepared by Leeson A, Johnson PC, Johnson RL, Vogel CM, Hinchee RE, Marley M, Peargin T, Bruce CL, Amerson IL, Coonfare CT, Gillespie RD, McWhorter DB for ESTCP, Arlington, VA, USA. http://www.serdp.org/content/advancedsearch/. Accessed August 5, 2011.

ESTCP. 2003. In-Situ Remediation of Methyl tert-Butyl Ether (MTBE) Contaminated Aquifers Using Propane Biosparging: Project ER-200515 Final Report. Prepared by Envirogen, Inc., Lawrenceville, NJ, USA. http://www.serdp.org/content/advancedsearch/. Accessed August 5, 2011.

ESTCP. 2005. Push-Pull Tests for Evaluating the Aerobic Cometabolism of Chlorinated Aliphatic Hydrocarbons. Project CU-9921Final Report. Prepared by Semprini L, Istok J, Azizian M, Kim Y; Oregon State University, Corvallis, OR, USA. http://www.serdp.org/content/advancedsearch/. Accessed August 5, 2011.

Johnson PC. 1998. An assessment of the contributions of volatilization and biodegradation to in situ air sparging performance. Environ Sci Technol 32:276–281.

Johnson RL, Johnson PC, McWhorter DB, Hinchee RE, Goodman I. 1993. An overview of in situ air sparging. Ground Water Monit Remediat 13:127–135.

Johnson PC, Johnson RL, Neaville C, Hansen CC, Stearns SM, Dortch IJ. 1997. An assessment of conventional in situ air sparging pilot tests. Ground Water 35:765–774.

Johnson PC, Das A, Bruce CL. 1999. Effect of flow rate changes and pulsing on the treatment of source zones by in situ air sparging. Environ Sci Technol 33:1726–1731.

Johnson PC, Johnson RL, Bruce CL, Leeson A. 2001a. Advances in in situ air sparging/biosparging. Bioremediation J 5:251–266.

Johnson PC, Leeson A, Johnson RL, Vogel CM, Hinchee RE, Marley M, Peargin T, Bruce CL, Amerson IL, Coonfare CT, Gillespie RD. 2001b. A practical approach for the selection, pilot testing, design, and monitoring of in situ air sparging/biosparging systems. Bioremediation J 5:267–281.

Johnson RL, Johnson PC, Amerson IL, Johnson TL, Bruce CL, Leeson A, Vogel CM. 2001c. Diagnostic tools for integrated in situ air sparging pilot tests. Bioremediation J 5:283–298.

Johnson RL, Johnson PC, Johnson TL, Leeson A. 2001d. Helium tracer tests for assessing contaminant vapor recovery and air distribution during in situ air sparging. Bioremediation J 5:321–336.

Johnson RL, Johnson PC, Johnson TL, Thomson NR, Leeson A. 2001e. Diagnosis of in situ air sparging performance using transient groundwater pressure changes during startup and shutdown. Bioremediation J 5:299–320.

Johnson PC, Bruce CL, Miller KD. 2003. In Situ Bioremediation of MTBE in Groundwater. NFESC Technical Report TR-2222-ENV. Prepared for ESTCP, Arlington, VA, USA.

Johnson PC, Johnson RL, Bruce CL. 2010. Air sparging for the treatment of chlorinated solvent plumes. In Stroo HF, Ward CH, eds, *In Situ* Remediation of Chlorinated Solvent Plumes. Springer Science + Business Media, New York, NY, USA, pp 455–480.

Johnston CD, Rayner JL, Patterson BM, Davis GB. 1998. Volatilisation and biodegradation during air sparging of dissolved BTEX-contaminated groundwater. J Contam Hydrol 33:377–404.

McCray JE, Falta RW. 1997. Numerical simulation of air sparging for remediation of NAPL contamination. Ground Water 35:99–110.

NIOSH (National Institute of Occupational Safety and Health). 2005. NIOSH Pocket Guide to Chemical Hazards. http://www.cdc.gov/Niosh/npg/.

Rutherford K, Johnson PC. 1996. Effects of process control changes on aquifer oxygenation rates during in situ air sparging in homogeneous aquifers. Ground Water Monit Remediat 16:132–141.

Salanitro JP, Johnson PC, Spinnler GE, Maner PM, Wisniewski HL, Bruce CL. 2000. Field-scale demonstration of enhanced MTBE bioremediation through aquifer bioaugmentation and oxygenation. Environ Sci Technol 34:4152–4162.

SERDP (Strategic Environmental Research and Development Program). 2003. Low-Volume Pulsed Biosparging of Hydrogen for Bioremediation of Chlorinated Solvent Plumes. ER-1206 Final Report. Prepared by Groundwater Services, Inc., Houston, TX, USA. http://www.serdp.org/content/advancedsearch/. Accessed August 5, 2011.

Thomson NR, Johnson RL. 2000. Air distribution during *in situ* air sparging: An overview of mathematical modeling. J Hazard Mater 72:265–282.

Tomlinson DW, Thomson NR, Johnson RL, Redman JD. 2003. Air distribution in the Borden aquifer during in situ air sparging. J Contam Hydrol 67:113–132.

U.S. Army. 2008. Engineering and Design: In Situ Air Sparging. EM1110-1-4005.192 p.

van Dijke MIJ, van der Zee SEATM. 1998. Modeling of air sparging in a layered soil: Numerical and analytical approximations. Water Resour Res 34:341–353.

Yu D-Y, Bae W, Kang N, Banks MK, Choi C-H. 2005. Characterization of gaseous ozone decomposition in soil. Soil Sediment Contam 14:231–247.

CHAPTER 9

INTRINSIC REMEDIATION IN NATURAL-GRADIENT SYSTEMS

Olaf A. Cirpka[1]

[1]Center for Applied Geoscience, University of Tübingen, Tübingen, Germany

9.1 INTRODUCTION

In recent years, monitored natural attenuation has been proposed as a cost-efficient alternative to the active remediation of contaminated sites (CGER, 2000). In most cases, contaminants in groundwater originate from highly polluted source zones, which may include nonaqueous phase liquids (NAPL) as continuous emitters. For degradation, the contaminants may be oxidized, as in the case of fuel compounds, or reduced, as is typical for highly chlorinated solvents. In both cases, the contaminants must mix with other reactants for degradation to take place.

The purpose of this chapter is to present efficient approaches of computing concentration distributions for mixing-controlled reactive transport. Emphasis is given on analytical methods which require simplifications regarding the geometric setup, the concentration distributions at the boundaries of the considered domain, and the reactive system. Of course, real field sites are always considerably more complex than the cases covered by analytical expressions so that one-to-one applications to the field are prohibited. Nonetheless, the idealized examples given below are informative about system behavior in mixing-controlled reactive transport, which would also retain in more complex settings. Key concepts presented may be modified to account for non-ideal conditions.

Consider for the moment the case of a former gas station, where diesel and/or gasoline had been leaking into the subsurface forming a NAPL spill (for illustration see Figure 9.1). The soluble fuel components, such as the small aromatic hydrocarbons – benzene, ethylbenzene, toluene, and total xylenes (BTEX) can be degraded by microbial oxidation. In the saturated zone, the most potent oxidants are dissolved compounds, namely oxygen and nitrate. A certain flux of oxidants will enter the source zone mainly by advection and readily react with the dissolved contaminants. In most cases, the dissolution rate of the contaminants is higher than the flux of oxidants entering the source zone. Thus, a plume evolves which is characterized by high concentrations of electron donors and essentially zero concentration of acceptors.

At the downstream front of the plume, mixing of dissolved electron donors and acceptors may take place due to longitudinal dispersion. However, BTEX compounds adsorb onto the soil matrix, whereas oxygen and nitrate do not. As a consequence, the invading front of the contaminants is slower than the receding front of dissolved oxidants, leading to chromatographic separation. That is, longitudinal mixing of dissolved compounds, which is restricted to the progressing front of the plume anyway, is not an efficient mechanism to facilitate natural degradation. In steady state, the only mechanism to bring dissolved electron acceptors to the plume core is by transverse dispersion (Cirpka et al., 1999; Thornton et al., 2001; Thullner et al., 2002; Huang et al., 2003; Ham et al., 2004).

P.K. Kitanidis and P.L. McCarty (eds.), *Delivery and Mixing in the Subsurface: Processes and Design Principles for In Situ Remediation*, doi: 10.1007/978-1-4614-2239-6_9, © Springer Science+Business Media New York 2012

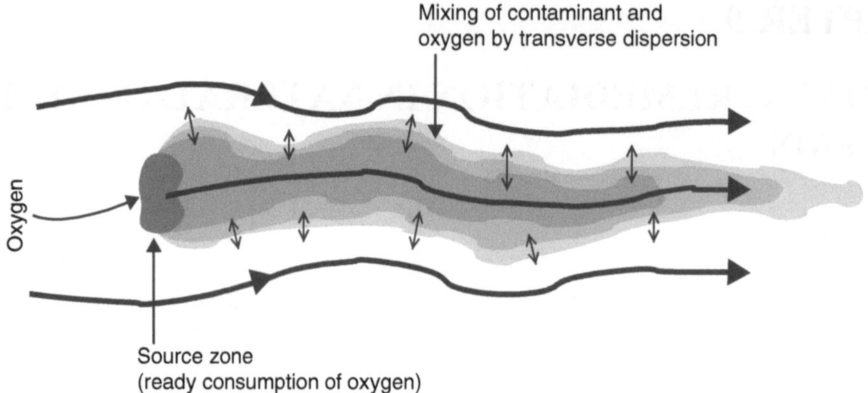

Figure 9.1. Situation of natural attenuation of a continuously emitted plume by reaction with a dissolved oxidant .Taken from Cirpka and Valocchi, 2007; © Elsevier Science Ltd.

While our focus is mixing, it is worth noting that dissolved compounds are not the only possible reaction partners for many contaminants, including BTEX compounds. Under anoxic conditions, ferric iron may be the dominant (and a sufficiently strong) oxidant. In natural aquifers, ferric iron exists mainly as iron oxides and hydroxides, that is, in solid form. For such compounds, dissolution of the solid phase and mass transfer from the dissolving minerals to the plume of dissolved contaminants may be the controlling processes. However, attenuation building on the dissolution of a solid phase cannot be sustained forever. After a sufficiently long exposure to the plume, the reactant in the solid phase will be completely depleted. That is, in the steady state the entire reservoir of accessible electron acceptors in the solid phase will have disappeared. Nonetheless, the reservoir may be large enough to oxidize the contaminants included in the entire source zone.

Another process disregarded in our simplified analysis is the fermentation of dissolved contaminants. In these biotic reactions, the contaminants are transformed to other chemically reduced compounds without requiring a reaction partner. The latter implies that the reduction equivalents are conserved, that is, one electron donor is transformed into another electron donor. The following discussion focuses more on the elimination of all electron donors, which requires mixing with suitable electron acceptors, rather than on the fate of an individual compound.

We consider a simplified setup of a plume in which a contaminant must mix with a compound from outside the plume in order to be degraded. We assume continuous injection of the contaminant without going into the details of the involved dissolution processes. We consider a semi-infinite domain starting directly downstream of the source zone. The contaminant is assumed to be introduced into the domain via a well-defined area at the inflow face of the domain. To obtain analytical expressions, we assume that the contaminant concentration is uniform within the plume entering the domain. At the inflow face, the contaminant concentration is zero outside of the plume. For the electron acceptors, the situation is reversed: At the inflow face the electron-acceptor concentration is zero within the plume, and has a uniform non-zero concentration outside of the plume. For this type of setup we will see that the steady-state length of the plume depends on the ratio of the concentrations entering the domain (weighted by their stoichiometric coefficients), the lateral extension of the source zone, the flow velocity, and the transverse dispersion coefficient.

9.2 ANALYTICAL SOLUTIONS FOR ZERO- AND FIRST-ORDER DECAY IN STEADY STATE

We start the discussion with known analytical expressions for steady-state transport in a uniform flow field accounting for first- or zero-order decay. These reaction laws depend only on a single concentration, namely that of the contaminant. That is, they cannot directly be applied to approximate reactive transport, in which the reaction depends on mixing with an originally separated compound. However, these reaction laws are commonly used in tools for rough estimation of natural attenuation. In Section 9.3 we will discuss the implications of choosing these reaction laws. In Sections 9.4 and 9.5, we will explicitly consider conditions in which analytical expressions for conservative transport can be used to assess mixing-controlled reactive transport.

Starting point is the steady-state advection-dispersion-reaction equation in uniform flow fields describing solute transport:

$$v\frac{\partial c}{\partial x} - D_\ell\frac{\partial^2 c}{\partial x^2} - D_{th}\frac{\partial^2 c}{\partial y^2} - D_{tv}\frac{\partial^2 c}{\partial z^2} = r(c) \qquad \text{(Eq. 9.1)}$$

in which c is the concentration of the contaminant, v is the seepage velocity, oriented in direction x; D_ℓ, D_{th}, and D_{tv} are the longitudinal, horizontal and vertical transverse dispersion coefficients, respectively; x, y, z are the longitudinal, transverse horizontal and transverse vertical dispersion coordinates; and $r(c)$ is the reaction rate. We will consider two common types of simplified reactions, the linear (first-order) and a nonlinear one (zeroth-order):

$$r(c) = \begin{cases} -\lambda c & \text{first order} \\ -r_0 H(c) & \text{zeroth order} \end{cases} \qquad \text{(Eq. 9.2)}$$

in which λ is the rate coefficient of linear decay, r_0 is the concentration-independent reaction rate of a zero-order reaction, and the so-called Heaviside function $H(c)$ takes value *1* when $c > 0$ and 0 otherwise.

The contaminant is introduced into the domain over a rectangle with width w and height h. That is, at the inflow boundary, the concentration of the contaminant takes the uniform value of c_0 within the rectangle, and zero outside of it. We choose the transverse coordinates y and z such that the origin of the coordinate system is in the center of rectangle. These assumptions result in the following boundary condition at $x = 0$:

$$c(x = 0,\ y,\ z) = \begin{cases} c_0 & \text{if } -\frac{w}{2} \le y \le \frac{w}{2} \text{ and } -\frac{h}{2} \le z \le \frac{h}{2} \\ 0 & \text{if } y < -\frac{w}{2} \text{ or } y > \frac{w}{2} \text{ or } z < -\frac{h}{2} \text{ or } z > \frac{h}{2} \end{cases} \qquad \text{(Eq. 9.3)}$$

All other boundaries are considered to be so far away that they don't influence the spatial concentration distribution.

In steady state, the concentration varies very gradually in the direction of flow (longitudinal direction) so that longitudinal dispersion may be neglected. Then the transport equation takes the simpler form:

$$\frac{\partial c}{\partial x} - \frac{D_{th}}{v}\frac{\partial^2 c}{\partial y^2} - \frac{D_{tv}}{v}\frac{\partial^2 c}{\partial z^2} = \frac{r(c)}{v} \qquad \text{(Eq. 9.4)}$$

with the following analytical solutions in three dimensions (3-D):

For first-order decay (Domenico, 1987):

$$c(x,y,z) = \frac{c_0}{4}\left(erf\left(\frac{y+\frac{w}{2}}{2\sqrt{\frac{D_{th}x}{v}}}\right) - erf\left(\frac{y-\frac{w}{2}}{2\sqrt{\frac{D_{th}x}{v}}}\right)\right) \times \left(erf\left(\frac{z+\frac{h}{2}}{2\sqrt{\frac{D_{tv}x}{v}}}\right) - erf\left(\frac{z-\frac{h}{2}}{2\sqrt{\frac{D_{tv}x}{v}}}\right)\right)\exp\left(-\frac{\lambda x}{v}\right) \quad \text{(Eq. 9.5)}$$

and for zero-order decay:

$$c(x,y,z) = \max\left(\frac{c_0}{4}\left(erf\left(\frac{y+\frac{w}{2}}{2\sqrt{\frac{D_{th}x}{v}}}\right) - erf\left(\frac{y-\frac{w}{2}}{2\sqrt{\frac{D_{th}x}{v}}}\right)\right)\left(erf\left(\frac{z+\frac{h}{2}}{2\sqrt{\frac{D_{tv}x}{v}}}\right) - erf\left(\frac{z-\frac{h}{2}}{2\sqrt{\frac{D_{tv}x}{v}}}\right)\right) - r_0 x, 0\right) \quad \text{(Eq. 9.6)}$$

In two dimensions (2-D), the factors involving the vertical coordinate z become unity. Extensions of Equations 9.5 and 9.6 to account for lateral boundary conditions (e.g., no dispersive flux across the top and bottom boundaries of a shallow aquifer or a fixed concentration along such a boundary) can be obtained by the method of imaging.

It may be worth noticing that the original solution of Domenico (1987) also includes also an approximation for the transient behavior. Various studies have elaborated on the validity of the solution, that is, whether it is appropriate to neglect longitudinal dispersion in the steady state, and whether the simplifications made in the derivation of the transient solution not shown here cause significant errors (West et al., 2007; Srinivasan et al., 2007). The common understanding is that the advection-dominated steady state is approximated well by Equations 9.5 and 9.6.

As already mentioned, zero- and fist-order degradation laws do not reflect the situation of mixing controlled reactions. In Equations 9.5 and 9.6 the reaction takes place all over the plume, whereas it should be restricted to the plume fringe if it depends on mixing of the plume with surrounding water. In the analysis of mixing-controlled reactions of Sections 9.4 and 9.5, however, we will make use of the expression for conservative transport. The latter is included in Equation 9.5 by setting the rate coefficient λ to zero, so that the exponential term at the very right becomes unity.

9.3 IMPLICIT ASSUMPTIONS OF ZERO- AND FIRST-ORDER DECAY

In the microbiological and chemical literature, first-order decay is the most common approximation of describing degradation behavior. These choices mainly stem from their simplicity and do not necessarily reflect situations in the field. Because the true reaction may depend in a non-linear way on the concentration of the contaminant, or may depend also on other constituents, the rate coefficients obtained in laboratory or field studies are often denoted as "pseudo first-order" ones. Using such coefficients without reflecting on the conditions under which reaction kinetics become linear, or concentration independent, may lead to erroneous results.

A process that can always be described by first-order rate laws is *radioactive decay*. In practically all other situations, observations of apparently first- or zero-order decay reflect limiting cases, some of which we will discuss in the following paragraphs.

We start by considering the chemical transformation of a compound by reaction with another reactant, such as an electron acceptor or donor. Then the reaction rate must depend at least on the concentrations of both reactants. Consider the one-way irreversible reaction of compound A and B forming the product C:

$$a_{\mathrm{A}}\mathrm{A} + a_{\mathrm{B}}\mathrm{B} \rightarrow a_{\mathrm{C}}\mathrm{C},$$

in which a_i is the stoichiometric coefficient of compound i. Then, the simplest kinetic model accounting for the concentrations of both reactants is bilinear:

$$r = -\frac{1}{a_{\mathrm{A}}}\frac{\partial c_A}{\partial t}\bigg|_{reac} = -\frac{1}{a_{\mathrm{B}}}\frac{\partial c_B}{\partial t}\bigg|_{reac} = \frac{1}{a_{\mathrm{C}}}\frac{\partial c_C}{\partial t}\bigg|_{reac} = \kappa c_A c_B \qquad \text{(Eq. 9.7)}$$

in which κ is the bilinear rate coefficient. If one of the two reactants, let's say B, is in excess, the concentration of B hardly changes in the course of the reaction, and the reaction rate may be expressed by a pseudo first-order rate coefficient $\lambda_A = \kappa c_B^{ini}$ with the initial concentration c_B^{ini} of compound B. Such a case, however, is not the typical mixing-controlled setup, in which the concentrations of A are high where those of B are low, or the concentrations of B are high where those of A are low.

If, in addition to the two reactants, a catalyst is needed, the reaction may be controlled by the concentration of the catalyst. The most important catalyst in contaminant hydrology is the biomass of active organisms. While the true mechanisms of all steps involved in biodegradation may require rather complicated models, it is common practice to apply simple laws derived for enzyme kinetics to an entire organism. These are Michaelis-Menten terms (e.g., Baveye and Valocchi, 1989):

$$r = -\frac{1}{a_{\mathrm{A}}}\frac{\partial c_A}{\partial t}\bigg|_{reac} = -\frac{1}{a_{\mathrm{B}}}\frac{\partial c_B}{\partial t}\bigg|_{reac} = \frac{1}{a_{\mathrm{C}}}\frac{\partial c_C}{\partial t}\bigg|_{reac} = r_{\max}\left(\frac{c_A}{K_A + c_A}\right)\left(\frac{c_B}{K_B + c_B}\right)c_{bio} \qquad \text{(Eq. 9.8)}$$

in which $r_{\max}$ is the maximum specific reaction rate, which is approached for large concentrations of A and B, whereas K_A and K_B are half-velocity concentrations, or Michaelis-Menten coefficients, indicating the concentration at which the reaction rate is reduced to half the maximum value. c_{bio} denotes the biomass concentration.

If the concentration c_{bio} of the biomass remains constant and the concentrations of reactants A and B are much larger than their respective half-velocity concentrations K_A and K_B, Equation 9.8 simplifies to a zero-order rate law with reaction rate r_0:

$$r_0 \approx r_{\max}\, c_{bio} \qquad \text{(Eq. 9.9)}$$

which does not depend on the concentrations of compounds A and B any more. That is, zeroth-order decay is a limiting case of Michaelis-Menten kinetics in which the reaction partners are in excess. In such cases, mixing of compounds A and B poses no limitation of the reaction.

If the concentration c_{bio} of the biomass remains constant and the concentration of reaction partner B is much larger than K_B, whereas the concentration of compound A is much smaller than K_A, Equation 9.8 simplifies to a first-order law for A with rate coefficient λ_A:

$$\lambda_A \approx \frac{r_{\max}}{K_A}\, c_{bio} \qquad \text{(Eq. 9.10)}$$

Again, pseudo first-order behavior is achieved in a system that has originally been described by Michaelis-Menten kinetics when one reaction partner is in excess so that mixing is not a controlling factor.

As last example, we may consider a system where a contaminant with high initial concentration reacts with a compound that is continuously emitted from the solid phase with a constant release rate. This situation may again be described by a quasi-zero-order law. Like in the limiting cases discussed above, this is not the situation of a mixing-controlled reaction.

9.4 GENERAL OUTLINE OF COMPUTING MIXING-CONTROLLED REACTIVE TRANSPORT

9.4.1 Direct Simulation of Coupled Systems

The most straightforward, but also tedious way of computing mixing-controlled reactive transport is by simulating the system of coupled reactive transport equations, here expressed for the reaction $a_{\mathrm{A}}\mathrm{A} + a_{\mathrm{B}}\mathrm{B} \rightarrow a_{\mathrm{C}}\mathrm{C}$:

$$\begin{aligned}
\frac{\partial c_A}{\partial t} + \mathbf{v} \cdot \nabla c_A - \nabla \cdot (\mathbf{D} \nabla c_A) &= -a_{\mathrm{A}}\, r \\
\frac{\partial c_B}{\partial t} + \mathbf{v} \cdot \nabla c_B - \nabla \cdot (\mathbf{D} \nabla c_B) &= -a_{\mathrm{B}}\, r \\
\frac{\partial c_A}{\partial t} + \mathbf{v} \cdot \nabla c_C - \nabla \cdot (\mathbf{D} \nabla c_C) &= a_{\mathrm{C}}\, r
\end{aligned} \qquad \text{(Eq. 9.11)}$$

subject to adequate initial and boundary conditions. For the reaction r, the corresponding rate law has to be substituted into above expressions (e.g., Equation 9.8 for Michaelis-Menten kinetics). In case of a microbially-mediated reaction, the growth and decay of biomass has to be included as well. Assuming that biomass is immobile and that biomass decay can be expressed by a first-order expression, the following ordinary differential equation has to be added to the system of equations:

$$\frac{\partial c_{bio}}{\partial t} = Yr - k_{dec}\, c_{bio} \qquad \text{(Eq. 9.12)}$$

where Y is the yield coefficient and k_{dec} is the decay rate of biomass. It may be noticed that Michaelis-Menten terms are called Monod terms when the transformation reaction is coupled to biomass growth.

In order to compute steady-state concentration distributions, transient reactive transport must be simulated until the concentrations don't change any more. As stated, this procedure is straightforward and allows for any modification of the underlying reaction laws. However, it comes at high computation costs.

9.4.2 Simulation Via Mixing Ratios

An alternative to direct simulation of the coupled system is based on the consideration of conservative components of the reactive system. For the simple system $a_{\mathrm{A}}\mathrm{A} + a_{\mathrm{B}}\mathrm{B} \rightarrow a_{\mathrm{C}}\mathrm{C}$, these are the so-called total concentrations c_A^{tot} and c_B^{tot}:

$$\begin{aligned}
c_A^{tot} &= c_A + \frac{a_{\mathrm{A}}}{a_{\mathrm{C}}} c_C \\
c_B^{tot} &= c_B + \frac{a_{\mathrm{B}}}{a_{\mathrm{C}}} c_C
\end{aligned} \qquad \text{(Eq. 9.13)}$$

or any linear combination of the two. The total concentrations do not change upon the reaction. Under the assumption that neither the seepage velocity $\mathbf{v}$ nor the dispersion tensor $\mathbf{D}$ depend on the particular compound, we can sum the weighted individual transport equations to obtain transport equations of the total concentration $c_A^{tot}$and c_B^{tot}:

$$\begin{aligned}
&\frac{\partial c_A^{tot}}{\partial t} + \mathbf{v} \cdot \nabla c_A^{tot} - \nabla \cdot \left(\mathbf{D} \nabla c_A^{tot}\right) = 0 \\
&\frac{\partial c_B^{tot}}{\partial t} + \mathbf{v} \cdot \nabla c_B^{tot} - \nabla \cdot \left(\mathbf{D} \nabla c_B^{tot}\right) = 0
\end{aligned} \qquad \text{(Eq. 9.14)}$$

which are advection-dispersion equations with zero source/sink term, reflecting that c_A^{tot} and c_B^{tot} are conservative components.

Because c_A^{tot} and c_B^{tot} are conservative, they mix linearly. That is, in the mixing of two uniform solutions, the total concentrations can be computed from the composition of the two end members and the volumetric ratio of one solution in the mixture with the other, denoted in the following mixing ratio $X(x,y,z,t)$. Consider the situation of the contaminant plume outlined above as the injection of the plume into ambient water. The total concentrations in the injected solution are denoted $c_A^{tot} = c_A^{in}$; $c_B^{tot} = c_B^{in}$, whereas the total concentrations in the ambient solution are $c_A^{tot} = c_A^{amb}$; $c_B^{tot} = c_B^{amb}$. Then linear mixing implies:

$$\begin{aligned}
c_A^{tot} &= X c_A^{in} + (1 - X) c_A^{amb} \\
c_B^{tot} &= X c_B^{in} + (1 - X) c_B^{amb}
\end{aligned} \qquad \text{(Eq. 9.15)}$$

in which $X(x,y,z,t)$ is the mixing ratio of the injected solution in the mixture with the ambient solution which depends on space and in case of transient transport also on time. The mixing ratio X undergoes conservative transport:

$$\frac{\partial X}{\partial t} + \mathbf{v} \cdot \nabla X - \nabla \cdot (\mathbf{D} \nabla X) = 0 \qquad \text{(Eq. 9.16)}$$

in which X equals unity along the boundary section where the plume enters the domain and X is zero at the remaining part of the inflow boundary. In this framework, the mixing ratio $X(x,y,z,t)$ may be interpreted as the probability that a solute particle observed at time t at location (x,y,z) originates from the part of the inflow section that belongs to the plume.

For steady-state transport under uniform flow conditions, the Domenico solution of Equation 9.5 with a rate coefficient λ of zero may be used to approximate the mixing ratio X. Subsequently, the total concentrations c_A^{tot} and c_B^{tot} can be computed by Equation 9.15. The extension to more complicated reactive systems, requiring more than two total concentrations, is straightforward. Also, it is possible to consider mixing involving more than two end members.

From the total concentrations, it is possible to compute the concentrations of the reactive species under certain circumstances discussed in the following section. The latter step may be addressed as the unique speciation problem and is addressed in Section 9.5.

In case of a unique relationship between mixing ratio X and concentration c_i of the reactive species i, we may evaluate the reaction rate by multiplying Equation 9.16 with the derivative $\partial c_i / \partial X$ at the location of interest:

$$\frac{\partial c_i}{\partial X}\frac{\partial X}{\partial t} + \frac{\partial c_i}{\partial X}\,\mathbf{v} \cdot \nabla X - \frac{\partial c_i}{\partial X} \nabla \cdot (\mathbf{D} \nabla X) = 0 \qquad \text{(Eq. 9.17)}$$

For the time derivative and the advective term, application of the chain rule of differentiation is straightforward. In case of the dispersive term, the following identity applies:

$$\begin{aligned}
\nabla \cdot (\mathbf{D} \nabla c_i) &= \nabla \cdot \left(\mathbf{D} \frac{\partial c_i}{\partial X} \nabla X\right) = \frac{\partial c_i}{\partial X} \nabla \cdot (\mathbf{D} \nabla X) + \left(\nabla \frac{\partial c_i}{\partial X}\right)^T \mathbf{D} \nabla X \\
&= \frac{\partial c_i}{\partial X} \nabla \cdot (\mathbf{D} \nabla X) + \frac{\partial^2 c_i}{\partial X^2} (\nabla X)^T \mathbf{D} \nabla X
\end{aligned} \qquad \text{(Eq. 9.18)}$$

Thus, the transport equation of the reactive species *i* reads as (De Simoni et al., 2005, 2007):

$$\frac{\partial c_i}{\partial t} + \mathbf{v} \cdot \nabla c_i - \nabla \cdot (\mathbf{D} \nabla c_i) = -\frac{\partial^2 c_i}{\partial X^2} (\nabla X)^T \mathbf{D} \nabla X \qquad \text{(Eq. 9.19)}$$

The right-hand side term of Equation 9.19 describes the reactive source/sink term. A few observations can be made:

- In an advective-dispersive system, mixing and thus reaction is controlled by (local) dispersion alone. For D = 0, the reactive source/sink term becomes zero.
- Areas with the strongest spatial gradient X of the mixing ratio are areas of strongest reaction.
- A compound that mixes linearly, i.e., $\partial c_i / \partial X = const.$, is conservative. The reactivity depends on the deviation from the linear behavior.

9.5 DETERMINING CONCENTRATIONS OF INDIVIDUAL REACTIVE SPECIES FROM TOTAL CONCENTRATIONS

The concept of total concentrations is common in chemical engineering (see "chemical equilibrium" on www.wikipedia.org). The simplest examples involve specific elements (e.g., the total concentration of all sulfur bearing compounds) or molecule groups (e.g., the sum of acetic acid and acetate) but other, chemically less intuitive examples are possible. In geochemical modeling, the total concentration related to a certain element in a particular valence state is often denoted master species (Appelo and Postma, 2005). Computing the distribution among various ions, complexes, etc. is a classical speciation problem. The speciation can be unique only when the relationship between the concentrations of the various species is a set of algebraic equations. This means, that the concentrations may be related to each other by fractions of polynomials and similar expressions, but not as differential equations. In the following, we will discuss the most common cases.

9.5.1 Chemical Equilibrium of Dissolved Compounds

Chemical equilibrium enforces that concentrations of reacting species are related to each other by an algebraic expression, namely the law of mass action. For the case of the reversible reaction $a_A A + a_B B \leftrightarrow a_C C$, the law of mass action is:

$$K = \frac{(\gamma_C C_C)^{a_C}}{(\gamma_A C_A)^{a_A} \times (\gamma_B C_B)^{a_B}} \qquad \text{(Eq. 9.20)}$$

in which K is the equilibrium constant related to the free energy ΔG of the reaction by $K = \exp(-\Delta G / RT)$, whereas γ_i is the coefficient of activity of compound i. The factors $\gamma_i^{a_i}$ can be included in the equilibrium constant K, leading to modified constant K_*. Then, making use of Equation 9.13, Equation 9.20 may be rewritten as:

$$K_* = \frac{C_C^{a_C}}{\left(c_A^{tot} - \frac{a_A}{a_C} C_C\right)^{a_A} \times \left(c_B^{tot} - \frac{a_B}{a_C} C_C\right)^{a_B}} \qquad \text{(Eq. 9.21)}$$

which may be solved for the concentration c_C of the reaction product C. In general, the latter step may require a numerical solution scheme. For the specific case of a 1:1:1 stoichiometry, that is, $a_A = a_B = a_C = 1$, we arrive at:

$$K_* \times \left(c_A^{tot} - c_C\right) \times \left(c_B^{tot} - c_C\right) = c_C \qquad \text{(Eq. 9.22)}$$

which is a quadratic equation with a valid and an invalid solution. The valid one is:

$$c_C = \frac{K_* \times \left(c_A^{tot} + c_B^{tot}\right) + 1 - \sqrt{K_*^2\left(c_A^{tot} - c_B^{tot}\right)^2 + 2K_*\left(c_A^{tot} + c_B^{tot}\right) + 1}}{2K_*} \qquad \text{(Eq. 9.23)}$$

The other solution is invalid, because it would lead to a negative concentration of either c_A or c_B after substitution of the solution into Equation 9.13 and rearrangement:

$$c_A = c_A^{tot} - \frac{a_A}{a_C} c_C$$
$$c_B = c_B^{tot} - \frac{a_B}{a_C} c_C \qquad \text{(Eq. 9.24)}$$

For the case of two uniform but different solutions introduced in parallel into the domain we now arrive at the following overall scheme to obtain the reactive species concentrations $c_A(x,y,z)$, $c_B(x,y,z)$, and $c_C(x,y,z)$ undergoing the reversible reaction $A + B \leftrightarrow C$:

1. Solve Equation 9.16 to obtain the spatial distribution of the mixing ratio $X(x,y,z)$. In case of uniform flow, steady-state transport and introduction of one solution via a rectangular section of the inflow boundary, the Domenico solution of Equation 9.5 with a rate coefficient λ of zero would hold to compute $X(x,y,z)$.
2. For the given mixing ratio $X(x,y,z)$ and total concentrations c_A^{in} and c_B^{in} in the plume respectively c_A^{amb} and c_B^{amb} in the ambient solution at the inflow boundary, compute the total concentrations $c_A^{tot}(x, y, z)$ and $c_B^{tot}(x, y, z)$ by Equation 9.15.
3. At each location (x,y,z), compute the concentrations of the reactive species, c_C, c_A, and c_B from $c_A^{tot}(x, y, z)$ and $c_B^{tot}(x, y, z)$ by Equations 9.23 and 9.24.

This scheme is considerably simpler than direct simulation of the coupled system of equations as outlined in Section 9.4.1.

Cirpka et al. (2006) considered a slightly more complicated case of mixing-controlled reactions involving local chemical equilibrium of dissolved compounds. The objective of their analysis was the assessment of transverse dispersion coefficients by measuring the length of a reactive plume in steady state. For this purpose, they injected an alkaline solution into acidic ambient water and added a pH indicator into both solutions to outline the mixing ratio at which the pH of the mixture was at the point where the indicator changed its color. For a general system of dissolved buffering compounds, they identified one total concentration per buffering system Buf_i in addition to the charge Q of all non-buffering counter ions. Because of electro-neutrality, the charge of the buffering compounds, including water itself, and the counter ions must cancel. Accounting for the law of mass action, this leads to:

$$Q = [\mathrm{Na}^{\oplus}] + [\mathrm{K}^{\oplus}] - [\mathrm{Cl}^{-}] \ \ldots = \frac{K_{\mathrm{H_2O}}}{[\mathrm{H}^{\oplus}]} + \sum_i^{n_{buf}} \frac{K_i[\mathrm{Buf}_i]_{tot}}{K_i + [\mathrm{H}^{\oplus}]} - [\mathrm{H}^{\oplus}] \qquad \text{(Eq. 9.25)}$$

here expressed without activity coefficients. Equation 9.25 can be solved for the concentration of the hydroxonium ion $[\mathrm{H}^{\oplus}]$ in an iterative manner:

$$[\mathrm{H}^{\oplus}] = \frac{1}{2}\left(\sqrt{Q^2 + 4 \times \left(K_{\mathrm{H_2O}} + \sum_i^{n_{buf}} \frac{K_i[\mathrm{H}^{\oplus}][\mathrm{Buf}_i]_{tot}}{K_i + [\mathrm{H}^{\oplus}]}\right)} - Q\right) \qquad \text{(Eq. 9.26)}$$

using the charge and the total buffer concentrations of the mixture:

$$Q_{mix} = (1 - X_{alk}) \times Q_{ac} + X_{alk} \times Q_{alk} \qquad \text{(Eq. 9.27)}$$

$$[\mathrm{Buf}_i]_{tot}^{mix} = (1 - X_{alk}) \times [\mathrm{Buf}_i]_{tot}^{ac} + X_{alk} \times [\mathrm{Buf}_i]_{tot}^{alk} \qquad \text{(Eq. 9.28)}$$

in which X_{alk} is the mixing ratio of the alkaline solution, that is, the volumetric fraction of the alkaline solution in the mixture with the acidic solution at a given point. $[\mathrm{Buf}_i]_{tot}^{ac}$ is the total concentration of the buffer compound i in the acidic solution, $[\mathrm{Buf}_i]_{tot}^{alk}$ in the alkaline solution, and $[\mathrm{Buf}_i]_{tot}^{mix}$ in the mixture.

In the example of Cirpka et al. (2006), obviously, there is a unique relationship between the mixing ratio X_{alk} and the resulting pH. Likewise, the mixing ratio meets Equation 9.16. Cirpka et al. (2006) performed their experiments in quasi two-dimensional sandboxes with uniform flow. The injection of the alkaline solution was via a line source of width w perpendicular to the direction of flow. Then, in steady state the two-dimensional variant of the Domenico solution, Equation 9.5, with a rate coefficient λ of zero holds to compute the distribution of the mixing ratio $X(x,y)$. Of particular relevance is the mixing ratio along the center line where the plume gets the longest:

$$X(x,\ y = 0) = \mathrm{erf}\left(\frac{w}{4\sqrt{x \times D_t / v}}\right) \qquad \text{(Eq. 9.29)}$$

Cirpka et al. (2006) used Equation 9.29 to estimate the length L of a plume. The general idea is that the pH indicator changes its color at a particular value of the mixing ratio $X_{\mathrm{pHind.}}$. The value of $X_{\mathrm{pHind.}}$ could either be computed by Equation 9.26 or obtained by a titration experiment. Then the length L of the plume is the longitudinal coordinate x along the center line for the given value of $X_{\mathrm{pHind.}}$:

$$L = \frac{v \times w^2}{16 D_t \times \left(\mathrm{inverf}(X_{\mathrm{pH\,ind}})\right)^2} \qquad \text{(Eq. 9.30)}$$

in which $\mathrm{inverf}(X_{\mathrm{pHind.}})$ is the inverse error function with argument X_{pHind}, that is, $A = \mathrm{inverf}(X_{\mathrm{pHind.}})$ is the solution to $\mathrm{erf}(A) = X_{\mathrm{pHind}}$.

In fact, Cirpka et al. (2006) rearranged Equation 9.30 to determine the transverse dispersion coefficient D_t from the length L and initial width w of the alkaline plume:

$$D_t = \frac{v \times w^2}{16 L \times \left(\mathrm{inverf}(X_{\mathrm{pH\,ind}})\right)^2} \qquad \text{(Eq. 9.31)}$$

For the setup, which is similar to that illustrated in Figure 9.1, the plume length L is inversely proportional to the transverse dispersion coefficient D_t, and proportional to the seepage velocity v and the initial width w squared. In a three-dimensional setup, X along the center line depends on the product of two error functions:

$$X(x,\ y = 0,\ z = 0) = \mathrm{erf}\left(\frac{w}{4\sqrt{x \times D_{th}/v}}\right)\mathrm{erf}\left(\frac{h}{4\sqrt{x \times D_{tv}/v}}\right) \qquad \text{(Eq. 9.32)}$$

which may be solved for x only by numerical methods.

The analysis of Cirpka et al. (2006) indicates that the length of a steady-state plume reacting with the ambient solution is inversely proportional to the bulk transverse dispersivity $\alpha_t = D_t/v$. In the two-dimensional setup, it also scales with w^2, the squared initial width. This result does not come as a surprise since the characteristic time scales of transverse dispersion and advection are w^2/D_t and L/x, respectively. At the plume boundary, these time scales must be in a fixed ratio, resulting in L proportional to $w^2 \times v/D_t$. For the purpose of estimating the transverse dispersion coefficient, measuring a mixing-controlled reactive plume length is advantageous, because typical values for α_t are in the range of 10^{-4} meters (m) and thus quite small. The smaller α_t, the larger is the plume length L, and the easier it can be measured.

For field applications, the proportionality with w^2 may be as important as the inverse proportionality to α_t. A wrong estimation of the initial plume width by a factor of 2 leads to an erroneous estimation of the plume length by a factor of 4.

Of course, the simple expressions given above relating the plume length to the transverse dispersion coefficient hold only for uniform flow fields. In heterogeneous aquifers, transverse mixing may considerably be enhanced by heterogeneity (Werth et al., 2006; Jankovic, 2009).

In this section, we have discussed mixing of two initially uniform solutions containing dissolved species undergoing reversible reactions that are locally in equilibrium. Acid–base reactions are good examples because they are quick and reversible. If the time scale of the chemical reaction is in the order of that of mixing, the kinetics of the reactions must be considered, and the simple approach of mapping mixing ratios to reactive-species concentrations may no longer be valid.

9.5.2 Chemical Equilibrium in the Presence of a Mineral Phase

In the examples given in the previous section, we have considered so-called homogeneous reactions in which all reactants are dissolved compounds. In heterogeneous reactions, by contrast, a second phase such as a mineral phase is involved. The simplest prototype of such a reaction is the precipitation/dissolution reaction of a cation $A^{\oplus}$ and an anion B^- and their precipitate AB:

$$A^{\oplus} + B^{-} \leftrightarrow AB_{\downarrow}$$

De Simoni et al. (2005) analyzed this system assuming local chemical equilibrium and restricting themselves to cases in which the precipitate AB is always present. By definition, the activity of a solid phase is unity, so that the law of mass action becomes the solubility product:

$$K = [A^{\oplus}] \times [B^{-}] \qquad \text{(Eq. 9.33)}$$

Figure 9.2 shows a plot of the two aqueous-phase concentrations in the presence of the mineral phase for 1:1 stoichiometry. The solid line marks equilibrium according to the solubility product of Equation 9.33. The gray shaded area underneath the solid line marks combinations of concentrations that are undersaturated with respect to the mineral phase. If a solution with this concentration combination comes into contact with the mineral, dissolution of the mineral occurs until the equilibrium line is met. By contrast concentration combinations in the white area of Figure 9.2 mark oversaturation with respect to the mineral enforcing precipitation to achieve equilibrium.

The interesting question is what happens if two solutions, both of them in equilibrium to the mineral phase, mix. This is a quite common problem in geochemical applications if water bodies of different origin, and thus different chemical composition, mix in the presence of minerals. As illustrated by the dotted line in Figure 9.2, the two end-member solutions, both of which are in equilibrium with the mineral phase, would mix linearly if there was no precipitation/

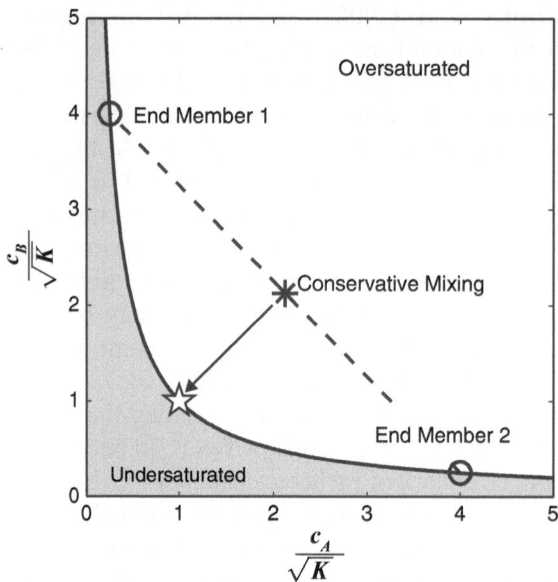

Figure 9.2. Mixing line of two compounds in the presence of a precipitate. *Solid line*: reactive mixing line, accounting for precipitation; *dashed line*: conservative mixing of the two end members marked by *circles*; *star*: 1:1 conservative mixing; *pentagram*: resulting composition after re-equilibration with the mineral phase.

dissolution of the mineral. For the simple system $A^{\oplus} + B^{-} \leftrightarrow AB_{\downarrow}$, however, the conservative mixture is always oversaturated with respect to AB, leading to precipitation. Re-equilibration thus requires that more precipitate is formed.

In the given example, the aqueous-phase concentrations, here denoted $[A^{\oplus}]$ and $[B^{-}]$ are not conservative. Also, the total concentrations as defined in Equation 9.13 would not be very useful because they involve the mineral phase, which does not undergo transport. However, as stated above, any linear combination of conservative components of a reactive system is also conservative; and thus the difference u of concentrations in the aqueous phase is a conservative component, too:

$$u = [A^{\oplus}] - [B^{-}] \qquad \text{(Eq. 9.34)}$$

which thus meets a standard conservative transport equation, or could be evaluated from the mixing ratio X if we consider the mixing of two uniform solutions, both of them in equilibrium with the mineral phase.

As long as the mineral phase remains present, the concentrations of the ions can be computed from the conservative component u by:

$$[A^{\oplus}] = \frac{u + \sqrt{u^2 + 4K}}{2}, \quad [B^{-}] = \frac{-u + \sqrt{u^2 + 4K}}{2} \qquad \text{(Eq. 9.35)}$$

In order to evaluate the aqueous-phase concentrations $[A^{\oplus}]$ and $[B^{-}]$ in a mixing-controlled reactive transport setup for the reaction $A^{\oplus} + B^{-} \leftrightarrow AB_{\downarrow}$ in local equilibrium, the following overall scheme applies:

1. Compute the conservative components u_1 and u_2 for the two end-member solutions.
2. Compute the mixing ratio $X(x,y,z,t)$ solving the advection-dispersion equation without sources or sinks and setting X to unity where solution 1 exists in the initial state, and to zero where solution 2 exists.

3. Compute the distribution of the conservative component $u(x,y,z,t)$ by linear mixing: $u(x,y,z,t) = X(x,y,z,t) \times u_1 + (1 - X(x,y,z,t)) \times u_2$
4. Evaluate the aqueous-phase concentrations according to Equation 9.35.

De Simoni et al. (2005) also computed the spatial and temporal distribution of the precipitation rate. Knowing how much precipitate is formed is of relevance, e.g., because the hydraulic conductivity of the porous medium may be reduced when too much precipitate is formed. In the analysis presented here, however, such feedbacks of the reactions onto flow and transport are not yet accounted for.

De Simoni et al. (2007) extended the previous analysis to systems involving several homogeneous and heterogeneous reactions in equilibrium, and demonstrated that the entire inorganic-carbon speciation in the presence of calcite can be evaluated by knowing the mixing ratio and the composition of end members. For such more complex systems, mixing of solutions that are in equilibrium to the mineral phase may lead to precipitation or dissolution. The latter is known as 'mixing corrosion' (e.g., Bögli, 1971). Sanchez-Vila (2007) extended the analysis to the case of non-equilibrium precipitation, using a series expansion about the equilibrium.

9.5.3 Instantaneous, Complete, Irreversible Reaction

The reactive systems discussed in Sections 9.5.1 and 9.5.2 involved reversible reactions in equilibrium, either considering only aqueous-phase compounds or including heterogeneous reactions with minerals. In contrast to that, we now discuss irreversible reactions in which the same principles of computing reactive-species concentrations apply as discussed above.

The reactive system of interest is an irreversible reaction of an electron donor with an electron acceptor catalyzed by microorganisms. Before discussing the case of kinetic bioreactions as expressed by dual Michaelis-Menten or Monod kinetics (see Equations 9.8 and 9.12 for the reaction laws), we consider a common simplification of bioreactive systems, in which it is assumed that the microbial reaction is significantly faster than mixing. Under this condition, one may treat the reaction as an instantaneous one (e.g., Borden and Bedient, 1986). For illustration, consider the situation of Figure 9.1. In the present conceptual model, oxygen and the organic contaminant cannot coexist. As soon as they mix, they react with each other so that one of the two concentrations becomes zero and the other concentration is reduced according to stoichiometry. This condition can be expressed as:

$$c_A \times c_B = 0 \tag{Eq. 9.36}$$

which clearly is an algebraic relationship. Substituting the definition of the total concentrations of Equation 9.13 into Equation 9.36, yields:

$$\left(c_A^{tot} - \frac{a_\mathrm{A}}{a_\mathrm{C}} c_C\right) \times \left(c_B^{tot} - \frac{a_\mathrm{B}}{a_\mathrm{C}} c_C\right) = 0 \tag{Eq. 9.37}$$

That is, either of the two factors must be zero. Considering that all concentrations are non-negative, the following analytical expressions hold:

$$\begin{aligned} c_A = 0,\ c_B = c_B^{tot} - \frac{a_\mathrm{B}}{a_\mathrm{A}} c_A^{tot},\ c_C = \frac{a_\mathrm{C}}{a_\mathrm{A}} c_A^{tot} \text{ if } \frac{c_A^{tot}}{a_\mathrm{A}} < \frac{c_B^{tot}}{a_\mathrm{B}} \\ c_A = c_A^{tot} - \frac{a_\mathrm{A}}{a_\mathrm{B}} c_B^{tot},\ c_B = 0,\ c_C = \frac{a_\mathrm{C}}{a_\mathrm{B}} c_B^{tot} \text{ if } \frac{c_A^{tot}}{a_\mathrm{A}} > \frac{c_B^{tot}}{a_\mathrm{B}} \end{aligned} \tag{Eq. 9.38}$$

For the case that the total concentrations are exactly in the stoichiometric ratio of the reaction, both c_A and c_B are zero:

$$c_A = c_B = 0 \text{ if } \frac{c_A^{tot}}{a_A} = \frac{c_B^{tot}}{a_B}. \qquad \text{(Eq. 9.39)}$$

In the setup of an injected solution, the latter condition corresponds to a particular critical mixing ratio X_{crit} of the injected solution in the mixture with the ambient solution:

$$X_{crit} = \frac{a_A\, c_B^{amb} - a_B\, c_A^{amb}}{\left(c_A^{in} - c_A^{amb}\right)a_B + \left(c_B^{amb} - c_B^{in}\right)a_A}, \qquad \text{(Eq. 9.40)}$$

which can be evaluated from Equations 9.15 and 9.37.

Figure 9.3 shows example calculations of reactive-species concentrations as a function of the mixing ratio X for an instantaneous reaction A + B → C. The solution representing $X = 1$ contains compound A at concentration $c_A^{in} = 0.33$ mmol/L and no compound B, whereas the solution representing $X = 0$ contains no compound A, and compound B has concentration $c_B^{amb} = 0.25$ mmol/L (c_B^{in} and c_A^{amb} are both zero). The stoichiometry of the reaction is 1:1:1. The dotted lines represent the total concentrations c_A^{tot} and c_B^{tot} according to Equation 9.15. The dashed lines show the reactive species concentrations of compounds A, B, and C for the case of an instantaneous, complete, irreversible reaction according to Equation 9.38. For the parameters given, the critical mixing ratio X_{crit} according to Equation 9.40 is 0.429. For values of $X < X_{crit}$ the concentration of compound A remains zeros if an instantaneous reaction is considered; in this regime, the concentration of compound B decreases linearly with X. The explanation is that $c_A^{tot} < c_B^{tot}$, so that compound A immediately vanishes, but the reaction also leads to a concentration decrease of compound B. For $X > X_{crit}$ we find that $c_A^{tot} > c_B^{tot}$,

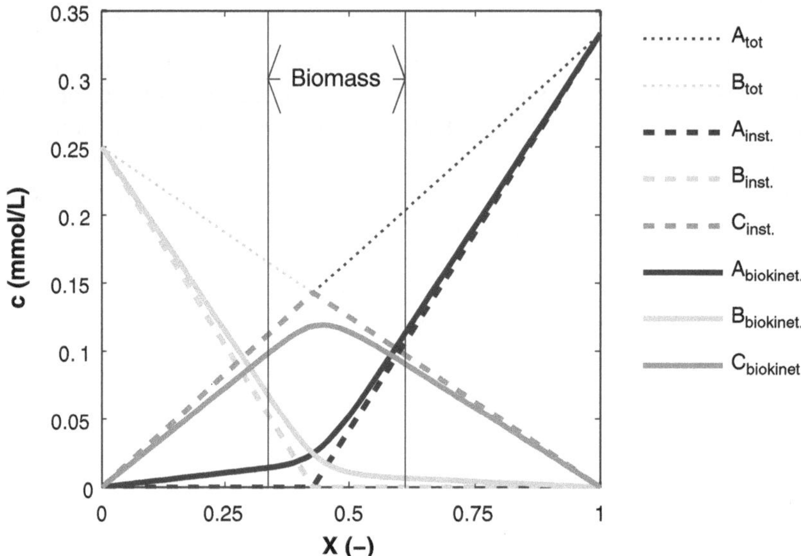

Figure 9.3. Dependence of compound-related concentrations on mixing ratio X for an irreversible reaction A + B → C. *Dotted lines*: total concentrations according to Equation 9.15; *dashed lines*: instantaneous, complete, irreversible reaction (see Section 9.5.3); *solid lines*: biokinetic reactions (see Section 9.5.4). *Black*: reactant A; *light gray*: reactant B; *dark gray*: reaction product C. Parameters are taken from Cirpka and Valocchi, (2007, 2009) and are listed in the text. Note: *mmol/L* millimole per liter.

resulting in instantaneous disappearance of compound B and a linear concentration increase of compound A with X. Figure 9.3 also includes reactive-species concentrations for the case of a biokinetic irreversible reaction in steady state which will be discussed in Section 9.5.4.

Figure 9.4 shows spatial distributions of steady-state concentrations for a line-injection perpendicular to uniform flow. The example, taken from Cirpka and Valocchi (2007), mimics a laboratory-scale experiment with a bulk transverse dispersivity D_t/v of 0.25 millimeters (mm), a seepage velocity of 1 meter per day (m/d) $= 1.16 \times 10^{-5}$ meters per second (m/s), a 1:1:1 stoichiometry, and concentrations c_A^{in} and c_B^{amb} as listed above. The mixing ratio X is computed by the two-dimensional version of the Domenico solution, Equation 9.5 with $\lambda = 0$. Subsequently the total concentrations c_A^{tot} and c_B^{tot} are computed by Equation 9.15. Finally, the reactive species concentrations are evaluated by Equation 9.38.

The solid line marks the boundary of the plume, which is identical to the contour line of the critical mixing ratio X_{crit} of 0.429. The width of the plume at the injection surface is 0.05 m, resulting in a plume length of 3.90 m according to Equation 9.30. Within the plume, the ambient compound B has zero concentration, whereas outside of the plume, the concentration of the injected compound is zero.

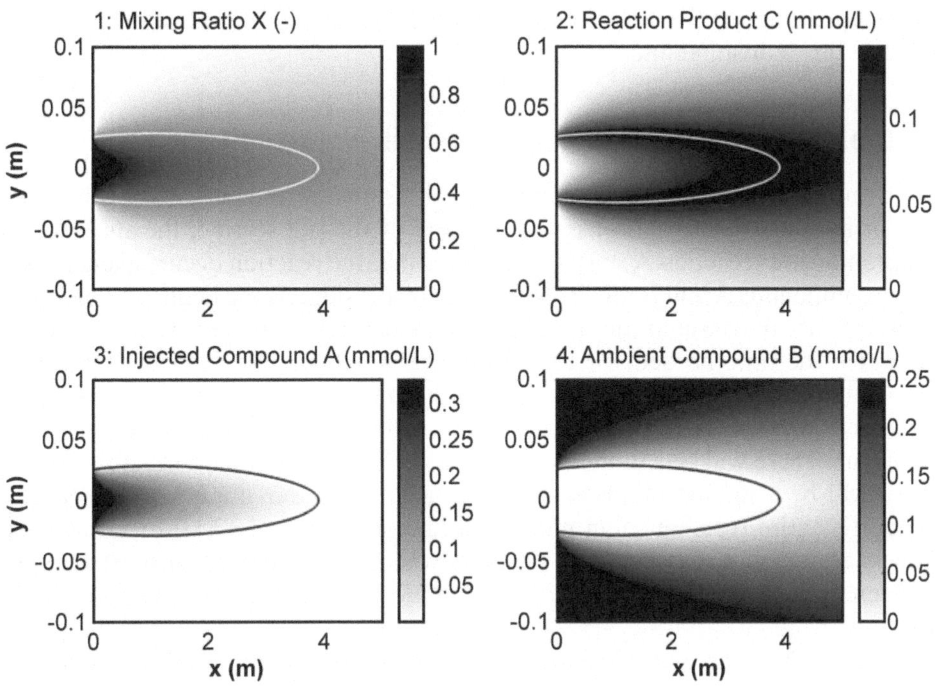

Figure 9.4. Mixing ratio X and distribution of reactive compounds for a complete, instantaneous, irreversible reaction A + B → C. *Solid line*: boundary of the plume of compound A. Taken from Cirpka and Valocchi, 2007; © Elsevier Science Ltd.

Expression 9.30 in conjunction with Equation 9.40 gives the convenient opportunity to assess the potential of natural attenuation by simple means. The length of the steady-state plume is given solely by the geometry of the injection, the concentrations in the inflow, the stoichiometry of the net reaction, and the bulk transverse dispersivity D_t/v. If a sensitive receptor was located downstream of the source at distances smaller than the computed plume length, natural attenuation must be ruled out as a management alternative. If such a receptor is considerably further away than the computed plume length, more detailed studies may be needed to confirm that natural attenuation is sufficient to control the mass flux released from the source.

The same reactive system has been considered by Ham et al. (2004), who analyzed a point-like injection, and by Liedl et al. (2005) who considered the injection of a contaminant over the entire thickness of an aquifer and the introduction of oxygen by diffusion through the capillary fringe. While the exact analytical expressions derived in these studies differ in details, they all show the same principal characteristic: (1) There is a definite length L of the plume which (2) is proportional to the squared width of the source, and (3) inversely proportional to the bulk transverse dispersivity $\alpha_t = D_t/v$.

At this point, we might come back to some theoretical considerations made at the end of Section 9.4.2 when we discussed Equation 9.19. At that point we observed that a linear dependence $c_i(X)$ of a reactive-species concentration c_i on the mixing ratio X indicates conservative behavior in mixing-controlled reactive transport. In Figure 9.3 we see that in the case of an instantaneous, irreversible reaction all reactive-species concentrations depend linearly on the mixing ratio X for all values of X except the critical one, X_{crit}, where the left-hand side gradient differs from the right-hand side one. Thus, in the two-dimensional example plotted in Figure 9.4, the compounds A, B, and C behave like conservative compounds except along the line of the critical mixing ratio X_{crit}. The entire reaction is spatially restricted to this individual line.

9.5.4 Biokinetic Irreversible Reaction in Steady State

The assumption of an instantaneous reaction as an approximation for a biotic reaction may be put into question. As discussed at the end of the last section, the analytical solution for an instantaneous reaction A + B → C implies that the reaction occurs exclusively along a line where compounds A and B mix by transverse dispersion. At the beginning of Section 9.5.3 we discussed that the assumption of an instantaneous reaction may be seen as the limiting case in which the time needed for the reaction is much smaller than the time needed for mixing. In this limiting case, the biomass is not computed explicitly, it is just assumed to be extremely quick in its transformations. In reality, the reaction A + B → C cannot be instantaneous. There will be a limited maximum specific transformation rate of the biomass, and there will be a limited biomass concentration. The latter implies that it is impossible to restrict the entire reaction to an individual line (or surface in 3-D) of an infinitesimally small thickness. Thus, it may be worth studying whether bioreactive transport using standard double-Monod expressions like those listed in Equation 9.8 gives qualitatively different results.

Chu et al. (2005) analyzed bioreactive transport in setups identical to the one considered here. The latter authors assumed that the biomass is concentrated in a narrow stripe, which is confirmed by experimental observations. Cirpka and Valocchi (2007, 2009) derived analytical expressions of the given problem under steady-state transport conditions. The starting point is the net growth rate of biomass, given by substituting Equation 9.8 into Equation 9.12:

$$\frac{\partial c_{bio}}{\partial t} = \left(r_{\max} Y \left(\frac{c_A}{K_A + c_A} \right) \left(\frac{c_B}{K_B + c_B} \right) - k_{dec} \right) c_{bio} \qquad \text{(Eq. 9.41)}$$

In steady state, Equation 9.41 must equal zero. There are two possible solutions to meet this condition. Either the biomass concentration c_{bio} equals zero, which is the trivial solution, or the term in the bracket before c_{bio} is zero. Substituting the definition of the total concentrations,

Equation 9.15, into Equation 9.41, requiring a net growth rate of zero, and disregarding the trivial solution, we arrive at (Cirpka and Valocchi, 2007):

$$r_{\max}Y\left(c_A^{tot}-\frac{a_A}{a_C}c_C\right)\left(c_B^{tot}-\frac{a_B}{a_C}c_C\right)=k_{dec}\left(K_A+c_A^{tot}-\frac{a_A}{a_C}c_C\right)\left(K_B+c_B^{tot}-\frac{a_B}{a_C}c_C\right)$$

(Eq. 9.42)

which is a quadratic expression of c_C. Cirpka and Valocchi (2007, 2009) derived the analytical expressions to compute the concentrations c_A, c_B, and c_C of the reacting species from the total concentrations, and thus from the mixing ratio. In the presence of biomass in steady state, the concentration of the reaction product c_C can be computed from the total concentrations c_A^{tot} and c_B^{tot} by (Cirpka and Valocchi, 2007):

$$c_C=\frac{-p_1-\sqrt{p_1^2-4p_0p_2}}{2p_2} \qquad \text{(Eq. 9.43)}$$

with the coefficients p_0, p_1, and p_2:

$$\begin{aligned}
p_0&=c_A^{tot}c_B^{tot}-\frac{k_{dec}}{Yr_{\max}}\left(K_A+c_A^{tot}\right)\left(K_B+c_B^{tot}\right)\\
p_1&=\frac{k_{dec}}{Yr_{\max}}\left(K_A+c_A^{tot}\right)\frac{a_B}{a_C}+\frac{k_{dec}}{Yr_{\max}}\left(K_B+c_B^{tot}\right)\frac{a_A}{a_C}-c_A^{tot}\frac{a_B}{a_C}-c_B^{tot}\frac{a_A}{a_C}\\
p_2&=\frac{a_Aa_B}{a_C^2}\left(1-\frac{k_{dec}}{Yr_{\max}}\right)
\end{aligned} \qquad \text{(Eq. 9.44)}$$

A particular difficulty lies in determining whether at a given location biomass can be retained at steady state or not. Cirpka and Valocchi (2009) argued that the transition $c_C(X)$ must be smooth between regions with and without biomass. This results in a minimum and a maximum mixing ratio $X_{\min}$ and $X_{\max}$, respectively, between which biomass can prevail in steady state. These values are implicitly defined by:

$$\begin{aligned}
c_C(X_{\min})-X_{\min}\left.\frac{\partial c_C}{\partial X}\right|_{X=X_{\min}}&=0\\
c_C(X_{\max})+(1-X_{\max})\left.\frac{\partial c_C}{\partial X}\right|_{X=X_{\max}}&=0
\end{aligned} \qquad \text{(Eq. 9.45)}$$

in which the derivatives of the relationship $c_C(X)$ are explicitly given by Cirpka and Valocchi (2009).

In regions where no biomass can exist in steady state, the concentration c_C of compound C must meet the conservative advection-dispersion equation, resulting in a linear dependence $c_C(X)$. Then, we arrive at the following expression for $c_C(X)$ for the entire range of the mixing ratio X:

$$\begin{aligned}
c_C(X)&=\begin{cases}X\left.\frac{\partial c_C}{\partial X}\right|_{X=X_{\min}} & \text{if } X<X_{\min}\\ c_C(X)\text{ according to Eqs. 9.43 \& 9.44} & \text{if } X_{\min}\le X\le X_{\max}\\ (X-1)\left.\frac{\partial c_C}{\partial X}\right|_{X=X_{\max}} & \text{if } X>X_{\max}\end{cases}\\
c_A&=c_A^{tot}-\frac{a_A}{a_C}c_C\\
c_B&=c_B^{tot}-\frac{a_B}{a_C}c_C
\end{aligned} \qquad \text{(Eq. 9.46)}$$

The solid lines in Figure 9.3 show the relationships $c_A(X)$, $c_B(X)$, and $c_C(X)$ for the biokinetic reaction A + B → C with 1:1:1 stoichiometry in steady state. The two end-member solutions are identical to the case of an instantaneous reaction (*dashed line* in Figure 9.3). In addition, the Monod coefficient of compound A is 8.33×10^{-5} moles per liter (mol/L), and that of compound B is 3.13×10^{-5} mol/L. The maximum specific reaction rate is 1 mol/g_{bio}/d, the rate coefficient of biomass decay is 0.1/d, and the specific yield is 1 g_{bio}/mol. Steady-state biomass is restricted to mixing-ratio values X in the range between $X_{min} = 0.337$ and $X_{max} = 0.611$. Within this range, we can observe a smooth transition from the linear set of relationships $c_i(X)$ of the range $X < X_{min}$ to that of the range $X > X_{max}$.

In contrast to the case of an instantaneous reaction, neither c_A nor c_B is zero, except for the end points $X = 0$ and $X = 1$. This implies incomplete transformation of the reacting compounds. From the slopes of $c_C(X)$ we can evaluate that in regions with mixing ratios $X < X_{min}$, that is outside of the plume, the concentration c_A of compound A is 13% of the value if there was no degradation at the plume fringe. In the plume core, defined here by $X > X_{max}$, the concentration c_B of compound B is 7% of the value if there was no degradation at the fringe.

The procedure summarized in Equations 9.43, 9.44, 9.45, and 9.46 yields a unique relationship between the mixing ratio $X(x,y,z)$ and the reactive-species concentrations $c_A(x,y,z)$, $c_B(x,y,z)$, and $c_C(x,y,z)$ for mixing-controlled bioreactive transport in steady state assuming dual Monod kinetics. Even though the reaction requires the presence of biomass, the equations presented above do not explicitly include a biomass concentration. This is so because, unlike the concentrations of the dissolved compounds A, B, and C, there is no unique mapping between mixing ratio X and biomass. In order to evaluate the distribution of biomass, the exact spatial setting must be analyzed. In regions where biomass can exist, requiring $X_{min} \leq X(x,y,z) \leq X_{max}$, the biomass concentration can be computed by postprocessing the reactive source term of the reaction product:

$$\frac{\partial c_C}{\partial t} + \mathbf{v} \cdot \nabla c_C - \nabla \cdot (\mathbf{D} \nabla c_C) = \frac{\partial^2 c_C}{\partial X^2} (\nabla X)^T \mathbf{D} \nabla X = a_C r_{max} \left(\frac{c_A}{K_A + c_A} \right) \left(\frac{c_B}{K_B + c_B} \right) c_{bio} \tag{Eq. 9.47}$$

Explicit terms of $c_{bio}(x,y)$ for two-dimensional steady-state transport in uniform flow have been presented by Cirpka and Valocchi (2007).

Figures 9.5 and 9.6 show results of steady-state bioreactive transport considering the same geometry, transport coefficients and inflow concentrations as used for the calculation of

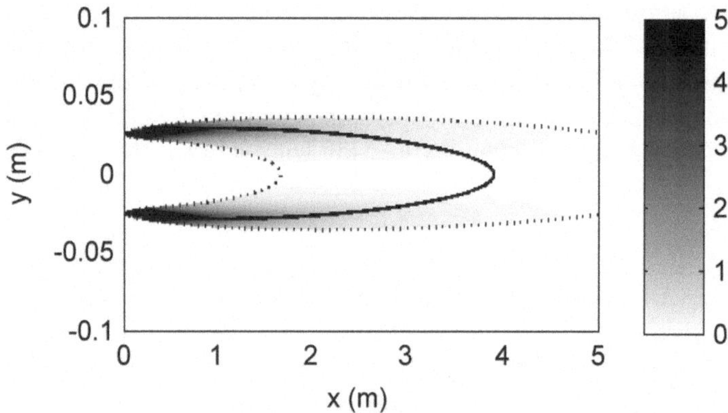

Figure 9.5. Distribution of steady-state biomass [mg/L] in the example calculation of mixing-controlled reactive transport with double-Monod kinetics. *Solid line*: outline of the plume if the reaction was instantaneous; *dotted lines*: boundaries of the zone containing steady-state biomass.

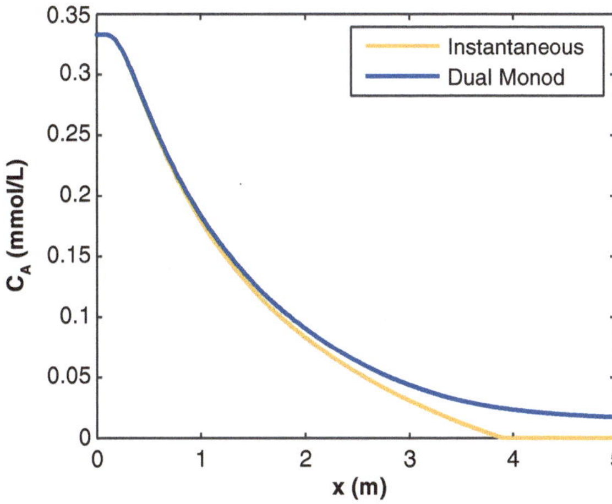

Figure 9.6. Longitudinal concentration profile of the contaminant along the center line in the example calculation of mixing-controlled reactive transport. *Black line*: instantaneous reaction; *gray line*: double-Monod kinetics.

concentrations in the case of an instantaneous reaction. The biokinetic coefficients have been presented in the discussion of Figure 9.3. The spatial concentration distributions of dissolved compounds for bioreactive transport are very similar to those of an instantaneous reaction shown in Figure 9.4. Figure 9.5 shows the distribution of biomass, which is focused along a narrow stripe along the hypothetical boundary of an instantaneously reacting compound. This observation, together with the similarity of dissolved-compound concentrations, indicates that the conceptual model of an instantaneous reaction is a good approximation of bioreactive transport described by double-Monod kinetics.

Figure 9.6 shows the longitudinal concentration profile of the contaminant along the center line of the plume for the cases of an instantaneous reaction and a biokinetic reaction, using the parameters given above. This plot marks an important difference between the two models. In the case of an instantaneous reaction, the contaminant plume has a definite boundary marked by the contour line of the critical mixing ratio X_{crit}. Outside of this line, the concentration of the contaminant is zero. The biokinetic model, by contrast, leads to incomplete degradation. At low concentrations of either the contaminant or the reaction partner from the ambient solution, the Monod terms become very small, resulting in almost negligible reaction rates. As demonstrated by Chu et al. (2005), a threshold concentration of the contaminant can be computed for a given concentration of the reaction partner. The threshold concentration depends on the maximum specific growth rate, $r_{\max}Y$, the two Monod coefficients, K_A and K_B, and on the rate coefficient of biomass decay k_{dec}. The faster the biomass decay, the more incomplete the degradation is in steady state.

Whether the remaining threshold concentration of the contaminant is too high to permit natural attenuation as a management option for a contaminated site depends on the actual value and the standards set for the specific compounds.

The analytical expressions of Cirpka and Valocchi (2007, 2009) have been implemented in the small Matlab program plume2D that allows for rapid assessment of the natural-attenuation potential, which is available from the author upon request (olaf.cirpka@uni-tuebingen.de). Figure 9.7 shows a screenshot of the graphical user interface. The user can choose between plots of the mixing ratio, the concentrations of compounds A, B, and C for instantaneous and biokinetic reactions and the biomass concentration. Plan views, longitudinal or transverse

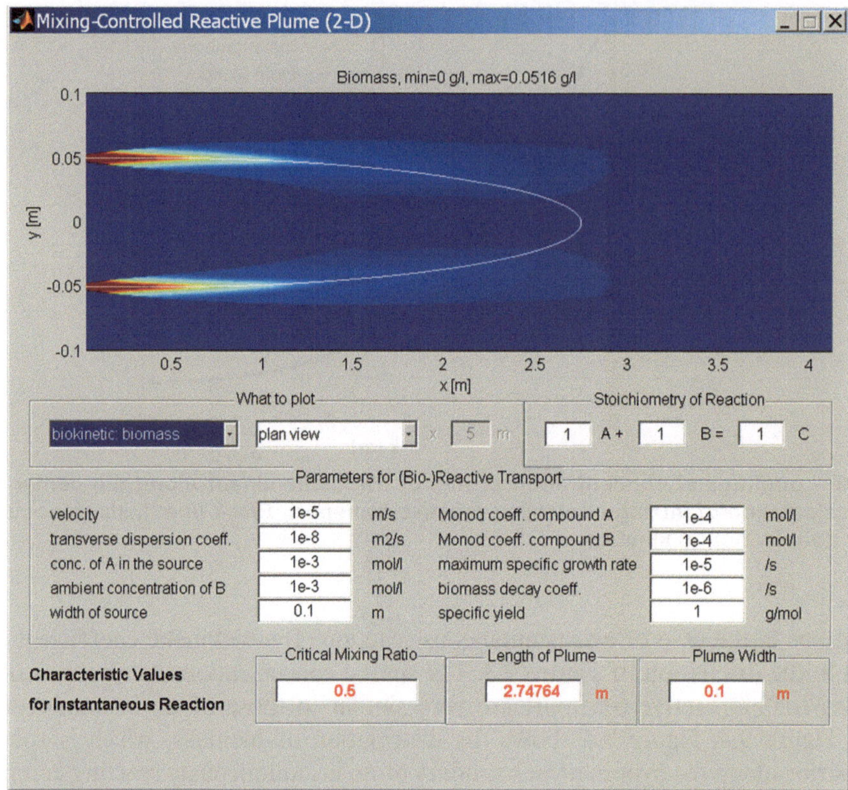

Figure 9.7. Graphical user interface of the Matlab program plume2D for the evaluation of reactive-species concentration distributions in 2-D mixing-controlled reactive transport in uniform flow fields.

profiles are plotted using the analytical expressions discussed in Sections 9.5.3 and 9.5.4. All parameters including the stoichiometry of the reaction can be changed interactively. The graph is replotted immediately after each change. In addition, the program yields the critical mixing ratio X_{crit} for the case of an instantaneous reaction, the corresponding length of the plume, and the widest width. This tool should be preferred over methods that are based on first-order decay, since the latter does not reflect a mixing controlled situation.

9.6 SUMMARY AND CONCLUSIONS

The key points of this chapter may be summarized as follows:

1. If a contaminant is continuously emitted into groundwater and if (bio)degradation requires a dissolved electron acceptor or donor, natural attenuation of the evolving plume will be controlled by transverse mixing between the plume and the ambient groundwater surrounding the plume, at least at late times.
2. The mixing of the plume with the ambient solution can be described by the mixing ratio X, which undergoes conservative transport. At any given location (x,y,z), the mixing ratio $X(x,y,z)$ expresses the volumetric fraction of the water originating from the plume at the inflow boundary at the domain in the mixture with water coming from outside the plume. For simple setups, such as uniform flow in semi-infinite domains, analytical solutions exist.

3. For each reactive system, it is possible to define conservative components that don't change upon the reaction. Prominent examples for conservative components are the so-called total concentrations, summing the concentrations of the original compound and its degradation product. Conservative components mix linearly. Total concentrations at any point within the domain can thus be computed from the mixing ratio $X(x,y,z)$ at that point and the composition of the end members.
4. One may directly evaluate the concentrations of reactive species from the mixing ratio if the relationship between the concentrations of reactive compounds is algebraic. Algebraic relationships do not contain derivatives in space or time. Such algebraic relationships between reactive-species concentrations exist for equilibrium reactions, instantaneous irreversible reactions, and biokinetic reactions in steady state. We denote the computation of the reactive-species concentrations from the conservative components a *speciation* problem, even though it may include cases beyond classical chemical speciation, such as the evaluation of organic and inorganic carbon concentration from that of total carbon.
5. For the case of unique speciation, the reaction rate is proportional to $(\nabla X)^T \mathbf{D} \nabla X$, in which $\mathbf{D}$ is the local dispersion tensor, and ∇X is the spatial gradient of the mixing ratio. This underlines the importance of local dispersion for mixing. The reaction rate is also proportional to the second derivative $\partial^2 c_i / \partial X^2$ of a reactive-species concentration with respect to the mixing ratio. This implies that a linear relationship $c_i(X)$ describes a situation without reaction.
6. The steady-state plume length of contaminants that instantaneously react with compounds from ambient groundwater is inversely proportional to the bulk transverse dispersivity D_t/v. In two dimensions, it also scales with initial plume width squared. The smaller the transverse dispersivity, the longer the plume is. Doubling the width of the source quadruples the length of the plume.
7. Biokinetic reactions described by double-Monod kinetics lead to concentration distributions similar to the case of an instantaneous reaction, but the degradation remains incomplete. The biomass is concentrated in a narrow strip along the line where an instantaneous reaction would take place.

REFERENCES

Appelo CAJ, Postma D. 2005. Geochemistry, Groundwater and Pollution, 2nd ed. A.A. Balkema Publishers, Leiden, The Netherlands. 649 p.

Baveye P, Valocchi AJ. 1989. An evaluation of mathematical models of the transport of biologically reacting solutes in saturated soils and aquifers. Water Resour Res 25:1413–1421.

Bögli A. 1971. Corrosion by mixing of karst waters. Trans Cave Res Group G.B 13:109–114.

Borden RC, Bedient PB. 1986. Transport of dissolved hydrocarbons influenced by oxygen-limited biodegradation: 1. Theoretical development. Water Resour Res 22:1973–1982.

CGER (Commission on Geosciences, Environment and Resources). 2000. Committee on Intrinsic Remediation: Natural Attenuation for Groundwater Remediation. National Academy Press, Washington, DC, USA. 292 p.

Cirpka OA, Valocchi AJ. 2007. Two-dimensional concentration distribution for mixing-controlled bio-reactive transport in steady state. Adv Water Resour 30:1668–1679, doi:10.1016/j.advwatres. 2006.05.022.

Cirpka OA, Valocchi AJ. 2009. Reply to comments on "Two-dimensional concentration distribution for mixing-controlled bioreactive transport in steady state" by H. Shao et al. Adv Water Resour 32:298–301.

Cirpka OA, Frind EO, Helmig R. 1999. Numerical simulation of biodegradation controlled by transverse mixing. J Contam Hydrol 40:159–182.

Cirpka OA, Olsson Å, Ju Q, Rahman MA, Grathwohl P. 2006. Determination of transverse dispersion coefficients from reactive plume lengths. Ground Water 44:212–221.

Chu M, Kitanidis PK, McCarty PL. 2005. Modeling microbial reactions subject to transverse mixing in porous media: When can the rates of microbial reaction be assumed to be instantaneous? Water Resour Res 41: W06002, doi: 10.1029/2004WR003495.

De Simoni M, Carrera J, Sánchez-Vila X, Guadagnini A. 2005. A procedure for the solution of7 multicomponent reactive transport problems. Water Resour Res 41:W11410, doi:10.1029/2005W R004056.

De Simoni M, Sanchez-Vila X, Carrera J, Saaltink MW. 2007. A mixing ratios-based formulation for multicomponent reactive transport. Water Resour Res 43:W07419, doi:10.1029/2006WR005256.

Domenico PA. 1987. An analytical model for multidimensional transport of a decaying contaminant species. J Hydrol 91:49–58.

Ham PAS, Schotting RJ, Prommer H, Davis GB. 2004. Effects of hydrodynamic dispersion on plume lengths for instantaneous bimolecular reactions. Adv Water Resour 27:803–813.

Huang WE, Oswald SE, Lerner DN, Smith CC, Zheng C. 2003. Dissolved oxygen imaging in a porous medium to investigate biodegradation in a plume with limited electron acceptor supply. Environ Sci Technol 37:1905–1911.

Jankovic I, Steward DR, Barnes RJ, Dagan G. 2009. Is transverse macrodispersivity in three-dimensional groundwater transport equal to zero? A counterexample. Water Resour Res 45:W08415, doi:10.1029/2009WR007741.

Liedl R, Valocchi AJ, Dietrich P, Grathwohl P. 2005. Finiteness of steady state plumes. Water Resour Res 41:W12501.

Sanchez-Vila X, Dentz M, Donado LD. 2007. Transport-controlled reaction rates under local non-equilibrium conditions. Geophys Res Lett 34:L10404, doi:10.1029/2007GL029410.

Srinivasan V, Clement TP, Lee KK. 2007. Domenico solution – Is it valid? Ground Water 45:136–146.

Thornton SF, Quigley S, Spence MJ, Banwart SA, Bottrell S, Lerner DN. 2001. Processes controlling the distribution and natural attenuation of dissolved phenolic compounds in a deep sandstone aquifer. J Contam Hydrol 53:233–267.

Thullner M, Mauclaire L, Schroth MH, Kinzelbach W, Zeyer J. 2002. Interaction between water flow and spatial distribution of microbial growth in a two-dimensional flow field in saturated porous media. J Contam Hydrol 58:169–189.

Werth CJ, Cirpka OA, Grathwohl P. 2006. Enhanced mixing and reaction through flow focusing in heterogeneous porous media. Water Resour Res 42:W12414, doi:10.1029/2005WR004511.

West MR, Kueper BH, Ungs MJ. 2007. On the use and error of approximation in the Domenico (1987) solution. Ground Water 45:126–135.

CHAPTER 10

SOURCE REMEDIATION CHALLENGES

L.M. Abriola,[1] J.A. Christ,[2] K.D. Pennell[1] and C.A. Ramsburg[1]

[1]Department of Civil and Environmental Engineering, Tufts University, Medford, MA, USA; [2]Department of Civil and Environmental Engineering, U.S. Air Force Academy, Colorado Springs, CO, USA

10.1 INTRODUCTION AND BACKGROUND

Widespread use of chlorinated solvents in textile dry cleaning and metal degreasing operations has resulted in extensive groundwater contamination by compounds such as perchloroethene (PCE; also termed perchloroethylene or tetrachloroethylene) and trichloroethene (TCE). Chlorinated solvents are often released into the environment as a separate, sparingly miscible, organic liquid phase, commonly known as a *dense nonaqueous phase liquid* (DNAPL). Although chlorinated solvents are the most common components of DNAPLs, it is not uncommon for these organic phases to contain substantial concentrations of oils or petroleum hydrocarbon compounds that are lighter than water constituents. Thus, it is important to make a distinction between a DNAPL constituent, that may be denser or lighter than water in its pure liquid form, and the DNAPL itself, that is a separate phase (potentially multicomponent) organic liquid. In this chapter, the term DNAPL is used to refer to the separate phase organic liquid.

When introduced into the subsurface, DNAPLs tend to migrate downward through the unsaturated zone and, depending upon DNAPL characteristics and release conditions (e.g., density, volume released, and release rate), can continue their downward migration across the water table, displacing water and infiltrating deep into the saturated zone. Under certain conditions, governed by local liquid and porous medium properties, vertical migration can proceed quite rapidly beneath the water table, accelerated by the onset of flow instabilities or 'fingering' (Zhang and Smith, 2002). During its downward migration, hysteretic capillary forces act to retain a portion of the DNAPL as discontinuous globules or ganglia within the pore structure (Chatzis et al., 1983; Lake, 1989). This retained DNAPL volume is commonly referred to as 'residual', and typically occupies between 10% and 35% of the pore volume (pv) in water-wet porous media (Pennell et al., 1993; Powers et al., 1992; Schwille, 1988). Substantial DNAPL mass can also be retained at textural interfaces, resulting in higher saturation DNAPL 'pools' that extend laterally above layers or lenses of lower-permeability (Feenstra et al., 1996). Subtle textural variations can create differences in capillary properties sufficient to induce DNAPL pooling even in apparently homogeneous units (Kueper et al., 1993). These processes tend to result in horizontal spreading over large distances relative to the vertical migration pathways, creating highly irregular spatial distributions of DNAPL mass. DNAPL distributions, thus, depend strongly on fine scale subsurface material textural and surface characteristics, as well as fluid properties, including density, viscosity, and interfacial tension with the pore water (Dekker and Abriola, 2000a; Mercer and Cohen, 1990; Miller et al., 1998). Given that DNAPLs tend to follow a tortuous migration pathway, only a small fraction of the subsurface volume at a contaminated site will actually contain the organic liquid and the location of the major part of this contaminant mass is often distant from the point of release. Within the contaminated

P.K. Kitanidis and P.L. McCarty (eds.), *Delivery and Mixing in the Subsurface: Processes and Design Principles for In Situ Remediation*, doi: 10.1007/978-1-4614-2239-6_10, © Springer Science+Business Media New York 2012

volume, the three dimensional distribution of the DNAPL will be highly irregular, with mass divided between entrapped ganglia and more highly-saturated "pools." The term 'architecture' is now commonly used to describe this complex distribution (i.e., extent, saturation, shape, location) of DNAPL mass within the subsurface (Sale and McWhorter, 2001; Moreno-Barbero et al., 2004).

The major characteristics of DNAPL infiltration beneath the water table are illustrated in Figure 10.1, which presents a photograph of a laboratory-scale aquifer cell experiment. The cell represents a quasi two dimensional release scenario with overall dimensions of 150 centimeters (cm) (length) by 48 cm (height) by 1.5 cm (thickness). Here, neat PCE (dyed red) was released into a water-saturated quartz sand media, with grain diameters that ranged from 0.05 to 0.85 millimeters (mm). Note that the finer-grained material, which appears as a lighter color in Figure 10.1a, acts as a capillary barrier to downward PCE migration; as the PCE migrated downwards under the influence of gravity, it encountered this finer material and spread laterally, pooling at the textural interfaces before cascading down to lower portions of the aquifer cell. Extensive pooling also occurred along the bottom of the cell, which serves as an impermeable barrier. Inspection of Figure 10.1a reveals the small thicknesses of the pools perched on the finer lenses and the presence of vertical DNAPL 'fingers', that are barely discernable downward paths to lower regions of the cell. The resulting PCE-DNAPL saturation distribution shown in Figure 10.1b was obtained using light transmission analysis. Note that saturation values (calculated over individual pixels, representing a volume of 0.15 mm^3) vary from a high of 0.45 within the pool to 0.01 in the ganglia region. Such variations in DNAPL saturation are consistent with field observations made in a heterogeneous source zone containing layers or lenses of fine-textured media (e.g., Kueper et al., 1993).

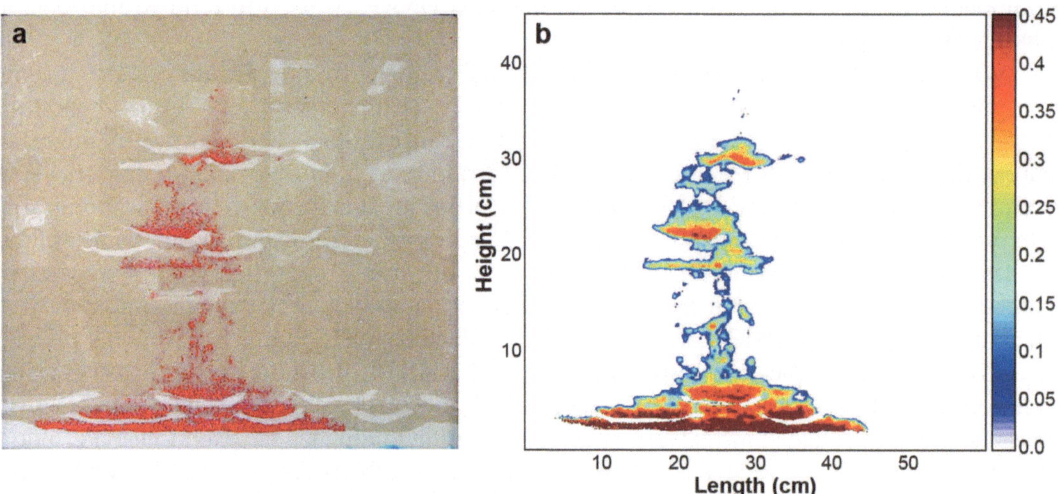

Figure 10.1. Photograph (a) and light transmission image (b) of a PCE-DNAPL saturation distribution, illustrating regions of ganglia, fingers, and high-saturation pools above lower-permeability media.

Once introduced, entrapped and pooled DNAPL mass tends to dissolve slowly into flowing groundwater, serving as a long-term source for downgradient plume contamination. It is now widely recognized that restoration of DNAPL-contaminated sites to pre-contamination concentration levels is rare, and may not be achievable within an acceptable time frame and cost using available technologies for many types of aquifer formations (e.g., USEPA, 2003; NRC, 2005). For purposes of DNAPL site management, it has become useful to distinguish between the highly contaminated region of the formation that 'feeds' the plume and the downgradient plume region. The former, is commonly referred to as the 'source zone' and is understood to incorporate all of the subsurface material that contains DNAPL. Although a number of distinct definitions for the source zone exist, the definition used herein is consistent with that proposed by USEPA (2003); for our purposes, the source zone is defined as "the closed region that encompasses all of the DNAPL mass present in the formation, as well as the subsurface material that was once in contact with DNAPL" (Figure 10.1). Note that, as defined here, the source zone also includes contaminant mass associated with the solid phase (e.g., sorption processes) and dissolved in groundwater. Thus, local concentrations (total mass concentrations) may vary substantially from point to point within a DNAPL source zone as a function of aquifer material properties, release history, and flow conditions.

The quantity and distribution of dissolved-phase contaminant mass emanating from a source zone are the primary factors controlling downgradient plume characteristics. Plume evolution from a source zone is illustrated in Figure 10.2, which presents results of a surfactant enhanced remediation experiment conducted in an aquifer cell. Here changes in the PCE-DNAPL source zone mass distribution, resulting from sequential injections of a solubilizing surfactant solution, are shown to have a substantial influence on the downgradient dissolved-phase plume concentrations (Suchomel and Pennell, 2006). In the effort to understand plume

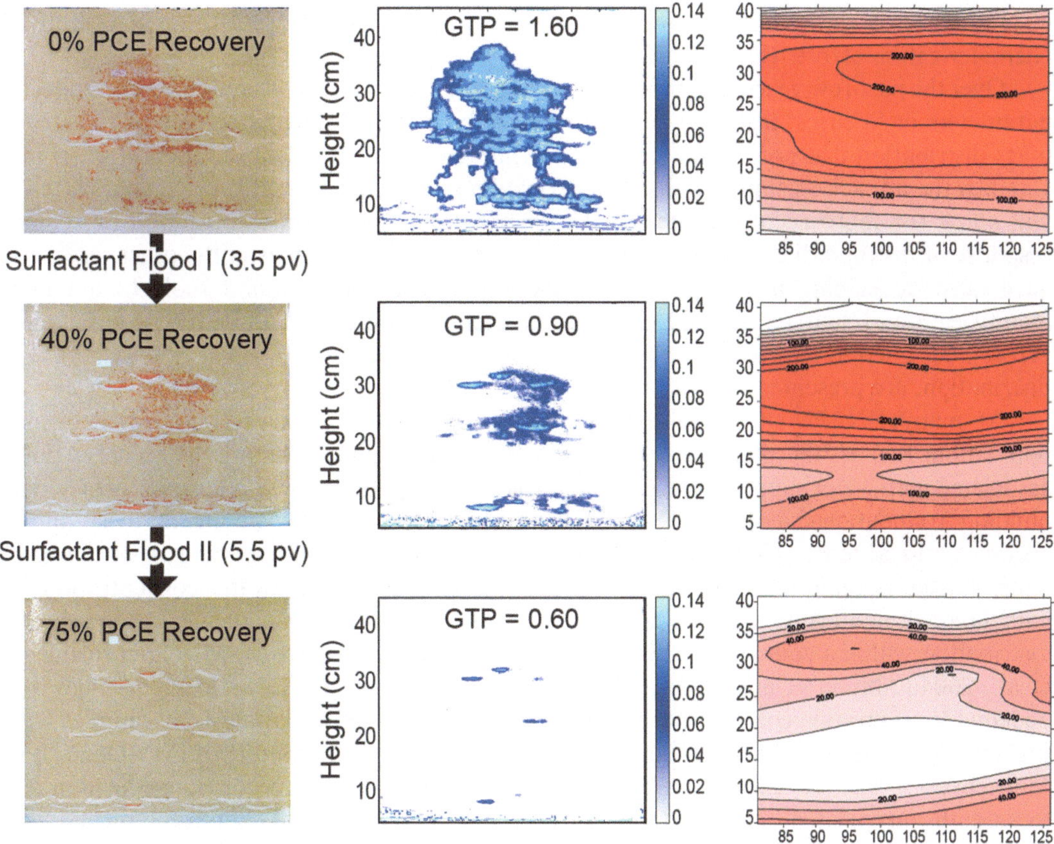

Figure 10.2. PCE-DNAPL source zone saturation distributions and corresponding downgradient dissolved-phase plume concentrations prior to surfactant flushing (*top row*), after one surfactant flood (*middle row*), and after a second surfactant flood (*bottom row*) (adapted from Suchomel and Pennell, 2006). Note: GTP – ganglia-to-pool.

longevity and to evaluate alternative plume management strategies, it, thus, becomes crucial to assess and quantify the relationship between DNAPL architecture and mass flux emanating from the source zone. Note that mass flux is defined as the mass (M) crossing a unit cross-sectional area of the formation (L^2), perpendicular to the average groundwater flow, per unit of time (t) (i.e., $M/L^2 \cdot t$). In field characterization work, it is important to differentiate between *mass flux* and the other commonly employed metric *mass discharge*. Mass discharge is the product of the mass flux and the cross-sectional area of interest and has units of mass per time (M/t). Thus, the spatial distribution of mass flux can be meaningfully delineated within a plume transect, but mass discharge will have only a single value for that transect (see Section 10.3.2 – for a more detailed discussion of mass flux and mass discharge). Mass flux is a complicated function of the subsurface flow field (and its spatial variability), DNAPL mass accessibility to the flow, and local DNAPL composition and constituent dissolution rates. Clearly, detailed knowledge of DNAPL mass distribution and composition within a source zone could provide invaluable information for site assessment. Unfortunately, such detailed knowledge is unavailable at virtually all sites, since exhaustive sampling of the subsurface is generally cost prohibitive, can pose risks (application of intrusive characterization methods may cause further DNAPL mobilization), and non-intrusive characterization methods lack the appropriate levels of spatial resolution and quantification. The challenge of source zone remediation is, thus, inextricably linked to the challenge of source zone characterization. In several cases, the failure of source zone treatment technologies to achieve remedial goals at DNAPL sites with complex hydrogeology can be traced to incomplete or inaccurate site characterization efforts (USEPA, 2003).

Over the past few years, a number of investigators have developed source zone architecture parameters or ‘metrics’ that may be used to describe particular characteristics of the DNAPL mass distribution that are associated with downgradient plume response. Their work is motivated by the implicit assumption that quantification of selected up-scaled parameters (metrics) will allow for more accurate, site-specific, prediction of the natural evolution or remedial response of downgradient contaminant flux in space and time. Concurrently, other researchers have been developing and refining site characterization methods and tools to aid in the quantification of source zone DNAPL mass distribution. Results of these recent and ongoing efforts, designed to improve our ability to predict and characterize DNAPL source zone mass distribution, are presented in Section 10.2.

At many DNAPL sites, the contaminant plume poses the largest risk to the general population; that is, the plume is responsible for transporting pollutants to receptor points, associated with water supply wells, discharges to surface water bodies, and vapor intrusion into building foundations/basements. Since the source zone feeds the plume at these sites, it becomes critical to be able to quantify and characterize the mass emanating from the source zone. Furthermore, it is now generally understood that local groundwater concentrations can vary quite dramatically within a source zone region. For example, the PCE concentration distribution within a source zone at the Bachman Road site in Oscoda, Michigan (Ramsburg et al., 2004; Abriola et al., 2005; Ramsburg et al., 2005) is illustrated in Figure 10.3. Here aqueous concentration contours have been developed based upon 26 multi-level samples within the domain, using the GSLIB sequential Gaussian simulation program (SGSIM) (Deutsch and Journel, 1998). The Gaussian assumption at the foundation of the interpolation algorithm tends to smooth interpolated concentrations within the source zone. This is a common drawback of most standard data interpolation algorithms in current use. Despite the smoothing effect, one can observe that concentrations in this small region are highly irregular and vary over more than three orders-of-magnitude.

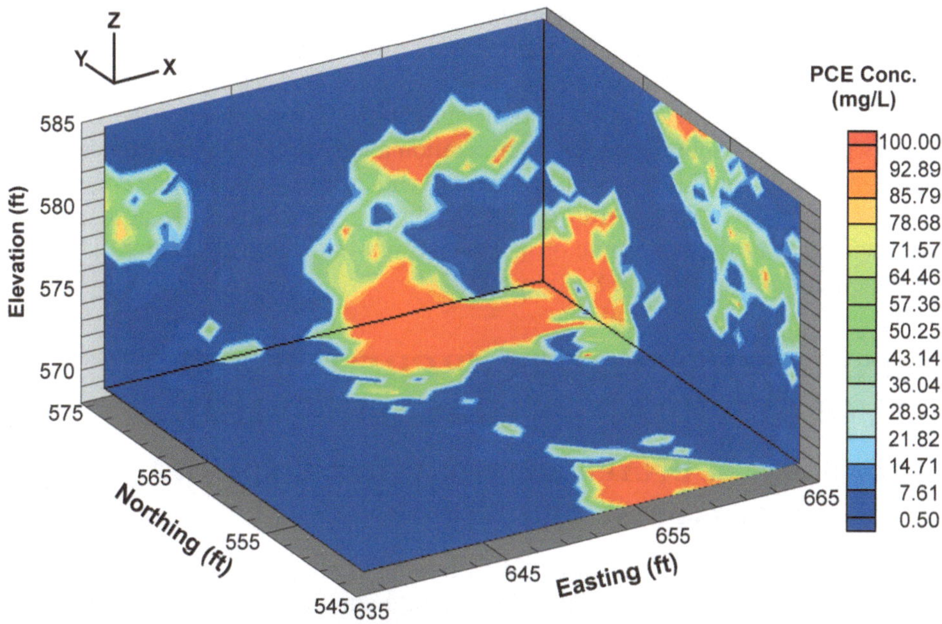

Figure 10.3. Representation of PCE concentrations (milligrams per liter [mg/L]) based upon measurements at 26 multilevel sampling points within the PCE-DNAPL source zone at the Bachman Road site. The volume shown here is a portion of the source zone, corresponding to the treated volume during a pilot-scale surfactant flushing test. The elevation of the water table at the Bachman Road site was about 585 feet (ft) above mean sea level when these concentrations were measured. The lower boundary of the figure represents the clay layer present at the Bachman Road site at approximately 568 ft above mean sea level.

Reduction of concentration levels throughout a source zone to meet maximum concentration level (MCL) standards is generally thought to be extremely difficult, if not infeasible (e.g., Sale and McWhorter, 2001). However, it is now recognized that aggressive mass removal within the source zone can lead to significant changes in the downstream mass flux (e.g., Rao and Jawitz, 2003; Lemke and Abriola, 2006; Suchomel and Pennell, 2006). Thus, mass flux response could be employed as an alternative metric to assess the effectiveness of a particular remediation effort.

Based on our improved understanding of source zone characteristics, increased attention has focused on the quantification of mass flux over the past few years. Indeed, a number of individuals have advocated that integrated downgradient mass flux or mass discharge be considered as a preferred remediation endpoint (e.g., an alternative to local dissolved-phase concentrations) (e.g., Einarson and Mackay, 2001; Rao et al., 2002). Unlike concentration, however, mass flux is not measured directly, but rather is inferred from other types of *in situ* measurements, such as passive flux meters (Hatfield et al., 2004) and integrated pump tests (Bockelmann et al., 2001; Basu et al., 2009). Typically, such measurements are distributed in space, as well as in time, and may have different scales of measurement support. Thus, quantification of source mass discharge at a site may be associated with substantial uncertainty. An overview of recent research directed towards quantifying the relationship between DNAPL architecture and mass flux and in developing tools for mass flux quantification and uncertainty analysis is provided in Section 10.3.

A number of *in situ* technologies, including surfactant and cosolvent flushing, steam injection, electrical resistive heating, air sparging, and chemical oxidation or reduction, have been developed to aggressively treat DNAPL source zones. Over the past decade, field-scale tests have demonstrated the ability of these technologies to recover or destroy substantial quantities of contaminant mass (e.g., Abriola et al., 2005; Falta et al., 1999; Heron et al., 2002, 2005; Jawitz et al., 2000; Londergan et al., 2001; Lowe et al., 2002; Ramsburg et al., 2005). In order to properly design and implement *in situ* remediation technologies, it is essential that the overall extent, accessibility, composition, and spatial distribution of the DNAPL source zone is known. Detailed source zone characterization and real time monitoring can provide the data necessary for targeted delivery of remedial agents, thereby minimizing costs and improving mass recovery. Possible benefits of partial mass removal and the potential for using combined remedies, at the same time or in series, to more effectively address DNAPL source zone management are explored in Section 10.4.

10.2 DNAPL SOURCE ZONE ARCHITECTURE: EVOLUTION AND CHARACTERIZATION

10.2.1 Influence of DNAPL and Subsurface Properties on Source Zone Architecture

As discussed above, the saturation distribution arising from a DNAPL spill will be influenced by a number of factors, including the rate and volume of the DNAPL release, the density, viscosity, and interfacial tension of the fluid phases, as well as the surface chemistry, permeability distribution, and initial water content of the porous medium (Dekker and Abriola, 2000a; Mercer and Cohen, 1990; Miller et al., 1998). In subsurface systems containing two liquid phases (i.e., water and DNAPL), the minimum height of a continuous DNAPL pool (*H*) that is required to overcome the entry pressure of a capillary barrier can be estimated as (McWhorter and Kueper, 1996):

$$H = 9.6\left(\frac{\rho_w}{\Delta\rho}\right)\left(\frac{\sigma_{nW}}{\sigma_{aw}}\right)\left(\frac{K}{\varepsilon}\right)^{-0.403} \quad \text{(Eq. 10.1)}$$

where ρ_w is the density of water, $\Delta\rho$ is the difference between the density of the DNAPL and water, σ_{nw} is the interfacial tension between the DNAPL and water, σ_{aw} is interfacial or surface tension between air and water, K and ε are the hydraulic conductivity and porosity, respectively, of the lower-permeability layer. Equation 10.1 indicates that DNAPLs with high densities relative to water, such as chlorinated solvents (e.g., PCE has a liquid density of 1.62 grams per milliliter [g/mL]), are more likely to penetrate a lower permeability layer and therefore less likely to form thick pools in comparison to DNAPLs with lower densities, such as coal tars (McWhorter and Kueper, 1996). Use of this relation is supported by observations of DNAPL behavior in the field, where pools of PCE-DNAPL ranging in thickness from a few millimeters to a few centimeters have been observed immediately above laminations and lenses of lower-permeability media (Kueper et al., 1993). Spill release characteristics will also influence the DNAPL architectural features. Hofstee et al. (1998a, b) investigated the infiltration and redistribution of PCE-DNAPL in quartz sands. Their observations suggested that continuous DNAPL spills are more likely to accumulate a pool height sufficient to penetrate horizontal fine layers and migrate vertically, in contrast to the behavior of short duration or intermittent releases that tend to flow laterally along capillary barriers. Thus, prolonged or continuous release of DNAPLs into heterogeneous aquifer formations containing low permeability media may result in a greater proportion of DNAPL mass existing as discrete ganglia at or near residual saturation.

The wettability, or tendency for a fluid to spread or "wet" a solid surface, of an aquifer material can also have a dramatic effect on DNAPL residual saturation and source zone architecture.

The angle formed by the tangent to the organic-water interface at the point where it contacts the solid surface is known as the contact angle of the liquid on a surface. Contact angle is often used as a measure of wettability; when the contact angle of an organic liquid drop in water is less than 75 degrees (°) (measured from the horizontal surface) the solid is considered to be water-wetting, while surfaces exhibiting contact angles of greater than 120° are considered to be nonaqueous phase liquid (NAPL) wetting, and intermediate contact angles are associated with neutral-wetting solids. The wettability of aquifer materials can vary both spatially and temporally, associated with grain mineralogy and surface coatings, and can be altered through the injection of remedial agents such as surfactants (Demond et al., 1994; Karagunduz et al., 2001), long-term exposure to NAPLs (Powers and Tamblin, 1995), and changes in aqueous geochemistry (Lord et al., 2000). Bradford et al. (1998) reported that spatial variations in contact angle that were correlated to intrinsic permeability resulted in higher DNAPL saturations, increased lateral spreading, and decreased infiltration depth. In two-dimensional aquifer cell experiments, layers of organic-wetting sand were shown to strongly retain PCE-DNAPL, increasing the residual saturation and inhibiting downward migration of DNAPL (O'Carroll et al., 2004).

As represented by Equation 10.1, the DNAPL-water interfacial tension (σ_{nw}) strongly influences the DNAPL entry pressure. Consistent with this expression, surfactant facilitated reductions in interfacial tension have been shown to substantially reduce the capillary resistance to vertical migration of PCE-DNAPL, allowing PCE to enter lenses of lower permeability not penetrated in the absence of surfactant (Rathfelder et al., 2003). Similarly, injection of cosolvent (100% ethanol) and surfactant (4% Aerosol MA/OT) solutions for source zone remediation have been demonstrated to mobilize entrapped DNAPLs, resulting in substantial DNAPL redistribution, formation of pools, and in some cases, penetration into lower confining units (van Valkanburg and Annable, 2002; Ramsburg and Pennell, 2001). The potential for reductions in interfacial tension to result in mobilization of entrapped DNAPL was explored by Pennell et al. (1996), who developed the total trapping number (N_T) concept to describe the combined influence of viscous, capillary, and gravitational forces on the entrapment and displacement of residual NAPLs;

$$N_T = \sqrt{N_{Ca}^2 + 2N_{Ca}N_B \sin\alpha + N_B^2} \quad \text{(Eq. 10.2)}$$

$$\text{where} \quad N_{Ca} = \frac{q_w \mu_w}{\sigma_{nw} \cos\theta} \quad \text{(Eq. 10.3)}$$

$$N_B = \frac{\Delta\rho g k k_{rw}}{\sigma_{nw} \, cos\theta} \quad \text{(Eq. 10.4)}$$

Here, N_{Ca} is the capillary number, N_B is the Bond number, q_w is the Darcy velocity of the aqueous phase (positive upward), α is the angle of flow relative to the horizontal, μ_w is the dynamic viscosity of the aqueous phase, θ is the contact angle, g is the gravitational constant, k is intrinsic permeability of the aquifer material, and k_{rw} is the relative permeability to the aqueous phase. Equation 10.2 reduces to $N_T = \sqrt{N_{Ca}^2 + N_B^2}$ for horizontal flow ($\alpha = 0°$) and $N_T = |N_{Ca} + N_B|$ for vertical flow ($\alpha = 90°$). In one-dimensional column experiments, the onset of PCE-DNAPL mobilization corresponded to N_T values ranging from 2×10^{-5} to 5×10^{-5}, while complete ganglia displacement was observed at N_T values approaching 1×10^{-3} (Pennell et al., 1996). In most DNAPL contamination scenarios, gravity (the Bond number) dominates N_T. Thus, low values of interfacial tension and large density DNAPLs (i.e., larger trapping numbers) will favor formation of pools and mobilization of DNAPL from coarser-grained into finer-grained subsurface media. Application of the trapping number concept to assessment of the risk of mobilization in the field, however, ultimately requires detailed knowledge of DNAPL architecture and its relation to variations in media texture (permeability).

As described above, the process of a non-wetting DNAPL infiltrating within a porous medium generally involves vertical migration until a finer-grained layer (capillary barrier) is encountered. Lateral spreading will occur until either a sufficient thickness of DNAPL accumulates to overcome the entry pressure or a pathway with lower entry pressure is encountered. In many instances, however, the presence of small scale textural variations (creating small-scale variations in entry pressure) favors the appearance of flow instabilities (Smith and Zhang, 2001), leading to the formation of narrow vertical preferential flow pathways, commonly known as fingers, that can serve as rapid conduits for DNAPL migration deep into the subsurface. Subtle textural variations that create differences in entry pressure sufficient to cause the onset of fingering can occur even in apparently homogeneous units (Kueper et al., 1993). Because fine-scale textural variability is not easily quantifiable in the field, these preferential pathways can appear to be somewhat random in their distribution (see for example, Glass et al., 2000; Rathfelder et al., 2003). The propagation of fingers is more common in coarse textured media and under conditions of low interfacial tension and high DNAPL density, since capillary forces will tend to oppose the formation of extensive fingers (Held and Illangasekare, 1995). Given their small lateral dimensions (typically less than a centimeter in width), fingers (and consequently, DNAPL migration pathways) are extremely difficult to locate in the field. Although fingering can be a very important propagation mechanism in DNAPL releases, mass contained within fingers will likely be at low saturations and represent a very small percentage of the total DNAPL mass within a source zone. Thus, fingers are expected to be less import (in comparison to ganglia and pools) in controlling longer term mass flux and source longevity.

10.2.2 Characterization Tools

Large technological strides in DNAPL source characterization have been made over the last two decades. Application of rapid screening tools and online, in-field, data analysis has accelerated interest in dynamic work plans during site characterization (Robbat et al., 1998; Pitkin et al., 1999; Costanza and Davis, 2000; Guilbeault et al., 2005). Several reviews present and assess the suite of noninvasive and invasive techniques now routinely employed to identify and delineate DNAPL source zones (ITRC, 2000, 2003; Kram et al., 2001, 2002; USEPA, 2003; NRC, 2005). Noninvasive methods include use of historical data, information on the regional geology, soil gas surveys, and surface geophysics. These characterization approaches generally provide qualitative lines of evidence to support delineation or characterization of NAPL source zones. Invasive technologies include direct push sampling, core analysis, downhole testing, pump tests, and partitioning interwell tracer tests. When employed correctly, most invasive techniques can provide meaningful, quantitative information. Invasive technologies tend to be more expensive to implement, and use of these technologies may alter the DNAPL architecture. Alteration of the DNAPL architecture is problematic from two perspectives: (1) any mobilized DNAPL may migrate downward through the preferential pathway created by the borehole casing or direct push rod, increasing the extent of the contaminated source region and (2) assessments of the architecture near the sampling location may not be representative of the overall source-zone architecture.

While extensive, fine-scale source zone DNAPL characterization may ultimately prove feasible, the cost and intrusiveness (risk) of existing characterization technologies have led to the consideration of alternative approaches to field characterization. A growing body of research and field data suggests that a detailed, fine-scale reconstruction of the spatial distribution of DNAPL saturations may not be necessary for order-of-magnitude remedial performance prediction. This realization has led to the emergence of interest in averaged characterization metrics that represent the salient features of the DNAPL architecture (e.g., Jawitz et al., 2005; Christ et al., 2006; Saenton and Illangasekare, 2007). The sections below review those technologies that hold potential for characterizing source zone architecture or associated source zone metrics. While each approach continues to evolve to meet the challenges of architecture assessment, the merits of technology integration should not be overlooked (Rossabi et al., 2000; Brusseau et al., 2007). A detailed treatment of all DNAPL source zone characterization methods is beyond the scope of this chapter. Interested readers are referred to the references cited above.

10.2.2.1 Surface-Based Geophysics

A number of geophysical methods are available for noninvasive characterization and monitoring of the subsurface environment (Benson, 2006). Because most methods are not currently able to detect and quantify DNAPL saturations (USEPA, 2003; NRC, 2005), applications of geophysical methods in DNAPL source zone characterization are typically limited to characterization of subsurface geology or hydrogeology. Detection and quantification of DNAPL at a spatial resolution appropriate to the scale of source zone architecture remains an open area of research. Recent studies related to seismic refraction and reflection, electrical impedance and resistivity, and ground penetrating radar suggest that these techniques may hold promise for surface-based remote sensing (see, for example, Temples et al., 2001; Grimm et al., 2005; Johnson and Poeter, 2007). Ground penetrating radar has been used to track DNAPL migration when site geology is well characterized prior to controlled spills. Static detection of DNAPL, however, is problematic because reflections are nonspecific (i.e., geologic and DNAPL features tend to appear similar) (Greenhouse et al., 1993). These studies are complemented by laboratory-scale investigations aimed at assessing the acoustic and electromagnetic properties of DNAPLs and soil systems containing water and DNAPL (Carcione et al., 2003; Ajo-Franklin et al., 2006, 2007). There is growing evidence that comparative temporal analysis using noninvasive techniques may prove beneficial in monitoring clean-up in the near surface environment (<15 meters (m) below ground surface), particularly when coupled or augmented with other characterization methods (Greenhouse et al., 1993; Newmark et al., 1997; Sneddon et al., 2000; Johnson and Poeter, 2005, 2007; Ajo-Franklin et al., 2007; Stewart and North, 2006).

10.2.2.2 Inverse Methods

The utility of inverse modeling techniques in multiphase flow systems has been long recognized by the petroleum engineering community. Inverse simulations of petroleum reservoir production seek to match production history (i.e., multiphase flow) through adjustment of parameters in constitutive relations (e.g., relative permeability- and capillary pressure-saturation relationships). Assessment of source architecture through the use of inverse methods, however, has received relatively little attention until recently (Saenton and Illangasekare, 2003; Maji et al., 2006; Newman et al., 2006; Sun et al., 2006). When applied to source zone architecture, inverse modeling makes use of spatially-distributed measurements (estimates) of mass flux (Saenton and Illangasekare, 2003). More reliable methods for quantifying mass discharge and mass flux have recently become available (see Section 10.3.2), and have thereby opened the possibility of employing inverse modeling for assessment of source architecture. Reviews of the various approaches for inverting mass flux data can be found in

Atmadja and Bagtzoglou (2001) and Michalak and Kitanidis (2004). High resolution mass flux data are, however, insufficient to provide a unique solution to such inverse simulations. The non-uniqueness of the inversion may require additional types of data, which may take the form of stratigraphic information or tracer response (James et al., 2000; Zhang and Graham, 2001; Datta-Gupta et al., 2002; Jawitz et al., 2003; Moreno-Barbero et al., 2004; Enfield et al., 2005; Johnson and Poeter, 2007; Sun, 2007).

10.2.2.3 Tracer Tests

Conservative, non-reactive tracer tests employing solutes that do not sorb, partition, or degrade are commonly used to evaluate important subsurface transport properties. However, such tracers cannot assess DNAPL saturations or architectural features. Tracer technologies for DNAPL source zone characterization are derivative of techniques first used to identify and assess petroleum reserves (Deans, 1971; Cooke, 1971; Tomich et al., 1973). Application of similar techniques for characterization of DNAPL source zones relies upon partitioning and sorption of selected solutes between the flowing aqueous and immobile organic phases (Jin et al., 1995; Annable et al., 1998a, b; Istok et al., 2002). Partitioning interwell tracer tests (PITTs) are typically implemented using a line-drive or five spot well pattern in a well-controlled flow field (Brooks et al., 2002). The test is conducted by introducing a pulse containing a suite of partitioning and non-partitioning tracers. Aqueous-phase concentration measurements for all tracers at specific observation points (e.g., extraction or multilevel wells) are used to construct breakthrough curves. These breakthrough curves are then analyzed to determine the extent of retardation associated with the partitioning tracers. Retardation of the partitioning tracers is typically assumed to be the result of equilibrium, linear partitioning, quantified by the NAPL-water partition coefficient (K_{NW}) of the tracer. The degree of retardation is then related to the volume of NAPL along the flow path (Jin et al., 1995; Annable et al., 1998a). PITTs have been successfully applied for characterization of numerous NAPL source zones (e.g., Annable et al., 1998a; Meinardus et al., 2002). These tests, however, provide estimates of saturation that are averaged over the volume interrogated by the test (Brusseau et al., 1999; Rao et al., 2000). Such integrated DNAPL saturations are essentially estimates of total DNAPL mass and do not quantify DNAPL architecture. Higher resolution estimation of the spatial distribution of DNAPL saturation may be possible using well-designed PITTs that incorporate acquisition of high-density temporal tracer concentration data at numerous locations within the source zone (James et al., 1997). The benefits of higher resolution implementations of PITTs, however, must be carefully considered in the context of the associated costs and the attendant risk of DNAPL mobilization. Thus, although PITTs can be a useful tool for estimating volume-averaged DNAPL mass between observation points, they are costly and complex to employ (Jackson and Jin, 2005). The increased monitoring and chemical analyses necessary for high spatial resolution may further constrain deployment of the PITT technique, which is already maligned as being only practicable at large federal facilities (USEPA, 2003).

Because PITT technologies are based upon parametric fitting of system responses (tracer breakthrough curves), they are also a form of inverse modeling and consequently, reliant on a number of assumptions and subject to some of the non-uniqueness problems described above. For example, PITT data analysis typically relies upon the assumption of linear equilibrium partitioning of the tracer species between phases, despite evidence suggesting that this partitioning may be nonlinear (Wise, 1999; Wise et al., 1999) and kinetic (Willson et al., 2000; Imhoff and Pirestani, 2004; Moreno-Barbero and Illangasekare, 2006). Jawitz et al. (2003) suggest that problems relating to nonlinearity may be overcome by careful selection of tracers and concentration levels, since in the dilute range, many tracers display near linear partitioning behavior

(Rao et al., 2000). Mass transfer limitations and the bypassing of aquifer regions containing higher NAPL saturations, however, require additional consideration during PITT design. Indeed, these effects may help explain discrepancies between actual and estimated NAPL in the blind PITT conducted by Brooks et al. (2002). Thus, spatially varying injection rates, residence time control, and assessment using kinetic models may prove necessary for accurate delineation of the source zone architecture when using PITTs (Dai et al., 2001).

10.2.3 Source Zone Architecture Metrics

A number of alternative metrics have been proposed to characterize source-zone DNAPL distributions, in place of fine-scale saturation delineation. These metrics generally represent averaged properties of the saturation distribution and include: domain-averaged saturations (i.e., statistical metrics) (e.g., Dekker and Abriola, 2000a; Jawitz et al., 2003); distribution moments (i.e., spatial metrics) (e.g., Lemke et al., 2004a; Christ et al., 2005a); stochastic-advective trajectory averaged saturations (e.g., Jawitz et al., 2003, 2005); and ganglia-to-pool (GTP) mass ratios (Lemke et al., 2004b; Christ et al., 2005a). Domain averaged saturation metrics are commonly reported at field sites (e.g., Annable et al., 1998a), but provide little information to link source zone architecture to plume response for a given remediation strategy (Mayer and Miller, 1996; Dekker and Abriola, 2000b; Lemke et al., 2004b). For example, consider Figure 10.4a and b that present hypothetical spill scenarios for two geologic formations

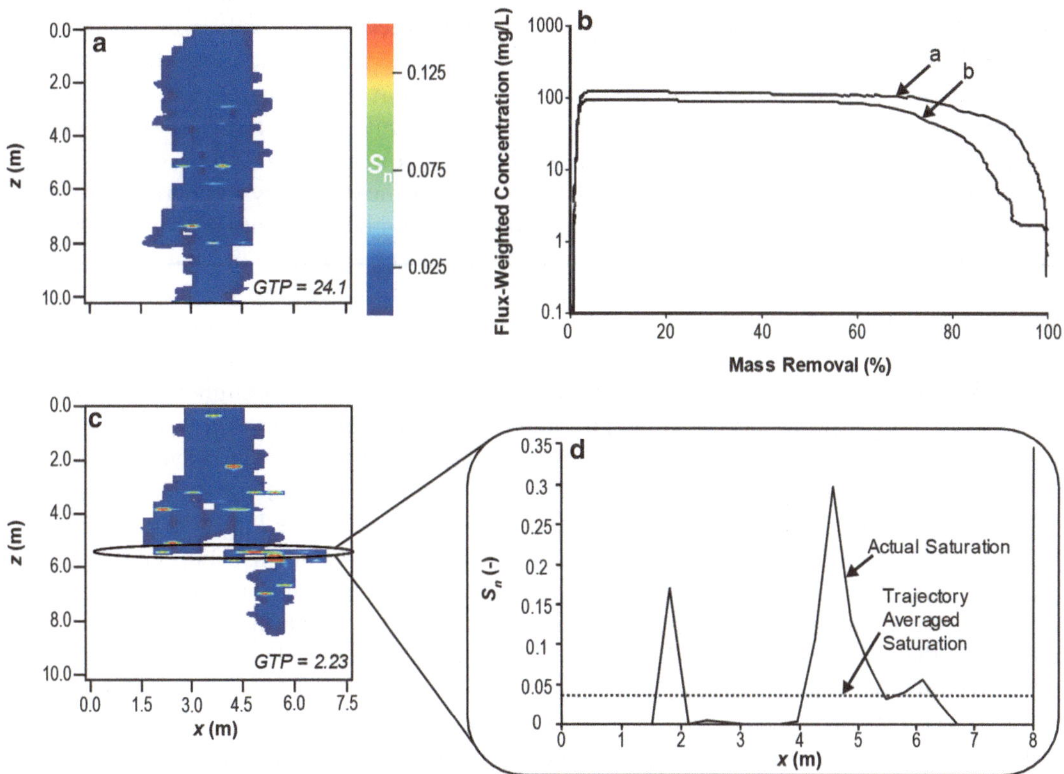

Figure 10.4. Two-dimensional source zone distributions with (a) a high GTP mass ratio and (b) a low GTP mass ratio. (c) Comparison of flux-weighted concentration as a function of % mass removal for scenarios depicted in (a) and (b). (d) Depiction of actual and trajectory-averaged NAPL saturation along a horizontal streamtube selected from scenario (b).

of identical mean hydraulic conductivity. Both spills are of the same magnitude and duration and thus, these two source zones have the same domain-averaged saturation value. Examination of Figure 10.4c reveals, however, that downstream flux response is distinctly different for each scenario (note the log concentration scale). Thus, domain-averaged saturation values provide little assistance for the prediction of system response and will not be discussed further herein.

Spatial moments (M_{ijk}) have been used widely as a metric for quantifying DNAPL saturation distributions in theoretical investigations, where the detailed saturation distribution is known (e.g., Lemke et al., 2004a):

$$M_{ijk} = \int_{-\infty}^{\infty}\int_{-\infty}^{\infty}\int_{-\infty}^{\infty} n\rho^n s_n(x,y,z)x^i y^j z^k dxdydz \quad \text{(Eq. 10.5)}$$

Here n is porosity, ρ^n is the DNAPL density, $s_n(x,y,z)$ is the saturation, x, y, and z are Cartesian coordinates and i, j, and k are corresponding moment orders. Written for a single dimension using a discrete number (l) of sample points (10.5) becomes:

$$M_i = \sum_l n^l \rho^n s_n^l x^i \quad \text{(Eq. 10.6)}$$

where all parameters are as given before. The zeroth moment (M_0) is a measure of the total mass in the system. The first moment (M_1) normalized by the zeroth moment (M_0) is the center of mass in the x-direction and the second normalized moment (M_2) quantifies the degree of spreading about that center of mass. Comparing the extent of spreading in the directions perpendicular to flow with the size of the down gradient control plane over which the flux is averaged provides an estimate of the amount of dilution in downgradient, flux-averaged concentration measurements. Estimation of spatial moments in the field, however, requires extensive sampling and has rarely been reported (Jawitz et al., 1998, 2000).

More recently, Jawitz et al. (2003, 2005) proposed a methodology to characterize source zone distributions using data from PITTs in conjunction with a streamtube (e.g., Lagrangian) conceptualization of the source zone flow field. Here conservative and partitioning tracer breakthrough data are analyzed (see description in Section 10.2.2.2) to provide a measure of the hydrodynamic heterogeneity within the source zone and the stochastic-advective, trajectory-averaged, DNAPL saturation. Because these measures (i.e., DNAPL saturation and hydrodynamic heterogeneity) rely upon moments of the tracer breakthrough curves, they are necessarily integrative along the flow path and provide no spatial resolution in the direction of flow (see Figure 10.4d). Higher order moment analysis, however, can provide information relating to the flow field heterogeneity and the mean and variance of the trajectory averaged DNAPL saturation (Jawitz et al., 2003). Metrics obtained using these Lagrangian methods are valuable because they can be expressed as a reactive travel time distribution (τ), a single statistical distribution (with a corresponding mean and variance) that integrates variability in both the source zone flow field and trajectory averaged saturation (Jawitz et al., 2005). Changing the variance in the τ-distribution then provides a simple method for predicting potential reductions in contaminant flux due to a given level of DNAPL mass removal (Jawitz et al., 2005). This methodology has been used successfully at several field sites, including Hill Air Force Base (AFB), Utah and Sage's dry cleaner site in Jacksonville, Florida (Jawitz et al., 2003). The employment of a flow-field specific PITT to quantify source zone characteristics, however, is subject to some limitations. Flow bypassing of high NAPL saturation zones, as a consequence of permeability reductions and/or rate-limited partitioning in these regions, may lead to underestimation of the NAPL saturation (Willson et al., 2000; Imhoff and Pirestani, 2004; Moreno-Barbero and Illangasekare, 2006). The use of low saturation estimates, particularly when trajectory averaged (e.g., Figure 10.4d), in combination with an equilibrium dissolution assumption will likely lead to overly optimistic streamtube model predictions of NAPL mass recovery during remediation.

Ganglia-to-pool (GTP) mass ratios have been suggested as an alternative source zone distribution metric, useful for quantifying plume response to NAPL mass removal. The GTP metric is based on the concept that the time evolution of mass flux from source zone distributions dominated by pools will differ from those dominated by low saturation ganglia (Christ et al., 2005a; Lemke and Abriola, 2006). GTP mass ratios quantify the distribution of NAPL between ganglia and pool regions according to:

$$GTP = \frac{\sum \rho^n s_n n \Delta x \Delta y \Delta z \; for\ all\ s_n \ < \ s_{nr}^{\max}}{\sum \rho^n s_n n \Delta x \Delta y \Delta z \; for\ all\ s_n \ \geq \ s_{nr}^{\max}} = \frac{\sum s_n \; for\ all\ s_n \ < s_{nr}^{\max}}{\sum s_n \; for\ all\ s_n \ \geq s_{nr}^{\max}} \qquad \text{(Eq. 10.7)}$$

Here, pooled regions are defined as source-zone regions with a saturation value greater than the maximum residual organic saturation ($s_{nr}{}^{max}$), the saturation above which the organic will be mobile in this medium for all release histories (Parker and Lenhard, 1987). In nonuniform media, $s_{nr}{}^{max}$ may vary spatially with medium composition and texture; however, for practical purposes, a single value associated with a reference medium can be used (e.g., $s_{nr}{}^{max} = 0.10$). Ganglia regions are defined as regions with DNAPL saturations at or below $s_{nr}{}^{max}$. Groundwater flow within pooled regions tends to be very slow, as a consequence of the low water saturations and relative permeability effects. Thus, water tends to flow more easily around these regions of the source zone, and there is, consequently, less DNAPL surface area exposed to the flowing water. Numerical simulations of hypothetical spill scenarios have shown that the plume evolution is related to the initial distribution of NAPL mass between high surface area ganglia and low surface area pools (Lemke et al., 2004b; Lemke and Abriola, 2006). This metric was employed successfully in an upscaled model to predict plume response for given levels of NAPL mass removal (Christ et al., 2006). However, like several of the other metrics described above, methodologies for its field characterization, in the absence of detailed saturation profiles, are still under development.

10.3 MASS FLUX FROM DNAPL SOURCE ZONES

10.3.1 Influence of Architecture on Mass Discharge and Plume Response

In the effort to quantify the potential benefits of DNAPL source zone mass removal, recent research has focused on the link between source zone characteristics, NAPL mass removal, and subsequent changes in contaminant mass discharge (USEPA, 2003; Falta et al., 2005a, b; Fure et al., 2006; Lemke and Abriola, 2006). Although field-scale experimentation with emplaced DNAPL source zones can provide meaningful insights (e.g., Frind et al., 1999; Broholm et al., 1999, 2005; Rivett and Feenstra, 2004), regulations prohibiting the introduction of contaminants into the subsurface generally preclude this line of research. Furthermore, exploring the influence of alternative DNAPL architecture configurations on downstream fluxes using such an approach would most certainly be cost-prohibitive. Therefore, numerical simulations and sophisticated bench-scale aquifer cell experiments have been employed as surrogates for full-scale testing.

Multiphase flow models have been used by a number of researchers to investigate DNAPL migration, entrapment and dissolution in nonuniform subsurface settings. Simulation studies have examined the influence of variations in spill characteristics (Kueper and Gerhard, 1995; Dekker and Abriola, 2000a), levels of solid media property heterogeneity (Essaid and Hess, 1993; Dekker and Abriola, 2000a; Lemke et al., 2004a), magnitude of capillary forces (Dekker and Abriola, 2000a; Lemke et al., 2004a; Phelan et al., 2004), and composition of the DNAPL (Bradford et al., 1998; Phelan et al., 2004) on the characteristics and extent of DNAPL contamination. The potential influence of local-scale mass transfer correlations or equilibrium assumptions on predictions of DNAPL dissolution has also been assessed in a variety of simulation scenarios (Mayer and Miller, 1996; Dekker and Abriola, 2000b; Zhu and Sykes, 2000; Sale and McWhorter, 2001; Rathfelder et al., 2001; Lemke et al., 2004b; Phelan et al., 2004; Jawitz et al., 2005). Although analysis of these simulation results has facilitated the identification of source zone architecture metrics and their connection to system parameters, conflicting conclusions have been drawn from these modeling exercises pertaining to the importance of source zone architecture in the determination of contaminant mass discharge (Stroo et al., 2003).

A modeling investigation by Sale and McWhorter (2001), in which a DNAPL source zone was idealized as a distribution of pools and fingers, predicted that contaminant mass discharge remains relatively constant until near complete mass removal, regardless of the source zone distribution. In contrast, Rao and Jawitz (2003) used an analytical stream tube model, incorporating a uniform NAPL source and nonuniform flow field, to demonstrate that significant changes in the mass discharge are predicted as source zone mass is depleted (i.e., the source zone architecture changes). These investigators have subsequently proposed trajectory averaged metrics that may be employed to quantify the link between source zone architecture and contaminant mass discharge (Jawitz et al., 2003, 2005). Successful use of trajectory averaged metrics in prediction of mass discharge has been reported for applications to laboratory experimental data (Fure et al., 2006) and to data generated by multiphase compositional simulations (Basu et al., 2008). However, other investigations have suggested trajectory averaged metric-based models do not perform as well as those that directly incorporate the changing DNAPL-water interfacial area as contaminant dissolves (Zhang et al., 2008). Other work has also explored the relationship between the source zone architecture and mass discharge. Lemke et al. (2004b) and Lemke and Abriola (2006) examined the relationship between the GTP mass ratio metric and source zone contaminant dissolution in heterogeneous porous media using a compositional, multiphase simulator and an ensemble of 200 source zone realizations. Their simulation results suggest that, during the initial stages of source evolution, contaminant mass discharge is controlled by dissolution of high surface area ganglia (Figure 10.5a). As the high surface area ganglia dissolve, the mass discharge is reduced and mass flux is characterized by disconnected "hot spot" areas emanating from persistent pools in the source region (Figure 10.5b). A recent bench-scale aquifer cell experiment supports these findings, suggesting that the source zone will evolve over time from architecture characterized by a high GTP ratio dominated by ganglia at residual saturation to that characterized by a low GTP ratio containing persistent, high-saturation DNAPL pools, with a corresponding reduction in the contaminant mass discharge (Suchomel and Pennell, 2006). On-going experimental (Fure et al., 2006; Brusseau et al., 2008; Kaye et al., 2008; Zhang et al., 2008) and field investigations (Brusseau et al., 2007; DiFilippo and Brusseau, 2008) are continuing to elucidate the linkage between source zone architecture and plume response.

Insights gained from these investigations are leading to the development of simplified, upscaled modeling tools for prediction of downstream mass discharge from source zone characteristics. Such tools are proposed for practitioner use for site assessment and screening of alternative remediation strategies. Typically, upscaled models account for spatial variability in NAPL architecture and flow by-passing through the use of a domain-averaged (upscaled)

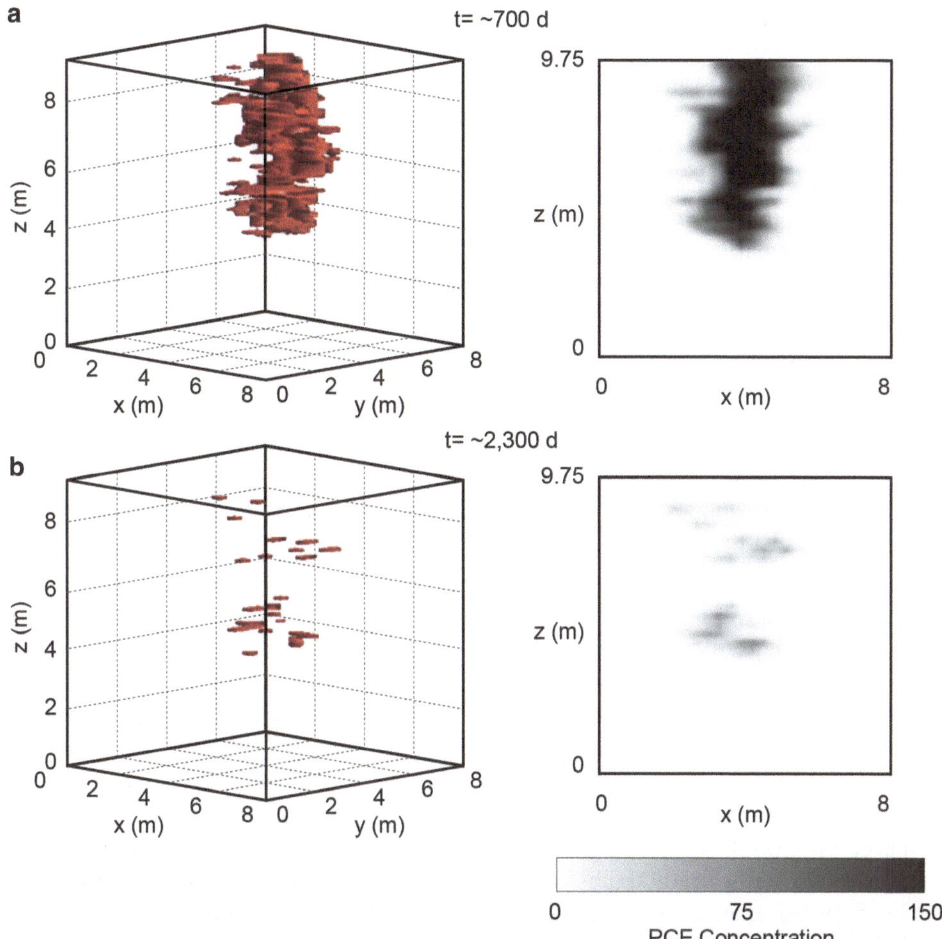

Figure 10.5. Temporal evolution of a 3-dimensional (3D) DNAPL source zone under natural gradient dissolution conditions (gradient = 0.01). 3D saturation distributions are depicted in the *left column* of the figure, while concentration profiles at the downgradient boundary are depicted on the *right*. Note, travel time from the center of the source zone to the down gradient boundary is approximately 25 days.

mass transfer coefficient (e.g., Parker and Park, 2004). In most studies, this domain averaged mass transfer coefficient has been parameterized by fitting a simplified one-dimensional solution of the advection-dispersion equation to experimentally (Saba and Illangasekare, 2000; Schaerlaekens and Feyen, 2004; Marble et al., 2008) or numerically generated results (Dekker, 1996; Parker and Park, 2004; Zhu and Sykes, 2004; Park and Parker, 2005):

$$\frac{\bar{C}(L)}{C^{eq}} = 1 - \exp\left(\frac{-\kappa_{eff} L}{\bar{q}}\right) \qquad \text{(Eq. 10.8)}$$

where $\overline{C}(x)$ is the flux-averaged downstream concentration at $x = L$, the downgradient boundary of the 1-D domain, C^{eq} is the equilibrium aqueous solubility of the selected component, and κ_{eff} is an upscaled mass transfer coefficient.

Such fitted models, however, are valid only for sites with source zone architecture characteristics consistent with those used to parameterize the coefficient (Parker and Park, 2004; Christ et al., 2006). Hence, recent work has focused on the incorporation of source zone architecture metrics into the upscaled mass transfer correlations to permit their use across a broader range of sites (Christ et al., 2006; Saenton and Illangasekare, 2007). Saenton and Illangasekare (2007) proposed an upscaled mass transfer correlation that integrates aquifer properties (variance of log hydraulic conductivity and correlation length) and source zone metrics (normalized second spatial moments) to facilitate model simulation of contaminant mass flux at scales larger than the numerical grid block. By extrapolation, this upscaled correlation could be used to predict mass discharge at the field scale (i.e., for the entire domain). In an alternative approach, Christ et al. (2006) integrated the GTP mass ratio into the upscaled mass transfer correlation proposed by Parker and Park (2004) to facilitate the prediction of field-scale mass discharge from site information. Application of the resulting model across a broad range of simulated subsurface conditions supports the utility of this approach for low to moderate levels of mass removal. Christ et al. (2006) note, however, that, at high levels of mass removal (>85%), the mass discharge and source longevity are controlled by persistent NAPL pools that are nearly impossible to capture using domain-averaged parameters such as the GTP mass ratio.

The simulation and modeling research described above has tended to focus on mass dissolution from DNAPL and has not explored mass elution from sorbed phase storage or as solute that has previously diffused into low permeability zones. However, the initial high solute concentrations found within a source zone can serve as a strong driving force for sorption and diffusion into lower permeability zones. While mass stored in these compartments probably represents only a small percentage of the total mass initially released in a DNAPL spill, over time this balance can shift, particularly in media with a large sorption capacity and/or significant regions of immobile water, such as fractured media (Parker et al., 1994). Furthermore, over long time scales, eluting mass from these compartments will account for an increasing fraction of the down gradient mass flux, as much of the DNAPL itself has dissolved (see Parker et al., 2008; Sale et al., 2008). Thus, long-term characteristics of elution (i.e., tailing) from these compartments will not be predicted by the upscaled models described above. Future efforts will need to be directed at refining these models for applications to such formations. It should be noted that these mass compartments are not only characteristic of source zones; mass sequestered in sorbed and immobile zones can also represent a large fraction of the mass stored within a contaminant plume.

10.3.2 Tools for Mass Flux Quantification

As illustrated above, near field contaminant plumes emanating from DNAPL source zones are characterized by heterogeneous distributions of dissolved-phase concentrations within a planar transect perpendicular to the average groundwater flow. Similarly, local mass flux values across this transect exhibit equivalent or even greater variability. These variations are a direct consequence of the coupling of nonuniform flow effects with the highly irregular spatial distribution of DNAPL within the upstream source zone. Total mass discharge across a downstream transect, also known as 'source strength,' is a popular metric that quantifies the contaminant mass contributed to a plume by the source. In the past 5 years, a great deal of attention has been directed towards quantification of source strength (Rao et al., 2002; USEPA, 2003; Stroo et al., 2003; ITRC, 2004; Falta et al., 2005a; Goltz et al., 2007). This metric is attractive, in that it provides a simple basis for comparing DNAPL contamination across sites of very different scales and geologic environments. The source strength concept has also been used in assessing natural attenuation, through comparisons of its value with that of the subsurface 'attenuation capacity' (USGS, 2003; Falta et al., 2005b), and in assessing the effectiveness of alternative remediation technologies, by tracking the evolution of its magnitude with treatment time or percent mass removal (Lemke et al., 2004a; Fure et al., 2006; Lemke and Abriola, 2006).

Source strength may be quantified as mass discharge [MT^{-1}] or, if normalized by a known area, as a spatially-averaged mass flux [$ML^{-2}\,t^{-1}$]. Note that the concept of a control *plane* is implicit in the definition and measurement of mass discharge or mass flux, in contrast to the concept of a compliance *point,* which is typically applied to concentration levels. Because the acquisition of meaningful data from within a control plane is economically and practically limited, it is important to consider the implications of employing mass discharge and mass flux metrics. Consider the heterogeneous distribution of local-scale contaminant flux shown in Figure 10.6 for a hypothetical plume transect. Definition of a control plane (defined by the white lines) permits integration of the local-scale flux of component i($\mathbf{N_i}$) over the area of the control plane (S) to obtain a mass discharge ($\dot{\mathbf{q}}_\mathbf{i}$) in the direction of a vector normal to the plane($\mathbf{n}$):

$$\dot{\mathbf{q}}_\mathbf{i} = \int_S \mathbf{N_i} \cdot \mathbf{n} dS = \int_S \mathbf{q} C_i \cdot \mathbf{n} dS \qquad \text{(Eq. 10.9)}$$

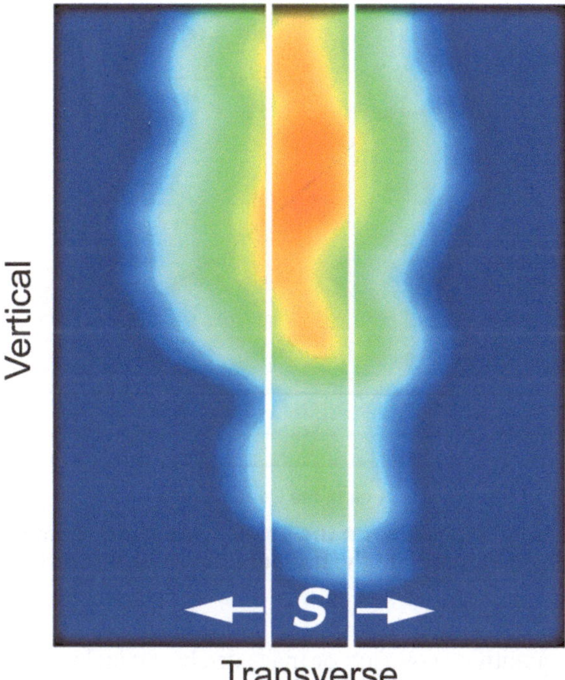

Figure 10.6. Representation of local-scale contaminant flux in a plume transect that is normal to the mean direction of flow (adapted from Li et al., 2007). *Warm colors* correspond to greater contaminant flux. *White lines* indicate the boundary of a control plane having area S. The effect of increasing S, as indicated by the *white arrows*, is shown in Figure 10.7 for this transect.

Here the local-scale contaminant mass flux is defined to be the product of the Darcy velocity ($\mathbf{q}$)and contaminant concentration(C_i). Normalization of the mass discharge by the area of the control plane produces a spatially-averaged mass flux of component i, $(\overline{\mathbf{N_i}})$, as shown in Equation 10.10.

$$\overline{\mathbf{N_i}} = \frac{\int\limits_S \mathbf{N_i} \cdot \mathbf{n} dS}{\int\limits_S dS} \qquad \text{(Eq. 10.10)}$$

The influence of the extent of the control plane on mass discharge and the spatially-averaged mass flux is shown in Figure 10.7. Increasing the size of the control plane beyond that defined by the plume boundaries (as indicated by the white arrows in Figure 10.6) affects the magnitude of the spatially-averaged mass flux, but does not affect the magnitude of the mass discharge. This has important implications when assessing source strength; control planes having areas greater than the areal extent of the plume will produce estimates of the spatially-averaged mass flux that are biased low. In practice, this will not be a substantial barrier to implementation of flux-based metrics for intra-site comparisons where the area used in the analysis remains constant (i.e., flux before and after aggressive source treatment). However, the above discussion emphasizes that extreme caution should be used for inter-site comparisons.

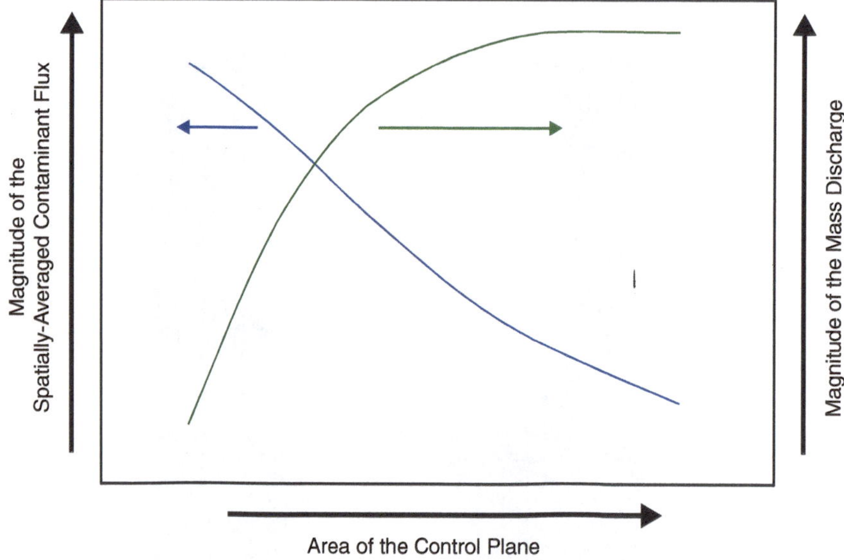

Figure 10.7. Effect of control plane size on the spatially-averaged contaminant flux (*blue*) and the contaminant mass discharge (*green*) for the transect shown in Figure 10.6. Based upon flux values found in Li et al., 2007.

Tools designed to quantify mass flux or mass discharge include integral pump tests, multi-level samplers, and passive flux meters. These techniques are described and discussed below. Because each of these three approaches makes use of different physical observations, perturbs the natural system in different ways, and is applied at different scales, it is important to identify the assumptions and uncertainties that underlie the application of each technique. Accurate estimation of uncertainty will be crucial to the use of mass discharge measurements in site assessment and remedial decisions.

Although scale effects, system perturbations, and measurement errors (all sources of estimate uncertainty) differ for each of these measurement techniques, uncertainty related to interpolation and aggregation is common to most applications. In practice, contaminant mass discharge across a control plane is typically computed as $\sum_i^n C_i q_i A_i$(e.g., Borden et al., 1997;

Einarson and Mackay, 2001), where q_i is the flow rate at location i; C_i is the corresponding contaminant concentration, and A_i is the weight for C_iq_i, which usually takes the value of the area associated with C_iq_i. Here C_i and q_i are local measurements. Note that C_i is inversely calculated in the integral pump test method and C_iq_i is directly measured in the passive flux meter method, as explained below. Thus, to obtain the mass discharge across the entire control plane, measurements must be interpolated and aggregated. A number of interpolation methods and aggregation sequences have been employed for this purpose (see Kübert and Finkel, 2006). Despite the high spatial variability of velocity at most sites, in the majority of cases, only C_i has been interpolated, with q_i assumed uniform in the transect (e.g., Semprini et al., 1995; Borden et al., 1997; King et al., 1999; Kao and Wang, 2001). The Thiessen polygon method has been used most commonly as an interpolation scheme (e.g., Borden et al., 1997; Einarson and Mackay, 2001; Kao and Wang, 2001), but other geostatistical approaches have also been employed (e.g., Semprini et al., 1995). Regardless of the method applied, the calculated mass discharge is always subject to uncertainty arising from interpolation and aggregation processes, even if measurement errors are negligible.

Although the uncertainty of field estimated mass discharge is beginning to be widely acknowledged (e.g., Jarsjö et al., 2005; Zeru and Schafer, 2005; Hatfield et al., 2004), there have been few attempts to quantify it. Using a classical Gaussian assumption, Willson et al. (2000) reported a few statistics (e.g., mean, 95% confidence interval) of variables (e.g., hydraulic conductivity, concentration) associated with their mass discharge estimation. However, such classical approaches may not be appropriate when data are correlated in space, as would be expected for this application. More recently, Li et al. (2007) presented a non-parametric approach, based upon a geostatistical stochastic simulation algorithm to estimate mass discharge and its associated probability distribution (uncertainty) from local measurements of concentration and hydraulic conductivity.

10.3.2.1 Integral Pump Tests

Integral pump tests employ short-term pumping of one or more fully-screened wells located in a plume transect (Holder et al., 1998; Teutsch et al., 2000; Bockelmann et al., 2001). Figure 10.8 depicts a typical application in which a plume emanating from a DNAPL source zone is fully encompassed by an extraction well capture zone. Pumping is initiated at a constant rate and the temporal evolution of effluent contaminant concentrations is recorded (Bockelmann et al., 2001). The change in concentration results from the time variation of the well capture zone, as illustrated in Figure 10.9. It is important to note that the circular isochrones (capture zones) depicted in Figure 10.9 are highly idealized and the consequence of several simplifying assumptions (confined, spatially homogeneous, isotropic aquifer having negligible natural gradient). Typical site conditions, however, will result in capture zones that are much more irregular and more elliptical in shape (Bauer et al., 2004). While the captured (integrated) mass discharge (M_d) may be measured directly using knowledge of the contaminant concentration and pumping rate, estimates of mass flux require inversion of the temporal evolution of the concentration signal (Schwarz et al., 1998; Bayer-Raich et al., 2003a, 2003b, 2004). These inversions may be based upon analytical or numerical solutions to the coupled groundwater flow and contaminant transport problem. Since inversions of contaminant transport for this pump test scenario are non-unique, results may be employed to provide a suite of possible realizations for the spatial distribution of concentration (or mass flux) within the plume (Bauer et al., 2004). Although such realizations of the plume structure may be beneficial in informing further site characterization efforts, the primary value of the integral pump test lies in its use in determining a spatially averaged value of contaminant flux (M_f):

$$M_f = \frac{M_d}{A} \qquad \text{(Eq. 10.11)}$$

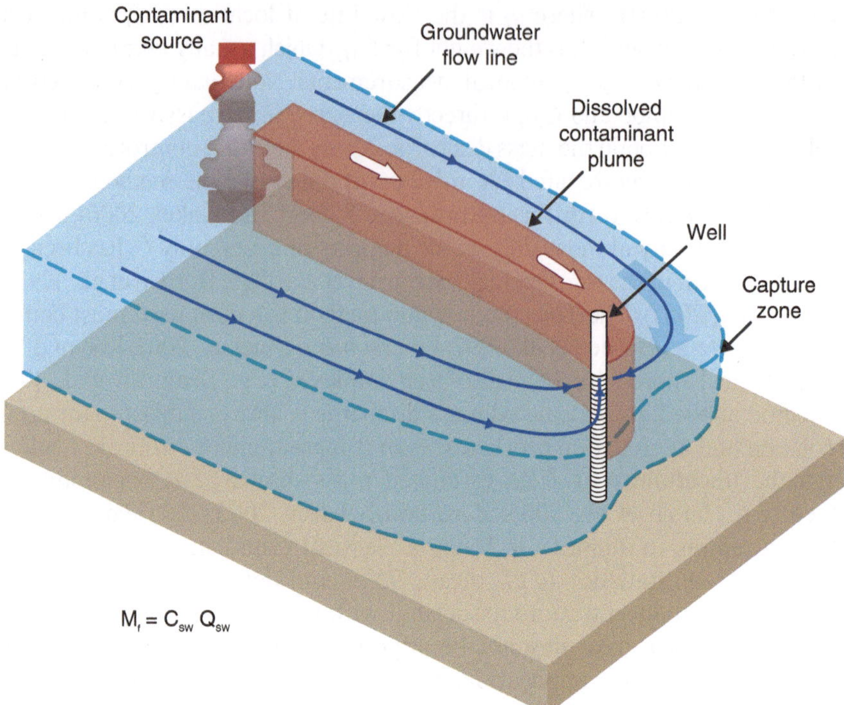

Figure 10.8. Diagram of source zone, contaminant plume, extraction well, and resultant capture zone when used for integral pump test method. Reproduced with permission from Nichols and Roth, 2004.

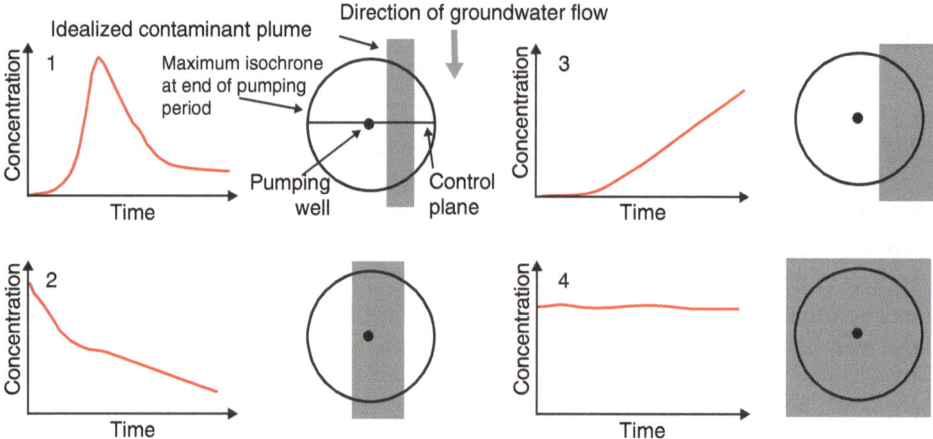

Figure 10.9. Illustrative examples of concentration signals during integral pumping tests. Reproduced with permission from Bauer et al., 2004.

where A represents the captured subsurface cross-sectional area of the plume (Goltz et al., 2007). The large sampling volume used during a pump test substantially reduces the influence of spatial-variability in contaminant concentration on the estimate of flux (Jarsjö et al., 2005). This reduction of estimate uncertainty, however, may be offset by the treatment and disposal costs associated with production of large volumes of contaminated water. Goltz et al. (2009) recently proposed an innovative variant of the integrated pump test, the tandem circulating well method that employs two recirculating wells, minimizing extracted volume.

The integral pump test has been deployed for use at several sites in Europe, and appears to be particularly well suited for application at large-scale industrial sites where there may be existing pumping wells and treatment facilities (e.g., Holder et al., 1998; Bockelmann et al., 2003; Bayer-Raich et al., 2003a; Bauer et al., 2004). Jarsjö and co-authors (2005) have developed a methodology to assess the uncertainty in average flux estimates generated using the integral pump test. Their work suggests that the largest uncertainty in the mass discharge estimate arises from the uncertainty in plume position relative to the pumping well. They also conclude that uncertainty bounds can be substantial (an order of magnitude) when integral pump tests are applied in strongly heterogeneous formations. In an experiment conducted in a relatively homogeneous artificial aquifer (9.5 m × 4.7 m × 2.6 m deep), Goltz et al. (2009) reported that a modified integral pump test method underestimated the flux by as much as 70%. This result was attributed to test conditions that deviated from the two dimensional, steady flow assumed for data analysis. In the same study, the tandem circulating well method performed well, estimating the flux to within 16%. It should also be noted that application of the integral pump test will perturb the natural flow field at a site. This perturbation will tend to change local flow directions and increase groundwater flow velocities. If the pump radius of influence intersects a DNAPL source zone or any zone of rate-limited mass transfer within the contaminant plume (e.g., areas of high sorption or diffusion into low permeability layers), mass transfer may be enhanced over that experienced under natural gradient conditions. Thus, integral pump test estimates of mass discharge could have an innate (high) bias and may not reflect the mass flux across the transect under natural gradient conditions and for rate-limited mass transfer scenarios.

10.3.2.2 Multilevel Sampling Arrays

Multi-level samplers installed in one or more plume transects may be used to estimate mass flux and mass discharge (Borden et al., 1997; Einarson and Mackay, 2001; King et al., 1999). Multi-level samplers are usually smaller-scale (0.5–2 ft screens) observation points that are nested within a selected plane to form a long-term sampling array typically oriented perpendicular to the contaminant plume (see Figure 10.10). In contrast to the integral pump test, the multilevel sampler approach yields spatially discrete measures of concentration (C_i). Information about the flow field (or hydraulic conductivity field) may be derived from discrete (e.g., cores or bore hole tracer tests) or integrated measurements (e.g., pump tests). Concentration measurements are then coupled with flow measurements (e.g., groundwater specific discharge, q_i) to produce a point estimate of mass flux ($M_{f,i}$):

$$M_{f,i} = C_i \cdot q_i \qquad \text{(Eq. 10.12)}$$

These flux estimates are then interpolated and integrated in the control plane to produce an estimate of the mass discharge and spatially-averaged mass flux. Experience with flux estimation using this method has suggested great sensitivity to sample spacing. In a field investigation, the estimate of mass flux in a transect was found to be more sensitive to vertical than horizontal resolution (Guilbeault et al., 2005). Recently, Li et al. (2007) developed a methodology to estimate mass flux and quantify its uncertainty using discrete concentration and hydraulic conductivity measurements in a transect (note that the methodology may also be applied to passive flux meters). For the spill scenarios examined, they found that greater sampling densities (up to 7% of the control plane) were required to accurately estimate the mass flux and its related uncertainty, as the concentration field became less uniform (either through mass removal or perhaps in more heterogeneous geologic settings). These results suggest the need for more densely sampled transects, to provide a basis for estimation of

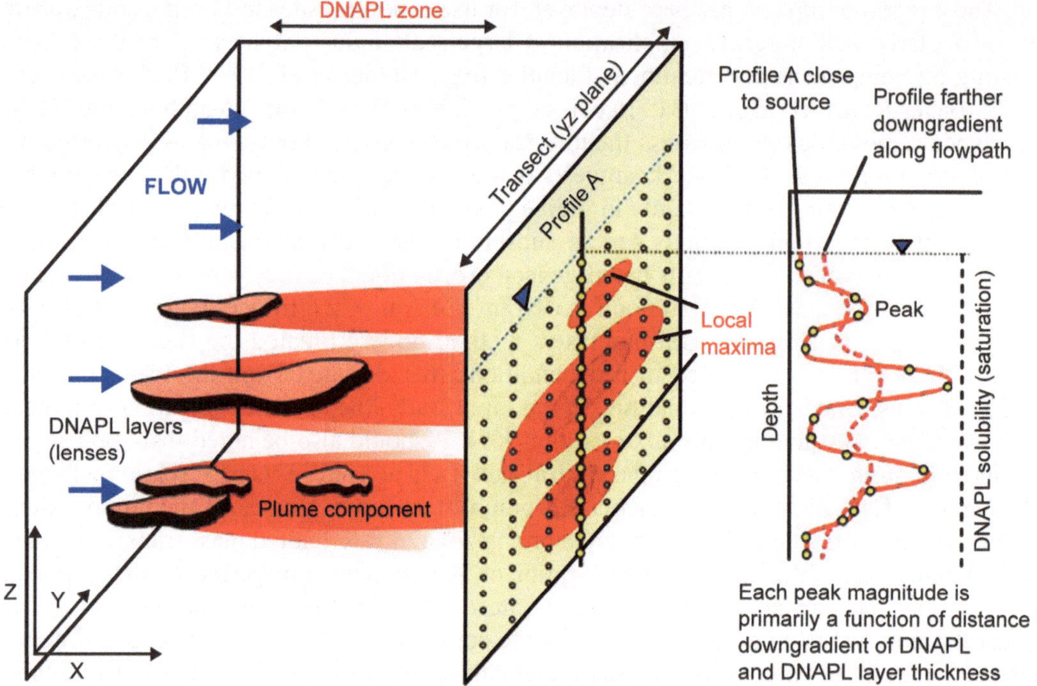

Figure 10.10. 3D depiction of contaminant plume eluting from DNAPL source zone intersected by multi-level sampling array for quantification of contaminant concentration, which may be used to compute the contaminant mass flux. Two concentration profiles depict smoothing effect as contaminant is transported down gradient of the source zone. Reproduced with permission from Guilbeault et al., 2005.

remediation effectiveness and/or risk reduction for downstream receptors. The requirement for greater resolution in the vertical direction is consistent with observed relationships between the integral scale (a.k.a. mean correlation length) in the vertical and horizontal directions for many well-characterized sites. Increased sampling densities present economic challenges for the application of this method, since greater sampling densities translate into greater installation and monitoring costs Mindful of these challenges, Li and Abriola (2009) recently developed and implemented a multi-stage spatial sampling algorithm, based upon geostatistical simulation and multiple criteria decision making theory, to select optimal sampling locations and determine minimal sampling density for accurate quantification of mass discharge uncertainty. Application of this staged methodology to numerically simulated plume transects produced promising results, a 50% reduction in required sampling density.

10.3.2.3 Passive Flux Meter

Passive flux meters are a novel method for estimating localized contaminant flux (Hatfield et al., 2002, 2004; Annable et al., 2005). In contrast to the integral pump test and multilevel sampling arrays, which measure contaminant concentrations, the passive flux meter provides a more direct measure of the local-scale, time-averaged contaminant flux. The flux meter depicted in Figure 10.11 is comprised of a permeable medium containing a sorbent which retains solutes as the groundwater migrates through the meter. Sorbents may be selected to accommodate organic or inorganic contaminants (Annable et al., 2005; Campbell et al., 2006). The self-contained flux meter is installed in an existing well for a predetermined duration ranging from

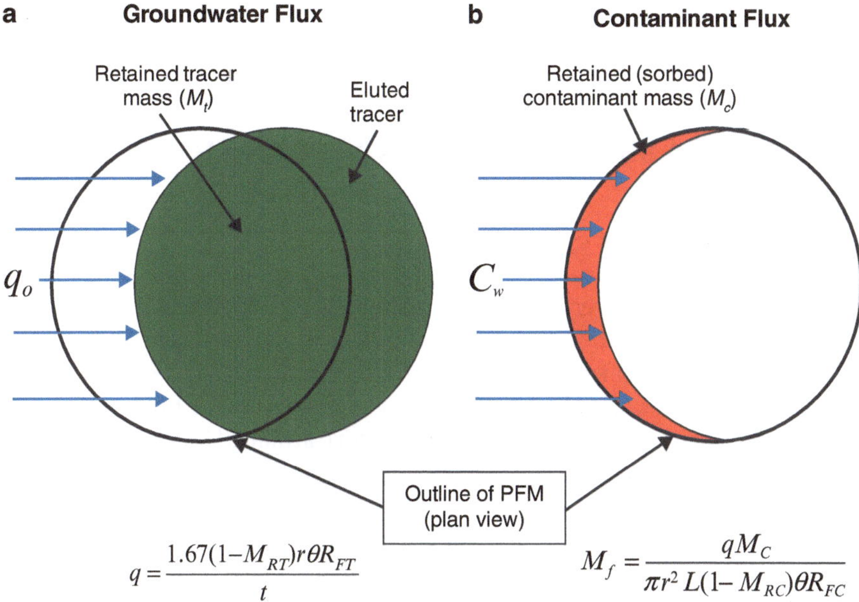

Figure 10.11. Plan view of passive flux meter used to determine (a) groundwater flux via the disappearance of a tracer with retardation factor (R_{FT}), and (b) contaminant flux as contaminant is sorbed to the meter.

days to weeks. Subsequent retrieval of the flux meter permits vertical sectioning and extraction of the permeable material. The amount of contaminant sorbed (Figure 10.11b) during the deployment period permits calculation of a time-averaged contaminant flux. Also present within the permeable medium of the flux meter are resident tracers. Groundwater flowing through the meter elutes these resident tracers (Figure 10.11a), the loss of which is used to compute a time-averaged volumetric flux of water (i.e., Darcy velocity). Calculation of the local mass flux ($M_{f,i}$) employs data collected during flux meter analysis (Annable et al., 2005):

$$M_f = \frac{qM_C}{\alpha\pi r^2 L(1 - M_{RC})\theta R_{FC}} \quad \text{(Eq. 10.13)}$$

where M_c is the mass of contaminant sorbed, L is length of the sorbent matrix for the vertical interval sampled, R_{FC} is the retardation factor of the contaminant on the sorbent, M_{RC} is the relative mass of a resident tracer retained for the exposure time period, where the tracer has the same retardation factor as R_{FC}, θ is the volumetric water content in the meter, q is the Darcy velocity of water through the meter, and α is the convergence or divergence of flow around the meter.

The passive flux meter provides estimates of time-averaged fluxes at specific vertical and horizontal locations. Thus, this technique requires interpolation and integration of values to obtain estimates of mass discharge and spatially-averaged mass flux. As is the case with measurements made from multilevel samplers, spatial non-uniformities in the flux field may limit the accuracy and certainty of these extrapolations. The advantage of the flux meter over the multilevel sampler array, however, is simultaneous, spatially discrete, co-located, measurement of time-averaged contaminant flux and groundwater flow. This allows contaminant fluxes to be assessed without detailed knowledge of the spatial variability of hydraulic conductivity or intrinsic permeability. Flux meters have been deployed at numerous sites throughout North America with promising results (Annable et al., 2005; Basu et al., 2006; EnviroFlux, 2007; Brooks et al., 2008). A recent comparison of flux meter and integral pump test results at Hill AFB, Utah and Ft Lewis, Washington indicated both flux estimation methods gave comparable results and were able to elucidate some of the variability in the downgradient flux signal (Brooks et al., 2008).

10.4 PARTIAL MASS REMOVAL AND COMBINED REMEDIES

10.4.1 Benefits of Partial Source Removal

Although DNAPL mass recovery or destruction efficiencies exceeding 90% have been reported for many *in situ* remediation technologies, it is now widely accepted that a portion of the initial DNAPL mass will persist following aggressive source zone treatment (Stroo et al., 2003). For example, up to 10% of the initial TCE-DNAPL mass remained in low-permeability layers following a surfactant flushing test at Camp Lejeune, North Carolina (Holzmer et al., 2000), while 35–38% of the initial PCE-DNAPL mass remained after cosolvent flushing of a former dry cleaning facility in Jacksonville, Florida (Jawitz et al., 2000). In the latter case, groundwater concentrations of PCE in the treated source zone were reduced by 92% despite the relatively modest mass recovery. Nevertheless, a greater than 2-log reduction in dissolved-phase concentrations is typically required to reach remediation goals that are based on drinking water standards (e.g., MCL = 5 micrograms per liter [μg/L]). Persistence of low concentration levels will also be facilitated by elution from mass sequestered in the sorbed phase or within immobile aqueous zones. Several exhaustive surveys of field sites indicate this reduction is unlikely (NAVFAC, 2004; McGuire et al., 2006). A survey of chemical oxidation, enhanced bioremediation, thermal treatment, or surfactant/co-solvent flushing at 59 chlorinated DNAPL sites indicated a median concentration reduction of 88%, well below the 99.9% reduction likely required (McGuire et al., 2006). Furthermore, rebound was commonly observed, particularly at the chemical oxidation and enhanced bioremediation sites (McGuire et al., 2006). Thus, plume concentrations downgradient from treated source zones will likely exceed dissolved-phase remediation goals for extended periods of time despite substantial mass removal (Sale and McWhorter, 2001; Soga et al., 2004; NAVFAC, 2004; McGuire et al., 2006).

Recent laboratory-scale studies have also shown that greater than 90% mass removal is necessary to achieve order-of-magnitude reductions in dissolved-phase concentrations (Suchomel and Pennell, 2006). Furthermore, the relationship between reductions in source zone mass and mass discharge was found to strongly depend on the DNAPL saturation distribution within the source zone. In order to realize the potential benefits of aggressive source zone treatment, including substantial mass removal and reduced source longevity, more comprehensive remediation strategies must be developed to address both the residual contaminant mass within the source and corresponding downgradient dissolved-phase plumes. One possible strategy is to couple an aggressive DNAPL mass removal technology with a longer-term treatment or "polishing" step in series, with the latter designed to treat any dissolved-phase contaminants emanating from treated source zones (Christ et al., 2005b). A second possible strategy is to directly combine two or more aggressive remediation technologies in parallel, with the goal of treating the DNAPL source zone to meet regulatory requirements. Either approach will require careful and prolonged monitoring of the downstream plume (or mass flux) to assess treatment effectiveness.

10.4.2 Combined Remedies

The use of combined remedies, applied in series or in parallel, to more effectively treat a DNAPL source zone has received considerable attention during the past several years. While this approach holds great promise, very few controlled studies have been performed to address technical issues related to treatment compatibility and to demonstrate added benefit with respect to mass removal/destruction and dissolved-phase plume management. Several combined remedies that are currently being evaluated at the laboratory and field scales are highlighted below to illustrate the potential benefits and challenges of these strategies.

10.4.2.1 Surfactant Flushing and Bioremediation

Although many surfactants exhibit toxicity toward soil biota, those selected for aquifer remediation are typically biodegradable and/or approved for food-grade applications. More specifically, surfactants considered for sequential biotreatment of DNAPL source zones should (a) provide reducing equivalents for reductive dechlorination of chlorinated ethenes and (b) be tolerated by relevant microbial populations (e.g., *Dehalococcoides* sp.). One promising candidate is polyoxyethylene (20) sorbitan monooleate (Tween 80), a food-grade nonionic surfactant that has been used as a solubilizing agent for treatment of DNAPL source zones (e.g., Pennell et al., 1993; Ramsburg and Pennell, 2001; Abriola et al., 2005). In a recent field test, a 6% weight (wt.) aqueous solution of Tween 80 was used to enhance the recovery of PCE-DNAPL from a shallow aquifer formation (Ramsburg et al., 2005). Post-treatment monitoring of the treated source zone revealed elevated concentrations of volatile fatty acids (e.g., acetate, generated from the fermentation of T80), *cis*-1,2-dichloroethene (*cis*-DCE), and vinyl chloride (VC) (Ramsburg et al., 2004). A subsequent laboratory study demonstrated that Tween 80, at concentrations ranging from 50 to 5,000 mg/L, did not inhibit reductive dechlorination of PCE and TCE to *cis*-DCE by selected pure and mixed cultures (Amos et al., 2007). Although the presence of Tween 80 prevented ethene formation and reduced *Dehalococcoides* cell numbers, activity was recovered when Tween 80 was removed from the aqueous phase. Additionally, McGuire and Hughes (2003) reported that 40% of Tween 80 in a 1% (wt.) solution was degraded by mixed anaerobic culture after 70 days, and that Tween 80 exhibited no adverse effect on PCE conversion to VC, although subsequent transformation of VC to ethene occurred more slowly than in surfactant-free controls. These results suggest that biotransformation of VC to ethene may not be readily achieved within a surfactant-flushed source zone, requiring favorable conditions downstream of the source region for complete transformation. Thus, downstream monitoring for products of microbial reductive dechlorination becomes a crucial component of this combined technology remediation approach.

10.4.2.2 Chemical Oxidation and Bioremediation

In situ chemical oxidation (ISCO) typically involves the injection of an aqueous solution or suspension containing an oxidant, such as hydrogen peroxide (H_2O_2) (and a catalyst) or potassium permanganate ($KMnO_4$) (Siegrist et al., 2001). Applications of ISCO to treat DNAPL source zones have achieved mixed results, which has been attributed to several factors, including the inability to deliver sufficient quantities of oxidant, permeability reductions due to precipitant formation, and the formation of oxide coatings at the DNAPL-water interface (e.g., Li and Schwartz, 2003; Nelson et al., 2001). Despite the expected pH reduction and increased oxidation state following ISCO, microbial reductive dechlorination has been considered as a polishing step to treat residual DNAPL contamination. Following permanganate flushing, which achieved rapid but incomplete treatment of TCE-DNAPL, Hrapovic et al. (2005) observed no biological activity after addition of electron donor, and minimal degradation of TCE activity in two of four columns bioaugmented with a PCE-ethene dechlorinating mixed culture (KB-1). In a similar study, Sahl et al. (2007) observed a rebound in PCE dechlorination by KB-1 after flushing with $KMnO_4$ when followed by up to eight pore volumes of sterile growth medium. These findings suggest that coupling ISCO with microbial reductive dechlorination faces significant challenges and will require careful control of subsurface redox conditions.

10.4.2.3 Thermal Treatment and Bioremediation

The term thermal treatment encompasses a number of remediation technologies, including steam flushing, electrical resistive heating (ERH) and conductive heating. In general, these technologies involve heating the subsurface to a temperature of at least 90–100 degrees Celsius (°C) in order to achieve mass transfer of contaminants into the gas phase, which is subsequently extracted from the subsurface (NRC, 1999). It has generally been assumed that heating the subsurface to these temperatures would be analogous to autoclaving, resulting in sterilization or a shift in microbial community structure to favor thermophilic bacteria. Richardson et al. (2002) reported that samples collected following steam treatment of a creosote-contaminated field site showed high levels of microbial activity, consistent with laboratory tests that revealed a rebound in activity after gradual cooling. More recent studies performed with a PCE-ethene dechlorinating culture (KB-1) indicated that reductive dechlorination of TCE did not occur at temperatures above 40°C (Friis et al., 2007). Likewise, a recent investigation using samples from the Ft Lewis, Washington site following ERH indicated a lack of bioactivity above 40°C (Costanza et al., 2009). However, favorable redox conditions and the release of organic matter (which could serve as an electron donor) following thermal treatment of subsurface materials collected from a TCE-contaminated site undergoing ERH suggest that bioaugmentation holds promise as a polishing step for thermal remediation sites (Friis et al., 2005).

10.4.2.4 Thermal Treatment and Chemical Oxidation

Chemical oxidants, applied during or after thermal treatment, hold considerable promise as a means to enhance *in situ* destruction or to treat residual contamination. In particular, persulfate ($S_2O_8^{-2}$), which can be activated at elevated temperature to produce sulfate radicals ($SO_4^{-}\bullet$), has the potential to rapidly degrade chlorinated ethenes (e.g., Huang et al., 2005; Liang et al., 2003; Waldemer et al., 2007). In aqueous solution, Waldemer et al. (2007) reported a PCE half-life on the order of minutes at 70°C. Similarly, Liang et al. (2003) reported a dramatic reduction in the half-life of TCE and 1,1,1-trichloroethane (TCA) in the presence of persulfate, which decreased from 385 to 55 hours (h) at 20°C to 0.15 and 1.25 h at 60°C, respectively. In soil slurry tests, however, soil organic matter exhibited strong competition for sulfate radicals, suggesting that larger persulfate doses and longer treatment times may be required for effective treatment in the field. It is also known that transition metals ions (e.g., Fe^{+2}) can play an important role in persulfate activation (Liang et al., 2004), and thus, it may be important to control their availability during fields applications.

10.5 CONCLUSION

Despite substantial research and technology development over more than two decades, DNAPL sites remain a distinct remedial and management challenge. From the information presented above, a number of themes emerge.

First, it is clear that source zone characterization is crucial to all aspects of site management – from the development of a management/remedial plan, to the selection of remediation metrics/targets, to the monitoring and assessment of remediation success, and to the potential influence of remediation on downstream plume evolution and risk. Thus, investments in the development and refinement of source zone characterization technologies will likely provide the greatest future benefit to site managers. Data collection technologies and characterization tools for each of the above aspects of site management must be carefully selected and targeted to extract the information critical to that task. This suggests that a staged approach to DNAPL site characterization, which integrates multiple data sets and provides opportunities for critical evaluation of collected data and refinement of sampling protocols, will likely become the future norm, as opposed to rapid or "accelerated" site characterization. Similarly, information gathered during active site remediation can be utilized to direct remedial efforts toward the regions containing the greatest and/or most persistent contamination and can inform the need for and selection of subsequent remediation technologies as a "polishing" step or to facilitate long-term plume control.

Second, although modeling, laboratory, and field research has repeatedly shown that knowledge of the distribution of contaminant mass is crucial to the successful design and prediction of remedial performance, recent work also suggests that a detailed spatial delineation of DNAPL saturation may not be necessary. Thus, more emphasis should be placed on characterization efforts that can identify the extent of the source zone (in three dimensions) and yield estimates of source zone metrics, such as GTP, that characterize mass distribution in an averaged sense. An understanding of subsurface formation material properties (degree of physical and chemical heterogeneity, lithology, and correlation structure) will also be crucial to successful site management. It is typically these material properties that control contaminant mass persistence and accessibility to remedial efforts. Upscaled models hold much promise for use in site assessment, but current models should be used with caution as predictive tools in all but relatively homogeneous geologic settings, unless field calibration is undertaken. For use of these models in long-term site management, it is likely that averaged metrics similar to the GTP will need to be identified that characterize the impact of sorbed and dissolved mass distributions on down gradient flux behavior.

Third, research suggests that there is no 'silver bullet' in DNAPL remediation. Sustained stewardship and ongoing monitoring will be required at most sites. Although many remediation and characterization technologies offer much promise, no single technology is without substantial shortcomings. Furthermore, site heterogeneity tends to exacerbate the influence of these technological limitations, with effective remedial design often requiring that technologies be selected and integrated on a site-specific basis. Just as combined remedies may offer the greatest promise for site restoration, the most promising advances in DNAPL source zone management will likely be associated with refined and novel approaches to the coupling of invasive and noninvasive subsurface characterization techniques.

Finally, inverse modeling and uncertainty estimation tools have been both under developed and underutilized in DNAPL site management. Given the high variability of source zone mass and concentration distributions, it is imperative to include the potential influence of variability in all characterization plans, remedial designs, and mass flux assessments. Recent research has demonstrated that over reliance on the use of mean mass metrics (e.g., mean saturation, source mass, mass discharge) in site management can have substantial and deleterious impacts on predictions of site longevity, remedial performance, and downstream plume development. Future refinements in inverse and uncertainty modeling tools are likely to produce great dividends with respect to both active DNAPL site remediation and longer term site stewardship.

ACKNOWLEDGEMENTS

This review was sponsored by the Strategic Environmental Research and Development Program (SERDP) under contracts ER-1293 and ER-1612. The content of this publication has not been subject to agency review and does not necessarily represent the views of the agency sponsor.

REFERENCES

Abriola LM, Drummond CD, Hahn EJ, Hayes KF, Kibbey TCG, Lemke LD, Pennell KD, Petrovskis EA, Ramsburg CA, Rathfelder KM. 2005. A pilot-scale demonstration of surfactant-enhanced PCE solubilization at the Bachman Road site: (1) Site characterization and test design. Environ Sci Technol 39:1778–1790.

Ajo-Franklin JB, Geller JT, Harris JM. 2006. A survey of the geophysical properties of chlorinated DNAPLs. J Appl Geophys 59:177–189.

Ajo-Franklin JB, Geller JT, Harris JM. 2007. Ultrasonic properties of granular media saturated with DNAPL/water mixtures. Geophys Res Lett 34:L07404.

Amos BK, Daprato RC, Hughes JB, Pennell KD, Löffler FE. 2007. Effect of the nonionic surfactant Tween 80 on microbial reductive dechlorination of chloroethenes. Environ Sci Technol 41:1710–1716.

Annable MD, Rao PSC, Hatfield K, Graham W D, Enfield GC. 1998a. Partitioning tracers for measuring residual NAPL: Field-scale test results. J Environ Eng 124:498–503.

Annable MD, Jawitz JW, Rao PSC, Dai DP, Kim H, Wood AL. 1998b. Field evaluation of interfacial and partitioning tracers for characterization of effective NAPL-water contact areas. Ground Water 36:495–502.

Annable MD, Hatfield K, Cho J, Klammler H, Parker B, Cherry JA, Rao PSC. 2005. Field-scale evaluation of the passive flux meter for simultaneous measurement of groundwater and contaminant fluxes. Environ Sci Technol 39:7194–7201.

Atmadja J, Bagtzoglou AC. 2001. State of the art report on mathematical models for groundwater pollution source identification. Environ Forensics 2:205–214.

Basu NB, Rao PSC, Poyer IC, Annable MD, Hatfield K. 2006. Flux-based assessment at a manufacturing site contaminated with trichloroethene. J Contam Hydrol 86:105–127.

Basu NB, Fure AD, Jawitz JW. 2008. Predicting dense nonaqueous phase liquid dissolution using a simplified source depletion model parameterized with partitioning tracers. Water Resour Res 44:W07414, doi:10.1029/2007WR006008.

Basu NB, Rao PSC, Poyer IC, Nandy S, Mallavarapu M, Naidu R, Davis GB, Patterson BM, Annable MD, Hatfield K. 2009. Integration of traditional and innovative characterization techniques for flux-based assessment of dense non-aqueous phase liquid (DNAPL) sites. J Contam Hydrol 105:161–172.

Bauer S, Bayer-Raich M, Holder T, Kolesar C, Müller D, Ptak T. 2004. Quantification of groundwater contamination in an urban area using integral pumping tests. J Contam Hydrol 75:183–213.

Bayer-Raich M, Baumann R, Ptak T. 2003a. Application of pumping tests to estimate contaminant mass fluxes in a multi-layered aquifer: A numerically simulated field-scale experiment at the SAFIRA-Bitterfeld site. In Thornton SF, Oswald SE, eds, Groundwater Quality: Natural and Enhanced Restoration of Groundwater Pollution. IAHS Publication 275, IAHS Press, Wallingford, Oxfordshire, UK, pp 257–263.

Bayer-Raich M, Jarsjö J, Holder T, Prak T. 2003b. Numerical estimations of contaminant mass flow based upon concentration measurements in pumping wells. ModelCare 2002: A Few Steps Closer to Reality, IAHS Publication 277, IAHS Press, Wallingford, Oxfordshire, UK, pp 10–16.

Bayer-Raich M, Jarsjö J, Liedl R, Prak T, Teutsch G. 2004. Average contaminant concentration and mass flow in aquifers from time-dependent pumping well data: Analytical framework. Water Resour Res 40:W08303.

Benson RC. 2006. Remote sensing and geophysical methods for evaluation of subsurface conditions. In Nielsen DL, ed, Practical Handbook of Environmental Site Characterization and Ground-Water Monitoring, 2nd ed. CRC Press, Taylor & Francis Group, Boca Raton, FL, USA, Chapter 4.

Bockelmann A, Ptak T, Teutsch G. 2001. An analytical quantification of mass fluxes and natural attenuation rate constants at a former gasworks site: Natural attenuation of organic pollutants in groundwater. J Contam Hydrol 53:429–453.

Bockelmann A, Zamfirescu D, Ptak T, Grathwohl P, Teutsch G. 2003. Quantification of mass fluxes and natural attenuation rates at an industrial site with a limited monitoring network: A case study. J Contam Hydrol 60:97–121.

Borden RC, Daniel RA, LeBrun LE, Davis CW. 1997. Intrinsic biodegradation of MTBE and BTEX in a gasoline-contaminated aquifer. Water Resour Res 33:1105–1115.

Bradford SA, Abriola LM, Rathfelder KM. 1998. Flow and entrapment of dense nonaqueous phase liquids in physically and chemically heterogeneous aquifer formations. Adv Water Resour 22:117–132.

Broholm K, Feenstra S, Cherry JA. 1999. Solvent release into a sandy aquifer. 1. Overview of source distribution and dissolution behavior. Environ Sci Technol 33:681–690.

Broholm K, Feenstra S, Cherry JA. 2005. Solvent release into a sandy aquifer. 2. Estimation of DNAPL mass based on a multiple-component dissolution model. Environ Sci Technol 39: 317–324.

Brooks MC, Annable MD, Rao PSC, Hatfield K, Jawitz JW, Wise WR, Wood AL, Enfield CG. 2002. Controlled release, blind tests of DNAPL characterization using partitioning tracers. J Contam Hydrol 59:187–210.

Brooks MC, Wood AL, Annable MD, Hatfield K, Cho J, Holbert C, Rao PSC, Enfield CG, Lynch K, Smith RE. 2008. Changes in contaminant mass discharge from DNAPL source mass depletion: Evaluation at two field sites. J Contam Hydrol 102:140–153, doi:10.1016/j.jconhyd.2008.05.008.

Brusseau ML, Nelson NT, Cain RB. 1999. The partitioning tracer method for in-situ detection and quantification of immiscible liquids in the subsurface. In Brusseau ML, Sabatini DA, Gierke JS, Annable MD, eds, Innovative Subsurface Remediation, Field Testing of Physical, Chemical, and Characterization Technologies. American Chemical Society, (ACS) Symposium Series 725, ACS, Washington DC, USA.

Brusseau ML, Nelson NT, Zhang Z, Blue JE, Rohrer J, Allen T. 2007. Source-zone characterization of a chlorinated-solvent contaminated Superfund site in Tucson, AZ. J Contam Hydrol 90:21–40.

Brusseau ML, DiFilippo EL, Marble JC, Oostrom M. 2008. Mass-removal and mass-flux-reduction behavior for idealized source zones with hydraulically poorly-accessible immiscible liquid. Chemosphere 71:1511–1521.

Campbell TJ, Hatfield K, Klammler H, Annable MD, Rao PSC. 2006. Magnitude and directional measures of water and Cr(VI) fluxes by passive flux meter. Environ Sci Technol 40:6392–6397.

Carcione JM, Seriani G, Gei D. 2003. Acoustic and electromagnetic properties of soils saturated with salt water and NAPL. J Appl Geophys 52:177–191.

Chatzis I, Morrow NR, Lim HT. 1983. Magnitude and detailed structure of residual oil saturation. Soc Petrol Eng J 23:311–326.

Christ JA, Lemke LD, Abriola LM. 2005a. Comparison of two- and three-dimensional simulations of dense non-aqueous phase liquids (DNAPLs): Migration and entrapment in a nonuniform permeability field. Water Resour Res 41:W01007, doi:10.1029/2004WR 003239.

Christ JA, Ramsburg CA, Abriola LM, Pennell KD, Löffler FE. 2005b. Coupling aggressive mass removal with microbial reductive dechlorination for remediation of DNAPL source zones – A review and assessment. Environ Health Perspect 113:465–476.

Christ JA, Ramsburg CA, Pennell KD, Abriola LM. 2006. Estimating mass discharge from dense nonaqueous phase liquid source zones using upscaled mass transfer coefficients: An evaluation using multiphase numerical simulations. Water Resour Res 42:W11420, doi: 10.1029/2006WR004886.

Cooke CE Jr. 1971. Method of determining residual oil saturation in reservoirs. U.S. Patent 3,590,923.

Costanza J, Davis WM. 2000. Rapid detection of volatile organic compounds in the subsurface by membrane introduction into a direct sampling ion trap mass spectrometer. Field Anal Chem Technol 4:246–254.

Costanza J, Fletcher KE, Löffler FE, Pennell KD. 2009. Fate of trichloroethene in heated Fort Lewis soil. Environ Sci Technol 43:909–914.

Dai D, Barranco FT Jr, Illangasekare TH. 2001. Partitioning and interfacial tracers for differentiating NAPL entrapment configuration: Column-scale investigation. Environ Sci Technol 34:4894–4899.

Datta-Gupta A, Yoon S, Vasco DW, Pope GA. 2002. Inverse modeling of partitioning interwell tracer tests: A streamline approach. Water Resour Res 38:10.1029.

Deans HA. 1971. Method of determining fluid saturations in reservoirs. U.S. Patent 3,623,842.

Dekker TJ. 1996. An assessment of the effects of field-scale formation heterogeneity on surfactant-enhanced aquifer remediation. Doctoral Dissertation, University of Michigan, Ann Arbor, MI, USA.

Dekker TJ, Abriola LM. 2000a. The influence of field-scale heterogeneity on the infiltration and entrapment of dense nonaqueous phase liquids in saturated formation. J Contam Hydrol 42:187–218.

Dekker TJ, Abriola LM. 2000b. The influence of field-scale heterogeneity on the surfactant-enhanced remediation of entrapped nonaqueous phase liquids. J Contam Hydrol 42:219–251.

Demond AH, Desai FN, Hayes KF. 1994. Effect of cationic surfactants on organic liquid-water capillary-pressure saturation relationships. Water Resour Res 30:333–342.

Deutsch CV, Journel AG. 1998. GSLIB: Geostatistical Software Library and User's Guide, 2nd ed. Oxford University Press, New York, NY, USA.

DiFilippo EL, Brusseau ML. 2008. Relationship between mass flux reduction and source-zone mass removal: Analysis of field data. J Contam Hydrol 98:22–35.

Einarson MD, Mackay DM. 2001. Predicting impacts of groundwater contamination. Environ Sci Technol 35:66A-73A.

Enfield CG, Wood AL, Espinoza FP, Brooks MC, Annable M, Rao PSC. 2005. Design of aquifer remediation systems: (1) Describing hydraulic structure and NAPL architecture using tracers. J Contam Hydrol 81:125–147.

EnviroFlux. 2007. www.enviroflux.com/deployments. Accessed August 17, 2011.

Essaid HI, Hess KM. 1993. Monte Carlo simulations of multiphase flow incorporating spatial variability of hydraulic properties. Ground Water 31:123–134.

Falta RW, Lee CM, Brame SE, Roeder E, Coates JT, Wright C, Wood AL, Enfield CG. 1999. Field test of high molecular weight alcohol flushing for subsurface nonaqueous phase liquid remediation. Water Resour Res 35:2095–2108.

Falta RW, Rao PSC, Basu N. 2005a. Assessing the impacts of partial mass depletion in DNAPL source zones I. Analytical modeling of source strength functions and plume response. J Contam Hydrol 78:259–280.

Falta RW, Basu N, Rao PSC. 2005b. Assessing the impacts of partial mass depletion in DNAPL source zones II. Coupling source strength functions to plume evolution. J Contam Hydrol 79:45–66.

Feenstra S, Cherry JA, Parker BL. 1996. Conceptual models for the behavior of dense non-aqueous phase liquids (DNAPLs) in the subsurface. In Pankow JF, Cherry JA, eds, Dense Chlorinated Solvents and Other DNAPLs in Groundwater. Waterloo Press, Guelph, Ontario, Canada, pp 53–88.

Friis AK, Albrechtsen H-J, Heron G, Berg PL. 2005. Redox processes and release of organic matter after thermal treatment of a TCE-contaminated aquifer. Environ Sci Technol 39:5787–5795.

Friis AK, Heimann AC, Jakobsen R, Albrechtsen H-J, Cox E, Berg PL. 2007. Temperature dependence of anaerobic TCE-dechlorination in a highly enriched Dehalococoides-containing culture. Water Res 41:355–364.

Frind EO, Molson JW, Schirmer M, Guiger N. 1999. Dissolution and mass transfer of multiple organics under field conditions: The Borden emplaced source. Water Resour Res 35:683–694.

Fure AD, Jawitz JW, Annable MD. 2006. DNAPL source depletion: Linking architecture and flux response. J Contam Hydrol 85:118–140.

Glass RJ, Conrad SH, Peplinski W. 2000. Gravity-destabilizing nonwetting phase invasion in macroheterogeous porous media: Experimental observations of invasion dynamics and scale analysis. Water Resour Res 36:3121–3137.

Goltz MN, Kim S, Yoon H, Park J. 2007. Review of groundwater contaminant mass flux measurement. Environ Eng Res 12:176–193.

Goltz MN, Close ME, Yoon H, Huang J, Flintoft MJ, Kim S, Enfield C. 2009. Validation of two innovative methods to measure contaminant mass flux in groundwater. J Contam Hydrol 106:51–61.

Greenhouse J, Brewster M, Schneider G, Redman D, Annan P, Olhoeft G, Lucius J, Sander K, Mazzella A. 1993. Geophysics and solvents: The Borden experiment. The Leading Edge 12:261–267.

Grimm RE, Olhoeft GR, McKinley K, Rossabi J, Riha B. 2005. Nonlinear complex-resistivity survey for DNAPL at the Savannah River Site A-014 outfall. J Environ Eng Geophys 10:351–364.

Guilbeault MA, Parker BL, Cherry JA. 2005. Mass and flux distributions from DNAPL zones in sandy aquifers. Ground Water 43:70–86.

Hatfield K, Annable MD, Kuhn S, Rao PSC, Campbell T. 2002. A new method for quantifying contaminant flux at hazardous waste sites. In Groundwater Quality: Natural and Enhanced Restoration of Groundwater Pollution, IAHS Publication 275, pp 25–31.

Hatfield K, Annable MD, Cho J, Rao PSC, Klammler H. 2004. A direct passive method for measuring water and contaminant fluxes in porous media. J Contam Hydrol 75:155–181.

Held RJ, Illangasekare TH. 1995. Fingering of dense nonaqueous phase liquids in porous media: 1. Experimental investigation. Water Resour Res 31:1213–1222.

Heron G, Gierke JS, Faulkner B, Mravik S, Wood L, Enfield CG. 2002. Pulsed air sparging in aquifers contaminated with dense nonaqueous phase liquids. Ground Water Monitor Remediat 22:73–82.

Heron G, Carroll S, Nielsen SG. 2005. Full-scale removal of DNAPL constituents using steam-enhanced extraction and electrical resistance heating. Ground Water Monitor Remediat 25:92–107.

Hofstee C, Oostrom M, Dane JH, Walker RC. 1998a. Infiltration and redistribution of perchloroethylene in partially saturated, stratified porous media. J Contam Hydrol 34:293–313.

Hofstee C, Walker RC, Dane JH. 1998b. Infiltration and redistribution of perchloroethylene in stratified water-saturated porous media. Soil Sci Soc Am J 62:13–22.

Holder TH, Teutsch GP, Schwarz R. 1998. A new approach for source zone characterization: The Neckar Valley study. Groundwater Quality: Remediation and Protection, IAHS Publication 250, pp 49–55.

Holzmer FJ, Pope GA, Yeh L. 2000. Surfactant-enhanced aquifer remediation of PCE DNAPL in low-permeability sand. In Wickramanayake GB, Gavaskar AR, Gupta N, eds, Treating Dense Nonaqueous-Phase Liquids (DNAPLs): Remediation of Chlorinated and Recalcitrant Compounds. Battelle Press, Columbus, OH, USA, pp 211–218.

Hrapovic L, Sleep BE, Major DJ, Hood ED. 2005. Laboratory study of treatment of trichloroethene by chemical oxidation followed by bioremediation. Environ Sci Technol 39:2888–2897.

Huang KC, Zhao ZQ, Hoag GE, Dahmani A, Block PA. 2005. Degradation of volatile organic compounds with thermally activated persulfate oxidation. Chemosphere 61:551–560.

Imhoff PT, Pirestani K. 2004. Influence of mass transfer resistance on detection of nonaqueous phase liquids with partitioning tracer tests. Adv Water Resour 27:429–444.

Istok JD, Field JA, Schroth MH, Davis BM, Dwarakanath V. 2002. Single-well "push-pull" partitioning tracer test for NAPL detection in the subsurface. Environ Sci Technol 36:2708–2716.

ITRC (Interstate Technology and Regulatory Council). 2000. Dense non-aqueous phase liquids (DNAPLs): Review of emerging characterization and remediation technologies.

ITRC. 2003. An introduction to characterizing sites contaminated with DNAPL.

ITRC. 2004. Strategies for Monitoring the Performance of DNAPL Source Zone Remedies.

Jackson RE, Jin M. 2005. The measurement of DNAPL in low-permeability lenses within alluvial aquifers by partitioning tracers. Environ Eng Geosience 11:405–412.

James AI, Graham WD, Hatfield K, Rao PSC, Annable MD. 1997. Optimal estimation of residual non-aqueous phase liquid saturations using partitioning tracer concentration data. Water Resour Res 33:2621–2636.

James AI, Graham WD, Hatfield K, Rao PSC, Annable MD. 2000. Estimation of spatially variable residual nonaqueous phase liquid saturations in nonuniform flow fields using partitioning tracer data. Water Resour Res 36:999–1012.

Jarsjö J, Bayer-Raich M, Ptak T. 2005. Monitoring groundwater contamination and delineating source zones at industrial sites: Uncertainty analyses using integral pumping tests. J Contam Hydrol 79:107–134.

Jawitz JW, Annable MD, Rao PSC. 1998. Using interwell partitioning tracers and the method of moments to estimate the spatial distribution of non-aqueous phase contaminants in aquifers. GQ 98 Conference Proceedings, Groundwater Quality: Remediation and Protection, IAHS Publication 250, pp 422–425.

Jawitz JW, Sillan RK, Annable MD, Rao PSC, Warner K. 2000. In-situ alcohol flushing of a DNAPL source zone at a dry cleaner site. Environ Sci Technol 34:3722–3729.

Jawitz JT, Annable MD, Demmy GC, Rao PSC. 2003. Estimating nonaqueous phase liquid spatial variability using partitioning tracer higher temporal moments. Water Resour Res 39:1192, doi:10.10 29/2002WR001309.

Jawitz JT, Fure AD, Demmy GC, Berglund S, Rao PSC. 2005. Groundwater contaminant flux reduction resulting from nonaqueous phase liquid mass reduction. Water Resour Res 41:W10408, doi:10.1029/2004WR003825.

Jin M, Delshad M, Dwarakanath V, McKinney DC, Pope GA, Sepehrnoori K, Tilburg CE, Jackson RE. 1995. Partitioning tracer test for detection, estimation and remediation performance assessment of subsurface nonaqueous phase liquids. Water Resour Res 31:1201–1211.

Johnson RH, Poeter EP. 2005. Interpreting DNAPL saturations in a laboratory-scale injection using one- and two-dimensional modeling of GPR data. Ground Water Monitor Remediat 25:159–169.

Johnson RH, Poeter EP. 2007. Insights into the use of time-lapse GPR data as observations for inverse multiphase flow simulations of DNAPL migration. J Contam Hydrol 89:136–155.

Kao CM, Wang YS. 2001. Field investigation of the natural attenuation and intrinsic biodegradation rates at an underground storage tank site. Environ Geol 40:622–631.

Karagunduz A, Pennell KD, Young MD. 2001. Influence of a nonionic surfactant on the water retention properties of unsaturated soils. Soil Sci Soc Am J 65:1392–1399.

Kaye AJ, Cho J, Basu NB, Chen X, Annable MD, Jawitz JW. 2008. Laboratory investigation of flux reduction from dense non-aqueous phase liquid (DNAPL) partial source zone remediation by enhanced dissolution. J Contam Hydrol 102:17–28.

King MWG, Barker JF, Devlin JF, Butler BJ. 1999. Migration and natural fate of a coal tar creosote plume: 2. Mass balance and biodegradation indicators. J Contam Hydrol 39:281–307.

Kram ML, Keller AA, Rossabi J, Everett LG. 2001. DNAPL characterization methods and approaches, Part 1: Performance comparisons. Ground Water Monitor Remediat 21:109–123.

Kram ML, Keller AA, Rossabi J, Everett LG. 2002. DNAPL characterization methods and approaches, Part 2: Cost comparisons. Ground Water Monitor Remediat 22:46–61.

Kübert M, Finkel M. 2006. Contaminant mass discharge estimation in groundwater based on multi-level point measurements: A numerical evaluation of expected errors. J Contam Hydrol 84:55–80.

Kueper BH, Gerhard JI. 1995. Variability of point source infiltration rates for two-phase flow in heterogeneous porous media. Water Resour Res 31:2971–2980.

Kueper BH, Redman D, Starr RC, Reitsma S, Mah M. 1993. A field experiment to study the behavior of tetrachloroethylene below the water table: Spatial distribution of residual and pooled DNAPL. Ground Water 31:756–766.

Lake LW. 1989. Enhanced Oil Recovery. Prentice-Hall, Englewood Cliffs, NJ, USA.

Lemke LD, Abriola LM. 2006. Modeling DNAPL mass removal in nonuniform formations: Linking source zone architecture and system response. Geosphere 2:74–82.

Lemke LD, Abriola LM, Goovaerts P. 2004a. DNAPL source zone characterization: Influence of hydraulic property correlation on predictions of DNAPL infiltration and entrapment. Water Resour Res 40:W01511, doi: 10.1029/2003WR001980.

Lemke LD, Abriola LM, Lang JR. 2004b. DNAPL source zone remediation: Influence of source zone architecture on predictions of DNAPL recovery and contaminant flux. Water Resour Res 40:W12417, doi:10.1029/2004WR003061.

Li KB, Abriola LM. 2009. A multistage multicriteria spatial sampling strategy for estimating contaminant mass discharge and its uncertainty. Water Resour Res 45:W06407.

Li DX, Schwartz FW. 2003. Permanganate oxidation schemes for the remediation of source zone DNAPLs and dissolved contaminant plumes. In Henry SM, Warner SD, eds, Chlorinated Solvent and DNAPL Remediation: Innovative Strategies for Subsurface Cleanup (ACS Symposium Series, No. 837). ACS, Washington, DC, USA, pp 73–85.

Li K, Goovaerts P, Abriola LM. 2007. A geostatistical approach for quantification of contaminant mass discharge uncertainty using multi-level sampler measurements. Water Resour Res 43:W06436, doi:10.1029/2006WR005427.

Liang CJ, Bruell CJ, Marley MC, Sperry KL. 2003. Thermally activated persulfate oxidation of trichloroethylene (TCE) and 1,1,1-trichloroethane (TCA) in aqueous systems and soils slurries. Soil Sediment Contam 12:207–228.

Liang CJ, Bruell CJ, Marley MC, Sperry KL. 2004. Persulfate oxidation for in situ remediation of TCE. I. Activated by ferrous ion with and without a persulfate-thiosulfate redox couple. Chemosphere 55:1213–1223.

Londergan JT, Meinardus HW, Mariner PE, Jackson RE, Brown CL, Dwarakanath V, Pope GA, Ginn JS, Taffinder S. 2001. DNAPL removal from a heterogeneous alluvial aquifer by surfactant-enhanced aquifer remediation. Ground Water Monitor Remediat 21:57–67.

Lord DL, Demond AH, Hayes KF. 2000. Effects of organic base chemistry on interfacial tension, wettability, and capillary pressure in multiphase subsurface waste systems. Tran Porous Media 38:79–92.

Lowe KS, Gardner FG, Siegrist RL. 2002. Field evaluation of in situ chemical oxidation through vertical well-to-well recirculation of NaMnO4. Ground Water Monitor Remediat 22:106–115.

Maji R, Sudicky EA, Panday S, Teutsch G. 2006. Transition probability/Markov chain analyses of DNAPL source zones and plumes. Ground Water 44:853–863.

Marble JC, DiFilippo EL, Zhang Z, Tick GR, Brusseau ML. 2008. Application of a lumped-process mathematical model to dissolution of non-uniformly distributed immiscible liquid in heterogeneous porous media. J Contam Hydrol 100:1–10.

Mayer AS, Miller CT. 1996. The influence of mass transfer characteristics and porous media heterogeneity on nonaqueous phase dissolution. Water Resour Res 32:1551–1568.

McGuire T, Hughes JB. 2003. Effects of surfactants on the dechlorination of chlorinated ethenes. Environ Toxicol Chem 22:2630–2638.

McGuire TM, McDade JM, Newell CJ. 2006. Performance of DNAPL source depletion technologies at 59 chlorinated solvent impacted sites. Ground Water Monit Remediat 26:73–84.

McWhorter DB, Kueper BH. 1996. Mechanisms and mathematics of the movement of dense nonaqueous phase liquids (DNAPLs) in porous media. In Pankow JF, Cherry JA, eds, Dense Chlorinated Solvents and Other DNAPLs in Groundwater: History, Behavior, and Remediation. Waterloo Press, Portland, OR, USA, pp 89–128.

Mercer JW, Cohen RM. 1990. A review of immiscible fluids in the subsurface: Properties, models, characterization and remediation. J Contam Hydrol 6:107–163.

Meinardus HW, Dwarakanath V, Ewing J, Hirasaki GJ, Jackson RE, Jin M, Ginn JS, Londergan JT, Miller CA, Pope GA. 2002. Performance assessment of NAPL remediation in heterogeneous alluvium. J Contam Hydrol 54:173–193.

Michalak AM, Kitanidis PK. 2004. Estimation of historical groundwater contaminant distribution using the adjoint state method applied to geostatistical inverse modeling. Water Resour Res 40:W08302.

Miller CT, Christakos G, Imhoff PT, McBride JF, Pedit JA, Trangenstein JA. 1998. Multiphase flow and transport modeling in heterogeneous porous media: Challenges and approaches. Adv Water Resour 21:77–120.

Moreno-Barbero E, Illangasekare TH. 2006. Influence of dense nonaqueous phase liquid pool morphology on the performance of partitioning tracer tests: Evaluation of the equilibrium assumption. Water Resour Res 42:W04408.

Moreno-Barbero E, Saenton S, Illangasekare TH. 2004. Potential use of partitioning tracer and mass flux emission data to characterize DNAPL source zone architecture. In Gavaskar AR, Chen ASC, eds, Proceedings of the Fourth International Conference on Remediation of Chlorinated and Recalcitrant Compounds. Battelle Press, Columbus, OH, USA, Paper 1B-08.

NRC (National Research Council). 1999. Groundwater & Soil Cleanup. The National Academies Press, Washington, DC, USA.

NRC. 2005. Contaminants in the Subsurface: Source Zone Assessment and Remediation. The National Academies Press, Washington, DC, USA.

NAVFAC (Naval Facilities Engineering Services Center). 2004. Assessing the feasibility of DNAPL source zone remediation: review of case studies. Contract Report CR-004-02-ENV.

Nelson MD, Parker BL, Al TA, Cherry JA, Loomer D. 2001. Geochemical reactions resulting from in situ oxidation of PCE-DNAPL by KMnO4 in a sandy aquifer. Environ Sci Technol 35:1266–1275.

Newman MA, Hatfield K, Hayworth J, Rao PSC, Stauffer T. 2006. Inverse characterization of NAPL source zones. Environ Sci Technol 40:6044–6050.

Newmark RL, Daily WD, Kyle KR, Ramirez AL. 1997. Monitoring DNAPL pumping using integrated geophysical techniques. U.S. Department of Energy Report UCRL-ID-122215.

Nichols E, Roth T. 2004. Flux redux – Using mass flux to improve cleanup decisions. L.U.S.T. Line, New England Interstate Water Pollution Control Commission, Bulletin 46, pp 6–9.

O'Carroll DM, Bradford SA, Abriola LM. 2004. Infiltration of PCE in system containing spatial wettability variations. J Contam Hydrol 73:39–63.

Park E, Parker JC. 2005. Evaluation of an upscaled model for DNAPL dissolution kinetics in heterogeneous aquifers. Adv Water Resour 28:1280–1291.

Parker BL, Gillham RW, Cherry JA. 1994. Diffusive disappearance of immiscible-phase organic liquids in fractured geologic media. Ground Water 32:805–820.

Parker BL, Chapman SW, Guilbeault MA. 2008. Plume persistence caused by back diffusion from thin clay layers in a sand aquifer following TCE source-zone hydraulic isolation. J Contam Hydrol 102:86–104.

Parker JC, Lenhard RJ. 1987. A model for hysteretic constitutive relations governing multiphase flow, 1. Saturation-pressure relations. Water Resour Res 23:2187–2196.

Parker JC, Park E. 2004. Field-scale DNAPL dissolution kinetics in heterogeneous aquifers. Water Resour Res 40:W05109, doi:10.1029/2003WR002807.

Pennell KD, Abriola LM, Weber WJ. 1993. Surfactant enhanced solubilization of residual dodecane in soil columns 1. Experimental investigation. Environ Sci Technol 27:2332–2340.

Pennell KD, Pope GA, Abriola LM. 1996. Influence of viscous and buoyancy forces on the mobilization of residual tetrachloroethylene during surfactant flushing. Environ Sci Technol 30:1328–1335.

Phelan TJ, Lemke LD, Bradford SA, O'Carroll DM, Abriola LM. 2004. Influence of textural and wettability variations on predictions of DNAPL persistence and plume development in saturated porous media. Adv Water Resour 27:411–427.

Pitkin SE, Cherry JA, Ingleton RA, Broholm M. 1999. Field demonstrations using the Waterloo Groundwater Profiler. Ground Water Monitor Remediat 19:122–131.

Powers SE, Tamblin ME. 1995. Wettability of porous media after exposure to synthetic gasolines. J Contam Hydrol 19:105–125.

Powers SE, Abriola LM, Weber WJ Jr. 1992. An experimental investigation of NAPL dissolution in saturated subsurface systems: Steady-state mass transfer rates. Water Resour Res 28:2691–2705.

Ramsburg CA, Pennell KD. 2001. Experimental and economic assessment of two surfactant formulations for source zone remediation at a former dry cleaning facility. Ground Water Monitor Remediat 21:68–82.

Ramsburg CA, Abriola LM, Pennell KD, Löffler FE, Gamache M, Amos BK. 2004. Stimulated microbial reductive dechlorination following surfactant treatment at the Bachman road site. Environ Sci Technol 38:5902–5914.

Ramsburg CA, Pennell KD, Abriola LM, Daniels G, Drummond CD, Gamache M, Hsu H, Petrovkis EA, Rathfelder KM, Ryder J, Yarkarski T. 2005. A pilot-scale demonstration of surfactant enhanced PCE solubilization at the Bachman road site: 2. System operation and evaluation. Environ Sci Technol 39:1791–1801.

Rao PSC, Jawitz JW. 2003. Comment on "Steady state mass transfer from single-component dense nonaqueous phase liquids in uniform flow fields" by Sale TC, McWhorter DB, Water Resour Res 39:1068, doi:10.1029/2001WR000599.

Rao PSC, Annable MD, Kim H. 2000. NAPL source zone characterization and remediation technology performance assessment: recent developments and applications of tracer techniques. J Contam Hydrol 45:63–78.

Rao PSC, Jawitz JW, Enfield CG, Falta RW, Annable MD, Wood AL. 2002. Technology integration for contaminated site remediation: clean-up goals and performance criteria. Groundwater Quality: Natural and Enhanced Restoration of Groundwater Pollution. IAHS Publication 275, pp 571–578.

Rathfelder KM, Abriola LM, Taylor TP, Pennell KD. 2001. Surfactant enhanced recovery of tetrachloroethylene from a porous media containing low permeability lenses: 2. Numerical simulations. J Contam Hydrol 48:351–374.

Rathfelder KM, Abriola LM, Singletary MA, Pennell KD. 2003. Influence of surfactant-facilitated interfacial tension reduction on chlorinated solvent migration in porous media: Observations and numerical simulation. J Contam Hydrol 64:227–252.

Richardson RE, James CA, Bhupathiraju VK, Alvarez-Cohen L. 2002. Microbial activity in soils following steam treatment. Biodegradation 13:285–295.

Rivett MO, Feenstra S. 2004. Dissolution of an emplaced source of DNAPL in a natural aquifer setting, Environ Sci Technol 39:447–455.

Robbat A Jr, Smarason S, Gankin Y. 1998. Dynamic work plans and field analytics, the keys to cost-effective hazardous waste site Investigations. Field Anal Chem Technol 2:253–265.

Rossabi J, Looney BB, Eddy-Dilek CA, Riha BD, Jackson DG. 2000. DNAPL Site Characterization: The Evolving Conceptual Model and Toolbox Approach. US Department of Energy Report WSRC-MS-2000-0183.

Saba T, Illangasekare TH. 2000. Effect of groundwater flow dimensionality on mass transfer from entrapped nonaqueous phase liquid contaminants. Water Resour Res 36:971–979.

Saenton S, Illangasekare TH. 2003. Determining the entrapment architecture in the DNAPL source zone using down-gradient mass flux measurements: A combined numerical and stochastic study. In Proceedings of MODFLOW and More 2003: Understanding through Modeling, Vol. 2, pp 615–619.

Saenton S, Illangasekare T. 2007. Upscaling of mass transfer rate coefficient for the numerical simulation of dense nonaqueous phase liquid dissolution in heterogeneous aquifers. Water Resour Res 43:WR004274.

Sahl JW, Munakata-Marr J, Crimi ML, Siegrist RL. 2007. Coupling permanganate oxidation with microbial dechlorination. Water Environ Res 79:5–12.

Sale TC, McWhorter DB. 2001. Steady state mass transfer from single-component dense nonaqueous phase liquids in uniform flow fields. Water Resour Res 37:393–404.

Sale TC, Zimbron JA, Dandy DS. 2008. Effects of reduced contaminant loading on down-gradient water quality in an idealized two-layer granular porous media. J Contam Hydrol 102:72–85.

Schaerlaekens J, Feyen J. 2004. Effect of scale and dimensionality on the surfactant-enhanced solubilization of a residual DNAPL contamination. J Contam Hydrol 71:283–306.

Schwarz R, Ptak T, Holder TH, Teutsch G. 1998. Groundwater risk assessment at contaminated sites: A new investigation approach. Groundwater Quality: Remediation and Protection, IAHS publication 250, pp 68–71.

Schwille F. 1988. Dense Chlorinated Solvents in Porous and Fractured Media. Translated by Pankow JF. Lewis Publishers, Chelsea, MI, USA.

Semprini L, Kitanidis PK, Kampbell DH, Wilson JT. 1995. Anaerobic transformation of chlorinated aliphatic hydrocarbons in a sand aquifer based on spatial chemical distributions. Water Resour Res 31:1051–1062.

Siegrist RL, Urynowicz MA, West OR, Crimi ML, Lowe KS. 2001. Principles and Practices of In Situ Chemical Oxidation Using Permanganate. Battelle Press, Columbus, OH, USA.

Smith JE, Zhang ZF. 2001. Determining effective interfacial tension and predicting finger spacing for DNAPL penetration into water-saturated porous media. J Contam Hydrol 48:167–183.

Sneddon KW, Olhoeft GR, Powers MH. 2000. Determining and mapping DNAPL saturation values from noninvasive GPR measurements, In Powers MH, Ibrahim A-B, Cramer L, eds, Symposium on the Applications of Geophysics to Engineering and Environmental Problems (SAGEEP 2000), Environmental and Engineering Geophysical Society (EEGS), pp 293–302.

Soga K, Page JWE, Illangasekare TH. 2004. A review of NAPL source zone remediation efficiency and the mass flux approach. J Hazard Mater 110:13–27.

Stewart M, North L. 2006. A borehole geophysical method for detection and quantification of dense, non-aqueous phase liquids (DNAPLs) in saturated soils. J Appl Geophys 60:87–99.

Stroo HF, Unger M, Ward CH, Kavanaugh MC, Vogel C, Leeson A, Marquesee JA, Smith BP. 2003. Remediating chlorinated solvent source zones. Environ Sci Technol 37:224A-230A.

Suchomel EJ, Pennell KD. 2006. Reductions in contaminant mass discharge following partial mass removal from DNAPL source zones. Environ Sci Technol 40:6110–6116.

Sun AY. 2007. A robust geostatistical approach to contaminant source identification. Water Resour Res 43:W02418, doi:10.1029/2006WR005106.

Sun AY, Painter SL, Wittmeyer GW. 2006. A robust approach for iterative contaminant source location and release history recovery. J Contam Hydrol 88:181–196.

Temples TJ, Waddell MG, Domoracki W. 2001. Non-invasive determination of the location and distribution of DNAPL using advanced seismic reflection techniques. Ground Water 39:465–474.

Teutsch G, Ptak T, Schwarz R, Holder T. 2000. Ein neues integrals verfahren zur quantifizierung der grundwasserimmission: I. Beschreibung der grundlagen. Grundwasser 4:170–175.

Tomich JF, Dalton RL, Deans HA, Shallenberger LK. 1973. Single-well tracer method to measure residual oil saturation. J Petrol Technol, February.

USEPA (U.S. Environmental Protection Agency). 2003. The DNAPL remediation challenge: Is there a case for source depletion? EPA/600/R-03-143. USEPA, Washington DC, USA.

USGS (U.S. Geological Survey). 2003. Methodology for Estimating Times of Remediation Associated with Monitored Natural Attenuation. Water Resources Investigation Report 03–4057.

Van Valkanburg ME, Annable MD. 2002. Mobilization and entry of DNAPL pools into finer sand media by cosolvents: Two-dimensional chamber studies. J Contam Hydrol 59:211–230.

Waldemer RH, Tratnyek PG, Johnson RL, Nurmi JT. 2007. Oxidation of chlorinated ethenes by heat-activated persulfate: Kinetics and products. Environ Sci Technol 41:1010–1015.

Willson CS, Pau O, Pedit JA, Miller CT. 2000. Mass transfer limitation effects on partitioning tracer tests. J Contam Hydrol 45:79–97.

Wise WR. 1999. NAPL characterization via partitioning tracer tests: Quantifying effects of partitioning nonlinearities. J Contam Hydrol 36:167–183.

Wise WR, Dai D, Fitzpatrick EA, Evans LW, Rao PSC, Annable MD. 1999. Non-aqueous phase liquid characterization via partitioning tracer tests: A modified Langmuir relation to describe partitioning nonlinearities. J Contam Hydrol 36:153–165.

Zeru A, Schafer G. 2005. Analysis of groundwater contamination using concentration-time series recorded during an integral pumping test: Bias introduced by strong concentration gradients within the plume. J Contam Hydrol 81:106–124.

Zhang C, Yoon H, Werth CJ, Valocchi AJ, Basu NB, Jawitz JW. 2008. Evaluation of simplified mass transfer models to simulate the impacts of source zone architecture on nonaqueous phase liquid dissolution in heterogeneous porous media. J Contam Hydrol 102:49–60.

Zhang Y, Graham WD. 2001. Spatial characterization of a hydrogeochemically heterogeneous aquifer using partitioning tracers: Optimal estimation of aquifer parameters. Water Resour Res 37:2049–2063.

Zhang ZF, Smith JE. 2002. Visualization of DNAPL fingering processes and mechanisms in water-saturated porous media. Trans Porous Media 48:41–59.

Zhu J, Sykes JF. 2000. The influence of NAPL dissolution characteristics on field-scale contaminant transport in subsurface. J Contam Hydrol 41:133–154.

Zhu J, Sykes JF. 2004. Simple screening models of NAPL dissolution in the subsurface. J Contam Hydrol 72:245–258.

APPENDIX A
LIST OF ACRONYMS AND ABBREVIATIONS

°	Degrees
°C	Degrees Celsius
°F	Degrees Fahrenheit
°K	Degrees Kelvin
(aq)	Aqueous phase
(g)	Gas phase
(s)	Solid phase
μg	Microgram(s)
μg/kg	Microgram(s) per kilogram
μg/L	Microgram(s) per liter
1,1-DCA	1,1-dichloroethane
1,1-DCE	1,1-dichloroethene
1,2-DCA	1,2-dichloroethane
1,2-DCE	1,2-dichloroethene
1,1,1-TCA	1,1,1-trichloroethane
1-D	One dimensional
2-D	Two dimensional
3-D	Three dimensional
ADE	Advection-dispersion equation
AFB	Air Force Base
AFCEE	Air Force Center for Engineering and the Environment (previously the Air Force Center for Environmental Excellence)
AMD	Acid mine drainage
API	American Petroleum Institute
ASCE	American Society for Civil Engineering
ASU	Arizona State University
atm	Atmosphere
BCEE	Board Certified Environmental Engineer
bgs	Below ground surface
BOF	Basic oxygen furnace
BTEX	Benzene, toluene, ethylbenzene, and total xylenes
CERCLA	Comprehensive Environmental Response, Compensation, and Liability Act
CGER	Commission on Geosciences, Environment and Resources
CHP	Catalyzed hydrogen peroxide
***cis*-DCE**	*cis*-1,2-dichloroethene
cm	Centimeter(s)
cm/s	Centimeter(s) per second
COC	Contaminant(s) of concern
cP	Centipoise
CSM	Conceptual site model
cSt	Centistoke
CSTR	Continuously stirred tank reactor
CT	Carbon tetrachloride
d	Day
D_a	Damkohler number
DAT	Diaminotoluene
DCE	Dichloroethene
DCM	Dichloromethane
DDC	Density Driven Convection system
DDMT	Dual-domain mass transfer
DDT	Dichlorodiphenyltrichloroethane
DERP	Defense Environmental Restoration Program
DNA	Deoxyribonucleic acid
DNAPL	Dense nonaqueous phase liquid
DNT	Dinitrotoluene
DOC	Dissolved organic carbon
DoD	Department of Defense
EDTA	Ethylenediaminetetraacetic acid
ERH	Electrical resistive heating
EISB	Enhanced *in situ* bioremediation
ESTCP	Environmental Security Technology Certification Program
ETH	Ethene

FD Finite difference
FE Finite element
FeRB Iron(III)-reducing bacteria
ft Feet
FV Finite volume
g Gram(s)
g/cm^3 Gram(s) per cubic centimeter
g/kg Gram(s) per kilogram
g/L Gram(s) per liter
g/mol Gram(s) per mole
GAC Granular activated carbon
gal Gallon(s)
GCW Groundwater circulating well
GEM Genetically engineered microorganism
GI Global implicit
GTP Ganglia-to-pool
H Henry's Law constant
hr(s) Hour(s)
in Inch(es)
ISCO *In situ* chemical oxidation
ISCR *In situ* chemical reduction
ISS *In situ* sparging
ITRC Interstate Technology & Regulatory Council
k_d Partition coefficient
K_{OW} Octanol-water partition coefficient
K_{sp} Solubility product constant
kg Kilogram(s)
kJ Kilojoule(s)
km Kilometer(s)
kmol Kilomole(s)
L Liter(s)
LEA Local equilibrium assumption
LEL Lower explosion level
LNAPL Light nonaqueous phase liquid
LPM Low permeability media
LTM Long term monitoring
m Meter(s)
M Mass, molar
MC Methylene chloride
MCL Maximum contaminant level
MCPP Mecoprop
mg Milligram(s)
mg/L Milligram(s) per liter
min Minute(s)
mL Milliliter(s)
mM Millimolar
mm Millimeter(s)
mmol Millimole(s)
MNA Monitored natural attenuation
mol Mole(s)
MSU Michigan State University
MTBE Methyl tertiary-butyl ether
NAPL Nonaqueous phase liquid
NAS Naval Air Station
NFESC Naval Facilities Engineering Service Center
NIH National Institutes of Health
NIOSH National Institute of Occupational Safety and Health
nM Nanomolar
NRC National Research Council
O&M Operation and maintenance
OHSU Oregon Health & Science University
ORP Oxidation-reduction potential
OS Operator splitting
P Poise
P&T Pump-and-treat
PAH Polynuclear aromatic hydrocarbons
PCB Polychlorinated biphenyl
PCE Perchoroethene (also termed perchloroethylene or tetrachloroethylene)
PCR Polymerase chain reaction
PDTC Pyridine-2,6-bis-thiocarboxylate
PE Professional engineer
PFR Plug flow reactor
PITT Partitioning interwell tracer test
POP Persistent organic pollutant
ppb Part(s) per billion
ppmv Part(s) per million by volume
PRB Permeable reactive barrier
PRG Preliminary remedial goal(s)
psi Pound(s) per square inch
pv Pore volume
PVC Polyvinyl chloride
RAO Remedial action objective
RCRA Resource Conservation and Recovery Act
R_e Reynold's number
REL Recommended Exposure Limit
RG Remedial goal(s)
ROI Radius of influence
s Second(s)
scfm Standard cubic feet per minute
SEAR Surfactant enhanced aquifer remediation

SER Steam enhanced remediation
SERDP Strategic Environmental Research and Development Program
SGSIM Sequential Gaussian simulation
SI Sequential iterative approach
SOD Soil oxidant demand
SOM Soil organic matter
SRB Sulfate-reducing bacteria
SVE Soil vapor extraction
TCE Trichloroethene
TDS Total dissolved solids
TNT 2,4,6-trinitrotoluene
TOC Total organic carbon
***trans*-DCE** *trans*-1,2-dichloroethene
TRW Tandem recirculating well
TTZ Target treatment zone
UIC Underground Injection Control
U.S. United States
USDOE U.S. Department of Energy
USEPA U.S. Environmental Protection Agency
USGS U.S. Geological Survey
UV Ultraviolet
UVB™ Unterdruck-Verdampfer-Brunnen
VC Vinyl chloride
VOC Volatile organic compound
wt. Weight

APPENDIX B
UNIT CONVERSION TABLE

Multiply	By	To Obtain
Acres	0.405	Hectares
Acres	1.56 E−3	Square miles (statute)
Centimeters	0.394	Inches
Cubic feet	0.028	Cubic meters
Cubic feet	7.48	Gallons (U.S. liquid)
Cubic feet	28.3	Liters
Cubic meters	35.3	Cubic feet
Cubic yards	0.76	Cubic meters
Feet	0.305	Meters
Feet per year	9.66 E−7	Centimeters per second
Gallons (U.S. liquid)	3.79	Liters
Hectares	2.47	Acres
Inches	2.54	Centimeters
Kilograms	2.20	Pounds (avoir)
Kilograms	35.3	Ounces (avoir)
Kilometers	0.62	Miles (statue)
Liters	0.035	Cubic feet
Liters	0.26	Gallons (U.S. liquid)
Meters	3.28	Feet
Miles (statue)	1.61	Kilometers
Ounces (avoir)	0.028	Kilograms
Ounces (fluid)	29.6	Milliliters
Pounds (avoir)	0.45	Kilograms
Square feet	0.093	Square meters
Square miles	640	Acres

APPENDIX C
GLOSSARY[1]

Abiotic - Occurring without the direct involvement of organisms.

Absorption - The uptake of water, other fluids, or dissolved chemicals by a porous material, a cell or an organism.

Activated carbon - A highly adsorbent form of carbon used to remove odors and/or toxic substances from liquid or gaseous emissions.

Activation - Chemical reaction where an agent reacts with an oxidant parent chemical (e.g., hydrogen peroxide [H_2O_2]) to yield a reactive species (e.g., hydroxyl free radical, OH^-).

Adsorption - A process that occurs when a gas or liquid solute accumulates on the surface of a solid or a liquid (adsorbent), forming a film of molecules or atoms (the adsorbate).

Advection - Transport of a substance by a fluid (e.g., groundwater) through the fluid's bulk motion in a particular direction (aka convection).

Aerobic - Environmental conditions where oxygen is present. Aerobic respiration by living organisms requires oxygen to generate energy.

Air sparging - Technology in which air or oxygen is injected into an aquifer to volatize or biodegrade contaminants.

Aldehyde - A broad class of organic compounds having the generic formula R-CHO and characterized by an unsaturated carbonyl group (C=O). They are formed from alcohols by either dehydrogenation or oxidation and thus occupy an intermediate position between primary alcohols and the acids obtained from them by further oxidation.

Aliphatic compounds - Any chemical compound belonging to the organic class in which the atoms are not linked together to form a benzene ring.

---

[1] This glossary is a compilation of definitions of terms synthesized by the volume editors and chapter authors from a variety of published and unpublished sources, including previous volumes in the SERDP/ ESTCP Remediation Technology Monograph Series.

Alkane - Non-aromatic saturated hydrocarbons with one or more carbon-carbon single bonds and having the general formula $C_nH_{(2n+2)}$.

Alkalinity - A measure of the ability of a solution to neutralize acids, equal to the stoichiometric sum of the bases in the solution. An expression of the buffering capacity of the solution.

Alkene - Unsaturated, open chain hydrocarbons with one or more carbon–carbon double bonds, having the general formula $C_nH_{(2n)}$.

Anaerobic - Environmental conditions where oxygen is absent. In groundwater, a dissolved oxygen concentration below 1.0 milligrams per liter (mg/L) is generally considered anaerobic. Anaerobic respiration is a means for a living organism to generate energy in the absence of oxygen.

Analytical model - A mathematical model, often based on simplifying assumptions, that has a closed form solution (i.e., the solution can be expressed in terms of known functions).

Anion - A negatively charged ion.

Anisotropy - In hydrology, the conditions under which one or more hydraulic properties (e.g., permeability) of an aquifer vary with respect to direction.

Anomalous transport - Non-Fickian transport. Scale effects that are space (or time) dependent in solute transport such that a constant dispersion coefficient or dispersivity in the advection-dispersion equation inadequately describes the solute transport.

Anoxic - "Without oxygen." Anoxic refers specifically to conditions of no dissolved oxygen but possibly with nitrate present.

Aquifer - An underground geological formation that stores and conducts water in significant amounts and can supply the water for wells or springs, etc.

Aquitard - A geological formation, usually a layer adjacent to an aquifer, of low hydraulic conductivity.

Aquiclude - A solid, practically impermeable area layer, that may underlying or overlying an aquifer.

Assimilative capacity - The capacity of a natural body of water to receive and degrade wastewaters or toxic materials.

Attenuation - Reduction of contaminant concentrations over space or time. Includes both destructive (e.g., biodegradation, hydrolysis) and non-destructive (e.g., volatilization, sorption) removal processes.

Attenuation rate - The rate of contaminant concentration reduction over time. Example units are milligrams per liter per year (mg/L/year).

Autotrophic - Self-sustaining or self-nourishing. Organisms that must synthesize their own food from inorganic materials, such as carbon dioxide and ammonium.

Bacterium - A single-celled organism of microscopic size (generally 0.3–2.0 micrometers [μm] in diameter). As opposed to fungi and higher plants and animals (*eukaryotes*), bacteria are *prokaryotes* (characterized by the absence of a distinct, membrane-bound nucleus or membrane-bound organelles and by deoxyribonucleic acid (DNA) that is not organized into chromosomes).

Baseline - A set of data representing ambient conditions that are collected before remediation is implemented. Compared with post-treatment data to evaluate the effectiveness of remediation.

Bedrock - The solid or fractured rock underlying surface solids and other unconsolidated material or overburden.

Bench-test - See "Treatability test."

Bentonite - An expandable clay mineral, subject to swelling when wet and shrinking when dry. Can be formed by chemical alteration of volcanic ash.

Bioaugmentation - Addition of microbes to the subsurface to improve the biodegradation of target contaminants. Microbes may be "seeded" from populations already present at a site or from specially cultivated strains or mixtures of microorganisms.

Bioavailability - The degree to which a compound is available for uptake or transformation by an organism.

Biobarrier - Same as Biowall. A remediation technology designed to intercept and biologically degrade a contaminant as it passes through a permeable subsurface barrier with groundwater flow in an aquifer. Biobarriers are created by installing wells or trenches across the width of a plume to deliver substrates required by microorganisms for contaminate degradation.

Biochemical - Produced by or involving chemical reactions of living organisms.

Biodegradation - Biologically mediated conversion of one compound to another.

Biofouling - Impairment of the functioning of wells or other equipment as a result of the growth or activity of microorganisms.

Biomarker - A biochemical within an organism that has a particular molecular feature that makes it useful for identifying a specific biological activity.

Biomass - Total mass of microorganisms present in a given amount of water or solid material.

Bioremediation - Use of microorganisms to control and destroy contaminants.

Biotransformation - Biologically catalyzed conversion of a chemical to some other product.

Biowall - See Biobarrier.

Breakthrough curve - The evolution of solute concentration measured at a fixed location such as in a column or a well.

Buffering capacity - A measure of a solution's ability to resist changes in pH upon addition of acid or base.

Capture zone - The three-dimensional region that contributes the groundwater extracted by one or more wells or drains.

Carboxylic acid - An organic acid characterized by one or more carboxyl groups (−COOH).

Catalyst - A substance that promotes a chemical reaction but does not itself enter into the reaction.

Catalyzed hydrogen peroxide (CHP) - An oxidant formulation consisting of hydrogen peroxide and a catalyst, generally ferrous iron. Also called Catalyzed Hydrogen Peroxide Propagations. Generally used interchangeably with Fenton's Reagent and Modified Fenton's reagent.

Cation - A positively charged ion.

Chelating agent - A compound, typically organic, that is capable of causing chelation.

Chelation - The formation or presence of two or more separate modes of binding between a ligand and a single central atom. The ligands are normally organic compounds such as ethylenediaminetetraacetic acid (EDTA) and can be called chelants, chelators, chelating agents, or sequestering agents.

Chemical equilibrium - In a closed system, the net reaction rate is zero; the forward and backward rates are equal and opposite such that the concentrations of the reacting substances do not change with time.

Chlorinated solvent - A hydrocarbon in which chlorine atoms substitute for one or more hydrogen atoms in the compound's structure. Chlorinated solvents commonly are used for grease removal in manufacturing, dry cleaning, and other operations. Examples include trichloroethene (TCE), perchloroethene (PCE), and trichloroethane (TCA).

Chloroethane - (Also ethyl chloride) a colorless, flammable gas, C_2H_5Cl, belonging to the family of organohalogen compounds. Used as a refrigerant, solvent, and anesthetic. At one time, used as a high-volume industrial chemical in the preparation of the gasoline additive tetraethyl lead.

Cleanup level - Used to describe the degree of remediation required with respect to achieving a certain concentration of contaminants of concern (COCs) in soil, groundwater or other media at a given site or within a particular target treatment zone (TTZ). Cleanup levels are commonly specified by regulatory authorities and programs and can include numeric values for specific media. Under some regulatory programs, cleanup levels may be used as remediation goals.

Co-contaminant - A contaminant that is present, but is not considered the contaminant of concern (COC) as a primary driver of remediation, due to relatively lower concentration or level of risk. May or may not be targeted by ISCO (or other technologies) when treating the primary COC.

Cometabolism - The metabolism of two compounds by an organism, in which the degradation of the second compound (the secondary substrate) depends on the transformation of the first compound (the primary substrate). For example, in the process of oxidizing methane, some bacteria can fortuitously degrade chlorinated solvents that they would otherwise be unable to attack.

Complete segregation - The assumption that all mixing occurs at the inlet and outlet of a reactor system rather than within the reactor itself.

Conceptual site model (CSM) - A hypothesis about how contaminant releases occurred at a site, the current state of the contaminant source, site conditions transport/fate pathways to receptors, and the current plume characteristics (plume stability).

Confined aquifer - An aquifer overlain by an impermeable layer such that the piezometric head rises above the top of the aquifer.

Conservative tracer - A solute that maintains its mass in a fluid volume.

Contaminant of concern (COC) - One or more contaminants present at a site that contribute to the risk and impact the nature and extent of remediation. They may be selected as the targets to be destroyed or otherwise removed during remediation.

Contaminant rebound - An increase in concentration over the course of a post-treatment monitoring period following an initial decrease in aqueous contaminant concentration immediately after site remediation.

Continuous-flow stirred-tank reactor (CSTR) - A perfectly mixed reactor that achieves compositional uniformity throughout its volume.

Coupling - A term used to describe the proactive combination of two or more remediation approaches or technologies. Also known as combined remedies.

Damkohler number (Da) - A dimensionless number that relates the chemical reaction time scale to the time scale of a relevant transport or mass transfer time scale. (The characteristic time is inverse to the rate.) In transport in porous media, the Damkohler number is typically the transport time scale divided by the reaction time scale.

Darcy scale - A scale that averages over volumes and areas that contain many pores, enough so that Darcy's law holds. Smaller than the field scale

Darcy's Law - The relationship discovered by Henri Darcy that states the average flow rate, defined by the specific discharge in a porous medium is proportional to the negative hydraulic-head gradient.

Darcy velocity - The Darcy velocity is the specific discharge (discharge through an area divided by the total area, both pores and solids) used in modeling at the Darcy scale.

Dechlorination - A reaction involving the removal of one or more chlorine atoms from a chlorinated organic compound.

Degradation - The transformation of a compound through biological or abiotic reactions.

Dehalogenation - Removal of one or more halogen atoms (e.g., chlorine, fluorine, or bromine) from an organic compound.

Dehalorespiration - Energy-yielding respiratory metabolism that encompasses the reductive removal of halogen atoms from a halogenated compound, such as chlorinated or brominated ethenes.

Delivery performance monitoring - Period of a remedy during which measurements are made to evaluate the effects of a chemical delivery process on the target treatment zone and its surroundings.

Dense nonaqueous phase liquid (DNAPL) - A liquid that is denser than water and does not dissolve or mix easily in water (it is immiscible). In the presence of water, it forms a separate phase from the water. Many chlorinated solvents, such as TCE, are DNAPLs.

Desorption - Opposite of sorption; the release of chemicals from solid surfaces.

Dichloroethene (DCE) - An ethene containing two chlorine atoms that may be used as a degreaser; a dechlorination break- down product of PCE and TCE.

Diffusion (molecular diffusion) - The flux of solute from areas of higher concentration to areas of lower concentration due to random molecular motion.

Dilution - The increase in volume that a given amount of solute occupies. Dilution tends to decrease solute concentration.

Direct push - A method of drilling in which a rod is advanced with percussive techniques. Colloquially referred to as geoprobe.

Dispersion - The spreading of a stream or discrete volume (plume) of solute in a flow field. It is also known as hydrodynamic dispersion. Dispersion is typically considered in terms of mechanical dispersion and molecular diffusion.

Dispersion coefficient - Parameter in the advection-dispersion equation (ADE) that characterizes the rate of dispersion. Possible types include: effective, ensemble, local, macro, and mechanical. Units are length squared over time $[L^2T^{-1}]$.

Dispersivity - A parameter in advection-dispersion transport models. Originally assumed to be a characteristic property of the porous medium, but has been shown to vary with the scale of the problem. Multiplied by the average linear velocity to yield the mechanical dispersion coefficient when using the linear parameterization of dispersion. Units are length [L].

Dual-porosity model - A mathematical model that conceptualizes solute transport as occurring in two overlapping domains: advection and hydrodynamic dispersion in the mobile domain, and rate-limited mass transfer between the mobile and immobile domains

Effective porosity - The porosity of a rock or soil that is actually connected to provide flow through the rock or soil.

Electron - A negatively charged subatomic particle that may be transferred between chemical species in chemical reactions (e.g., oxidation-reduction reactions). Every chemical contains electrons and protons (positively charged particles).

Electron acceptor - Substance that receives electrons (and therefore is reduced) in an oxidation-reduction reaction, which may be abiotic or biotic. Common electron acceptors in the subsurface are oxygen, nitrate, sulfate, ferric iron, and carbon dioxide. Chlorinated solvents (e.g., TCE) can serve as electron acceptors under anaerobic conditions.

Electron donor - Substance that donates electrons (and therefore is oxidized) in an oxidation-reduction reaction, which may be abiotic or biotic.

Emulsified vegetable oil - A formulation in which an edible oil (such as soybean oil) is dispersed into water (e.g., through stirring or use of homogenizers) to form a mixture of oil droplets in water. Emulsifying an oil can greatly improve its ability to disperse in the subsurface.

Emulsion - A suspension of small globules of one liquid in a second liquid with which the first will not mix (e.g., oil and water).

Enhanced *in situ* bioremediation (EISB) - See "*In situ* bioremediation."

Enzyme - A protein created by living organisms to use in transforming a specific compound. The protein serves as a catalyst in the compound's biochemical transformation.

Equilibrium reaction - A chemical or biological reaction that is reversible and will tend to move in either a forward or reverse direction towards a state of chemical equilibrium.

Ex situ - Latin term referring to the removal of a substance from its natural or original position, such as the treatment of contaminated groundwater aboveground.

Fenton's reagent - A solution consisting of hydrogen peroxide and an iron catalyst used to oxidize contaminants. The reagent was discovered by H.J.H. Fenton in the 1890s.

Fermentation - Oxidation or reduction of an organic compound occurring without the uses of an external electron acceptor.

Ferrous salt - A salt containing iron in the plus two valuent state.

Fick's Law - A mathematical equation that quantifies the diffusive mass flux as proportional to the negative concentration gradient.

First-order reaction - Chemical reaction in which the rate is proportional on the concentration of a reactant.

Fluvial - Of, relating to, or happening in a river.

Free radical - See "Radicals."

Full scale - Implementation of a remediation technology at a scale that is intended to represent what would be deployed to treat the entirety of a target treatment zone.

Geochemical - Produced by or involving non-biochemical reactions in the subsurface.

Growth substrate - An organic or inorganic compound used as an electron donor and upon which bacteria can obtain energy for growth.

Half-life - Time required to reduce the concentration of a constituent to half of its initial value.

Heterogeneity - The occurrence of variability of properties in space.

Heterogeneous reaction - A chemical or biological transformation in which the reactants occur in different phases (i.e. between solid and aqueous phases as in a sorption reaction, or between aqueous and gaseous phases as in an aerobic oxidation).

Homogeneous reaction - A chemical or biological transformation in which the reactants exist in a single phase.

Hydraulic conductivity - A measure of the ability of a porous medium to transmit a liquid (typically water) when subjected to a difference in hydraulic head. Has units of length over time $[LT^{-1}]$ (e.g., meters/day). It depends on the permeability of the porous medium and the viscosity of the fluid.

Hydraulic fracturing - Method used to create fractures that extend from a borehole into the surrounding subsurface formations. Fractures are typically maintained by a proppant, a material such as grains of sand or other material that prevent the fractures from closing. Used to increase or restore the ability of the subsurface to transmit fluids.

Hydraulic gradient - Change in hydraulic head per unit distance; a vector that points in the direction of maximum increase of the head.

Hydraulic head - Elevation of a water body above a particular datum level. Specifically, the energy possessed by a unit weight of water at any particular point; measured by the level of water in a manometer at the laboratory scale or by water level in a well, borehole, or piezometer in the field. Under constant density, water flows from points of larger hydraulic head to points of lower head.

Hydraulic residence time - The average time water spends within a specified region of space, such as a reactor or a treatment zone within the subsurface.

Hydrocarbons - Chemical compounds that consist entirely of carbon and hydrogen.

Hydrogen peroxide (H_2O_2) - An unstable compound used especially as an oxidizing and bleaching agent, antiseptic, and as a propellant.

Hydrolysis - The splitting of an organic compound by interaction with water which is also split.

Hydrophilic - Having a strong affinity for water. Hydrophilic compounds tend to be found in the aqueous phase.

Hydrophobic - "Water-fearing." Hydrophobic compounds, such as oils and chlorinated solvents, have low solubilities in water and tend to form a separate nonaqueous phase.

Hydroxyl (^{-}OH) - A chemical group that consists of one atom of hydrogen and one of oxygen and is neutral or negatively charged.

Hydroxyl radical - The neutral form of the hydroxide ion (OH–). Hydroxyl radicals are highly reactive and consequently short-lived. Hydroxyl radicals are produced from natural processes and engineered reactions.

Hypoxic - A condition of "low" or "deficient" oxygen content.

Hysteresis - A retardation of an effect when the forces acting upon a body are changed. For example, the relationship between moisture content and water potential generally differs depending on whether a porous media is being wetted or dried. Similarly, sorption and desorption of a compound may occur at different rates.

Immiscibility - The inability of two or more substances or liquids to readily dissolve into one another, such as oil and water.

Impermeable - Not easily penetrated. The property of a porous media or soil that does not allow, or allows only with great difficulty, the movement or passage of water.

Infiltration gallery - Horizontal wells or trenches that are installed in the unsaturated zone for the purpose of receiving reagents or fluids, which are injected into the gallery so that fluids can percolate downward into a treatment zone.

In situ - Latin term meaning "in place" – in the natural or original position, such as the treatment of groundwater in the subsurface.

***In situ* air stripping** - Treatment system that removes or "strips" volatile organic compounds from contaminated groundwater or surface water by forcing an air stream through the water, causing the compounds to volatilize.

***In situ* bioremediation** - The use of microorganisms to degrade contaminants in place with the goal of producing harmless chemicals as end products. Generally, *in situ* bioremediation is applied to the degradation of contaminants in saturated soils and groundwater, although bioremediation in the unsaturated zone can occur.

***In situ* chemical oxidation (ISCO)** - Technology that oxidizes contaminants in place by adding strong oxidants, such as potassium permanganate or hydrogen peroxide, resulting in detoxification or immobilization of the contaminants.

***In situ* chemical reduction (ISCR)** - Technology that reduces contaminants in place by addition of chemical reductants, such as zero-valent iron, resulting in detoxification or immobilization of the contaminants.

***In situ* thermal treatment** - Treatment system that generates high temperatures to remove and destroy contaminants in place. In practice, three types of technologies have been used – steam injection, electrical resistance heating (generating heat by applying an electrical current) and thermal conductive heating (using electrical subsurface heaters to radiate heat outwards through the solid matrix).

Influent - Water, wastewater, or other liquid that flows into a reservoir, basin, or *in situ* target treatment zone.

Injection well - A well installed for the purpose of injecting remediation agents into an aquifer.

Inorganic compound - A chemical that does not contain a reduced form of carbon, that is a carbon atom with a valence state less than plus four.

Instantaneous reaction - A reaction that occurs very rapidly. As a consequence, for irreversible reactions, the reacting compounds cannot coexist at the same location.

Interfacial tension - The force at the interface between two immiscible liquids (such as a DNAPL and water) that results from the attractive forces between the molecules in the different fluids. Generally, the interfacial tension of a given liquid surface is measured by finding the force across any line on the surface divided by the length of the line segment (so that interfacial tension is expressed as force per unit length, equivalent to energy per unit surface area).

Intrinsic bioremediation - A type of intrinsic remediation that uses the innate capabilities of naturally occurring microbes to degrade contaminants without requiring engineering steps to enhance the process.

Intrinsic remediation - *In situ* remediation that uses naturally occurring processes to degrade or remove contaminants without using engineering steps to enhance the process. Also known as natural attenuation and if process monitoring is carried out, monitored natural attenuation.

Ionization - The physical process of converting an atom or molecule into an ion by adding or removing charged particles, such as electrons or other ions.

Isoconcentration - More than one sample point exhibiting the same concentration. An isoconcentration line or surface is characterized by the same concentration.

Isotope - Any of two or more species of an element in the periodic table with the same number of protons. Isotopes have nearly identical chemical properties but different atomic masses and physical properties. For example, the isotopes chlorine 37 (^{37}Cl) and chlorine 35 (^{35}Cl) both have 17 protons, but ^{37}Cl has two extra neutrons and thus a greater mass.

Isotope fractionation - Selective degradation of one isotopic form of a compound over another isotopic form. For example, microorganisms can transform the ^{35}Cl isotopes of perchlorate more rapidly than the ^{37}Cl isotopes.

Karst - Geologic formation containing irregular limestone deposits with sinks, underground streams, and caverns.

Kinetics - The study of rates; in chemical kinetics it is the study of the rate of a chemical process.

Kinetically controlled reaction - A chemical or biological transformation that occurs at a rate faster than that of the other fate and transport processes governing a particular species.

Lactate - A salt or ester of lactic acid.

Leachate - Solution formed when a fluid (e.g., water) percolates through a permeable medium. When passing through contaminated media, the leachate may contain contaminants in solution or in suspension.

Light nonaqueous phase liquid (LNAPL) - A nonaqueous phase liquid with a specific gravity less than 1.0. Because the specific gravity of water is 1.0, most LNAPLs float on top of a groundwater table. Most common petroleum hydrocarbon fuels and lubricating oils are LNAPLs.

Liquid chromatography - A chemical separation technique in which a mobile phase (a liquid) passes over or through a stationary phase.

Local equilibrium assumption (LEA) - An assumption that reactions of interest (such as NAPL dissolution or sorption) are "sufficiently fast" so that local equilibrium can be assumed. In the case of dissolution, it could be that the dissolution occurs rapidly compared to flow through the porous media so that the dissolved concentration immediately adjacent to the NAPL is equal to the NAPL's solubility. For sorption, it means that sorption and desorption occur rapidly relative to flow through the porous media so that any change in concentration is immediately accompanied by a corresponding change in sorbed mass.

Local dispersion - Dispersion at the pore-to-Darcy scale that does not account for large-scale heterogeneities, which are accounted for in macrodispersion.

Log K_{ow} - Logarithmic expression of the octanol-water partition coefficient (K_{ow}); a measure of the equilibrium concentration of a compound between octanol and water.

Long-term monitoring (LTM) - Monitoring conducted after a remedial measure achieves its objectives, to ensure continued protection and performance.

Longitudinal dispersion - Dispersion in the direction of bulk flow.

Low permeability media (LPM) - A region of low permeability within the subsurface. Can act as a localized barrier to groundwater flow or NAPL migration. Can initially act as a sink for dissolved and sorbed contaminant mass, and later as a secondary source via back diffusion.

Macroscopic - Large enough to "homogenize" or "average over" smaller-scale processes. For example, the Darcy scale is macroscopic in relation to the pore scale.

Macrodispersion - Spreading of solute due to variability in local advective velocity at Darcy or field scales.

Magnetite - Common mineral of black iron oxide. Capable of reducing chlorinated solvents in groundwater.

Mass balance - An accounting of the total inputs and outputs to a system. For dissolved plumes, it refers to a quantitative estimation of the mass loading to a dissolved plume and the mass attenuation capacity within the affected subsurface environment.

Mass discharge - The rate of mass flow across an entire plume at a given location. Also referred to as "total mass flux" or "integrated mass flux." Expressed in units of mass per time (e.g., grams per day [g/day]), mass discharge essentially integrates several individual mass flux measurements (expressed as mass/area/time, such as grams per square meter per day [g/m^2/day]).

Mass flux - The rate of mass flow across a unit area (typically measured in g/m2/day). Typically calculated by integrating measured groundwater contaminant concentrations across a transect. Often incorrectly used interchangeably with mass discharge or mass loading (expressed in g/day) to describe the mass emanating from a source zone or the mass passing a given transect across the plume.

Mass transfer - The general term for the physical processes involving molecular and advective transport of atoms and molecules within physical systems. In this context, the term refers to the movement of solute mass between different locations such as occurs in absorption, evaporation, precipitation, and distillation. Here, a location usually means a stream, phase, domain, fraction, or component.

Maximum contaminant level (MCL) - Standards set by the U.S. Environmental Protection Agency (USEPA) or state equivalent for drinking water quality that provide for a legal threshold limit on the amount of a hazardous substance that is allowed in drinking water under the Safe Drinking Water Act. The limit is usually expressed as a concentration in milligrams or micrograms per liter of water.

Mechanical dispersion - Transport phenomena due to the variations in local velocity, both in magnitude and direction along the tortuous flow paths and between adjacent flow paths as a result of the velocity distribution, that causes a solute (tracer) mass to spread and occupy an ever-increasing volume of the porous media.

Media - Groundwater, porous media, soil, air, surface water, or other parts of an environmental system that can contain contaminants and be the subject of regulatory concern and remediation activities.

Metabolism - The chemical reactions in living cells that are necessary to maintain life, including reproduction and the conversion of food sources to energy and new cell mass.

Metabolite - The intermediates and products of metabolism.

Metal chelators - Chemicals that form multiple bonds with a single metal ion to produce soluble, complexed molecules. Used to enhance solubility and uptake of metals or to inhibit production of precipitates or scale.

Methanogen (methanogenic archaea) - A microorganism that exists in anaerobic environments and produces methane as the end product of its metabolism. Some methanogens use carbon dioxide as an electron acceptor to produce methane while others obtain energy by splitting acetate into carbon dioxide and methane.

Methanogenesis - Process of producing methane during biological metabolism.

Methanotroph (methanotrophic bacteria) - A microorganism that is able to oxidize methane for energy.

Micelle - An aggregate of surfactant molecules dispersed in a liquid colloid. A typical micelle in aqueous solution forms an aggregate with the hydrophilic "head" regions in contact with surrounding solvent, sequestering the hydrophobic single-tail regions in the micelle center.

Microcosm - A laboratory vessel established to resemble the conditions of a natural environment.

Microemulsion - Clear, stable, isotropic liquid mixtures of oil, water and surfactant, frequently in combination with a cosurfactant. The aqueous phase may contain salt(s) or other ingredients; the "oil" may actually be a complex mixture of different hydrocarbons and olefins. Microemulsions form upon simple mixing of the components and do not require the high shear conditions generally used in the formation of ordinary emulsions. The two basic types of microemulsions are direct (oil dispersed in water) and reversed (water dispersed in oil).

Microorganism (microbe) - An organism of microscopic or submicroscopic size. Bacteria are microorganisms.

Mineral - A naturally occurring solid formed through geological processes that has a characteristic chemical composition, a highly ordered atomic structure, and specific physical properties. A rock, by comparison, is an aggregate of minerals and/or mineraloids and need not have a specific chemical composition.

Mineralization - The complete degradation of an organic chemical to carbon dioxide, water, and possibly other inorganic compounds or elements.

Miscible - Two or more liquids that can be mixed and will remain mixed under normal conditions.

Mixing - A process by which two or more substances are joined together such as occurs with the overlap of plumes containing different substances.

Mixing ratio - The space and time (for transient transport) dependent ratio of an injected solution in a mixture with an ambient (background) solution.

Mobile porosity - Ratio of the pore space filled by mobile fluids to the total pore space.

Mole fraction - The number of moles of a component of a solution divided by the total number of moles of all components.

Molecular diffusion - see “Diffusion.”

Monitored natural attenuation (MNA) - Refers to the reliance on natural attenuation processes (within the context of a carefully controlled and monitored site cleanup approach) to achieve site-specific remediation objectives.

Monitoring well - A well installed for the purpose of monitoring groundwater quality in an aquifer, and not used to facilitate the injection of remediation agents.

Monod kinetics - Equation similar to the Michaelis-Menten equation for enzyme kinetics that relates a microbial culture’s specific growth rate to the concentration of rate limiting substrates.

Monte Carlo simulation - A problem-solving technique used to approximate the probability of certain outcomes by running multiple trial runs, called simulations, using random variables. Monte Carlo methods allow evaluation of complex situations involving random behavior, such as games of chance, and can help reduce uncertainty in estimating future outcomes in areas such as risk assessment or actuarial analyses.

Mudstones - A fine-grained sedimentary rock whose original constituents were clays or muds – hardened mud; a mix of silt and clay-sized particles.

Nanoscale - Generally deals with structures of the size 100 nanometers (nm) or smaller. For example, reactive iron produced in this size range is referred to as nanoscale iron.

Natural attenuation - Reduction in the mass, toxicity, mobility, volume, or concentration of contaminants in soil or groundwater caused by natural processes that act without human intervention. These *in situ* processes include biodegradation, dispersion, dilution, sorption, volatilization, radioactive decay, and chemical or biological stabilization, transformation, or destruction of contaminants.

Natural organic matter (NOM) - A form of naturally occurring organic matter that has been broken down to some base-level compounds (such as cellulose, chitin, protein, lipids, etc.). NOM provides nutrients to insects, bacteria, fungi, fish, and other organisms at the base of the food chain.

Natural oxidant demand (NOD) - Refers to one or more chemical reactions that can occur between an oxidant (typically permanganate) and naturally occurring substances in the subsurface (e.g., NOM, reduced metals, minerals). The oxidant consumed during these reactions is unavailable for reaction with the target COCs.

Nonaqueous phase liquid (NAPL) - An organic liquid that does not mix easily with water, and thus maintains itself as a separate phase from water.

Non-wetting DNAPL - Wettability is a measure of a liquid's relative affinity for a solid. Where two liquid phases are present, the "wetting" fluid will preferentially spread over the solid surface at the expense of the "non-wetting" fluid. Wettability is depicted by the concept of a Contact Angle. Since wettability conventionally refers to the nonaqueous phase, the angle is measured through the aqueous phase. A majority of DNAPL contaminants are non-wetting (water occupies the smaller pore spaces and preferentially spreads across solid surfaces while the DNAPLs are restricted to the larger openings).

Numerical model - A mathematical model that uses a numerical time-stepping procedure to estimate behavior of a system over time (as opposed to an analytical model). The mathematical solution is represented by a generated table and/or graph. Numerical models require greater computing power, but they can allow more realistic simulations of complex systems.

Octanol-water partition coefficient (Kow) - Ratio of the concentration of a chemical in octanol and in water at equilibrium and at a specified temperature. Octanol with the chemical formula $CH_3(CH_2)_6CHOH$ is an organic solvent used as a surrogate for NOM. This parameter is used in many environmental studies to help determine the fate of chemicals in the environment. Inversely related to aqueous solubility (a high Kow indicates a compound will preferentially partition into an organic phase rather than into water).

Operation and maintenance (O&M) - Activities conducted at a site to ensure a technology or approach is effective and operating properly. The term O&M covers a wide range of activities, from overseeing the proper functioning of a system to conducting monitoring to evaluate the effectiveness of an action.

Organic - Referring to or derived from living organisms. In chemistry, organic compounds contain reduced forms of carbon in combination with hydrogen and other elements.

Oxic - Containing oxygen or oxygenated. Often used to describe an environment, a condition, or a habitat in which oxygen is present.

Oxidant - A chemical compound that gains electrons in a chemical reaction. An oxidant can also be referred to as an oxidizing agent. As a result of the reaction, the oxidizing agent becomes reduced.

Oxidant concentration - The concentration (mass/volume) of an oxidant in a liquid oxidant solution.

Oxidant dose - See "Oxidant loading rate". Often incorrectly used as a synonym for oxidant concentration.

Oxidant loading rate - A design parameter that is the ratio of the mass of oxidant applied to the mass of subsurface solids in the target treatment zone, usually expressed in units of g/kg or mg/kg.

Oxidant persistence - Refers to the ability of an oxidant (usually applied to CHP, persulfate and ozone) to remain present and reactive over time in the subsurface after its initial delivery via an injection well, probe, or other method.

Oxidation - Transfer (loss) of electrons from a substance, such as an organic contaminant. Oxidation of compounds can supply energy that microorganisms use for growth and reproduction. Often but not always, oxidation results in the addition of an oxygen atom and/or the loss of a hydrogen atom.

Oxidation-reduction potential (ORP) - The tendency of a solution to either gain or lose electrons when it is subject to change by introduction of a new species. A solution with a higher (more positive) reduction potential than the new species will have a tendency to gain electrons from the new species (to be reduced by oxidizing the new species); a solution with a lower (more negative) reduction potential will have a tendency to lose electrons to the new species (to be oxidized by reducing the new species). A positive ORP indicates the solution is oxidizing, while a negative ORP indicates reducing conditions are dominant.

Ozone (O_3) - A simple triatomic molecule, consisting of three oxygen atoms. An allotrope of oxygen that is much less stable than the diatomic oxygen (O_2). A powerful oxidizing agent. Unstable at high concentrations, decaying to ordinary diatomic oxygen.

Partition coefficient (Kd) - Ratio of the concentrations of a substance in a liquid phase in contact with a solid phase. Measure of the sorption potential, whereby a contaminant is distributed between the solid and water phase.

Partitioning interwell tracer testing (PITT) - Method to quantify the volume of NAPL in a contaminated aquifer by injecting and recovering a tracer that will partition into the NAPL phase. Provides information about the NAPL volume distribution in a relatively large-scale area.

Passivation - Process of making a material "passive" in relation to another material. Often used to refer to the formation of a hard non-reactive surface film on many reactive or corrosive materials (such as aluminum, iron, zinc, magnesium, copper, stainless steel, titanium, and silicon) that inhibits further reactivity.

Passive injection (passive treatment) - Remediation approach involving additions of amendments to the subsurface on a one-time or very infrequent basis.

Passive treatment - *In situ* bioremediation approach in which amendments are added to the subsurface on a one-time or infrequent basis. Passive treatment relies on the use of slow-acting materials, which can be injected into the subsurface or placed in trenches or wells.

Pathogen - Microorganisms (e.g., bacteria, viruses, or parasites) that can cause disease in humans, animals, and plants.

Peclet number (Pe) - A dimensionless quantity that expresses the relative importance of advection and diffusion or dispersion of solutes.

Percarbonate - Any of a family of perhydrates of carbonate compounds, such as sodium percarbonate ($2Na_2CO_3$–$3H_2O$). Percarbonate compounds can undergo chemical reactions under certain environmental conditions to yield free radicals.

Perchlorate - A salt derived from perchloric acid ($HClO_4$), with the chlorine atom present at an oxidation state of +7. May occur naturally in small concentration, but is a potent oxidizer that has been manufactured and used for solid rocket propellants, explosives, and in road flares.

Perchloroethene (PCE, perchloroethylene, tetrachloroethene, tetrachloroethylene) - A colorless, nonflammable organic solvent, $Cl_2C{=}CCl_2$, used in dry-cleaning solutions and as an industrial solvent.

Percolation - The movement and filtering of fluids through porous materials.

Permanganate - General name for a chemical compound containing the manganate (VII) ion, (MnO_4^-). Because manganese is in the +7 oxidation state, the manganate (VII) ion is a strong oxidizing agent.

Permeability - A measure of the ability of a material, such as soil or aquifer porous media, to transmit fluids such as water. It is the measure of the relative ease of fluid flow under unequal pressure. Units of measurement are length squared [L^2].

Permeable reactive barrier (PRB) - A permeable wall or vertical zone containing reactive media or creating a set of reaction conditions and oriented to intercepting and remediating a contaminant plume as groundwater migrates through the wall or zone.

Peroxone - A combination of ozone and hydrogen peroxide yielding a product not requiring a catalyst and used to treat contaminated soil and water.

Persulfate - A strong oxidant containing the persulfate anion such as sodium persulfate ($Na_2S_2O_8$). Persulfate compounds can be activated by transition metals, heat, or elevated pH to yield sulfate free radicals.

pH - Equals the negative logarithm of the hydrogen ion concentration, and used to express the intensity of the basic or acid condition of a liquid; may range from 0 to 14, where 0 is the most acid and 7 is neutral. Natural waters usually have a pH between 6.5 and 7.5.

Photolysis - The splitting of molecules by means of light energy.

Physical equilibrium - At a location and time of interest, a condition in which enough time has occurred for diffusive forces to counterbalance advective forces.

Phytoremediation - The use of plants and in some cases the associated rhizosphere (root zone) microorganisms for *in situ* remediation of contaminants.

Pilot-scale - A scale of demonstration, testing or evaluation under laboratory or field conditions that can incorporate certain features and processes that are representative of a full-scale system. A pilot-scale study is often used to investigate the design and performance of a full-scale system. See "Full-scale" and "Pilot test".

Pilot test - A trial run of a remediation technology implemented at the field scale. Performed to assess the feasibility of the remediation technology and/or to collect field-scale data on which to base full-scale design. Generally conducted at smaller scale than full- scale treatment.

Plug-flow reactor (PFR) - A reactor in which substances are transported downstream in the reactor as a "plug". Every particle stays in the company of particles of the same age and there is no mixing between particles introduced earlier or later.

Plume - A zone of environmental media containing contaminants. As applied to groundwater, it usually originates from a contaminant source zone and extends under the effects of momentum, diffusion, etc.

Pneumatic fracturing - Injection of gas into the subsurface at pressures exceeding the natural *in situ* pressures and at flow volumes exceeding the natural permeability of the subsurface. Creates a network of artificial fractures in a geologic formation that can facilitate removal of contaminants out of the geologic formation; may be used to introduce remedial agents.

Polychlorinated biphenyls (PCBs) - Organic compounds composed of 1–10 chlorine atoms attached to two joined benzene rings (biphenyls) and used in electrical transformers and capacitors for insulating purposes and in gas pipeline systems as lubricant. The sale and new use of these chemicals were banned by U.S. law in 1979.

Polycyclic aromatic hydrocarbon (PAH) - Chemical compound that consists of fused aromatic rings and does not contain heteroatoms or carry substituents. PAHs occur in oil, coal, and tar deposits, and are produced as byproducts of fuel burning (whether fossil fuel or biomass). As a pollutant, they are of concern because some compounds have been identified as carcinogenic, mutagenic, and teratogenic.

Polymerase chain reaction (PCR) - Technique to amplify a single or few copies of a specific DNA sequence by several orders of magnitude. Allows detection of a target gene or parts of a gene, even when present at low concentrations in soil or groundwater, for example. PCR relies on thermal cycling, consisting of cycles of repeated heating and cooling of the reaction for DNA melting and enzymatic replication.

Polyvinyl chloride (PVC) - A tough, environmentally indestructible plastic formed through polymerization of vinyl chloride that releases hydrochloric acid when burned.

Pore volume (PV) - The volume of void space within a porous medium (e.g., soil). Used to determine a design metric termed the number of pore volumes that is the ratio of the volume of injected reagents to the volume of pore space in a target treatment zone.

Pore scale - The scale of the pore. The pore velocity is the velocity measured at this scale.

Porosity - The fraction of the subsurface volume filled with pores or cavities through which water or air can move.

Potassium permanganate ($KMnO_4$) - A chemical oxidant commonly used for ISCO; characterized by its purple to pink color in solution.

Potassium persulfate ($K_2S_2O_8$) - A chemical oxidant commonly used for ISCO.

Potentiometric surface map - A contour map that represents the top of the groundwater surface in an aquifer.

Precipitate - The formation of a solid from a solution or suspension by chemical or physical change.

Pressure transducer - A sensor device that converts pressure into an analog electrical signal, allowing measurement.

Primary substrates - The electron donors and electron acceptors that are essential to ensure the growth of a given microorganism. These compounds can be viewed as analogous to the food and oxygen that are required for human growth and reproduction.

Propagation reaction - Chemical reactions involving free radicals in which the total number of free radicals remains constant.

Pseudo first-order reaction - A second-order reaction in which one of the reactants is present in such great amounts that its effect is not seen and the reaction thus behaves as first-order.

Pseudokinetic - Rates of transformation that appear kinetically controlled but are associated with (multiple) equilibrium reactions in heterogeneous media.

Pump-and-treat (P&T) - A remediation approach in which groundwater is extracted from the subsurface using a network of pumping wells and treated *ex situ* to remove COCs. P&T also can be applied as a containment strategy.

Pyrite - An iron sulfide mineral with the formula FeS_2. The most common of the sulfide minerals. Also called fool's gold.

Radicals - Atoms, molecules, or ions with unpaired electrons, which are highly reactive. Chemical oxidants like H2O2 can be activated during use in ISCO and yield one or more types of radicals (often call free radicals), which serve as the primary oxidizing agents.

Radioactive decay - The process by which an atomic nucleus of an unstable atom decays to form other atoms and in the process yields energy.

Radius of influence (ROI) - The radial distance from the center of an injection point or well to the point where there is no significant impact from the injected material.

Raoult's Law - Relates the vapor pressure of components to the composition of the solution. If the components are sufficiently similar, the vapor pressure of the solution will depend on the vapor pressure of each chemical component and the mole fraction of the component present in the solution. Used to predict the soluble concentrations of each compound in a mixture of similar compounds (e.g., benzene, toluene, ethyl benzene and xylenes [BTEX] in gasoline) that is in equilibrium with the aqueous phase, based on the mole fraction of each compound in the mixture.

Rebound - see "Contaminant rebound."

Recharge - Process by which water is added to a zone of saturation, usually by percolation from the ground surface (e.g., the recharge of an aquifer via precipitation and infiltration). Also, the amount of water added.

Recirculation wells - A groundwater well that is specially designed so groundwater enters and exits the well and causes a spherical recirculation pattern in the groundwater formation. While groundwater is within the well, treatment can be achieved (e.g., by air stripping or sorption processes) and *in situ* treatment in the groundwater formation can also be enabled if amendments are added and carried out as groundwater exits the well (e.g., bionutrients, oxidants).

Redox reactions - Reduction/oxidation reactions are those in which atoms have their oxidation number changed. For example, carbon may be oxidized by oxygen to yield carbon dioxide or reduced by hydrogen to yield methane. The redox potential (ORP) reflects the tendency of a chemical species to acquire electrons and thereby be reduced. In a redox reaction, one chemical species – the reductant or reducing agent – loses electrons and is oxidized, and the other – the oxidant or oxidizing agent – gains electrons and is reduced.

Reducing - Environmental conditions that favor a decrease in the oxidation state of reactive chemical species (e.g., reduction of sulfates to sulfides).

Reduction - Transfer of electrons to a substance such as oxygen; occurs when another substance is oxidized.

Reductive dechlorination - Reaction involving removal of one or more chlorine atoms from an organic compound and their replacement with hydrogen atoms. A subset of reductive dehalogenation. Key reaction for anaerobic degradation of chlorinated solvents.

Reductive dehalogenation - The process by which a halogen atom (e.g., chlorine or bromine) is replaced on an organic compound with a hydrogen atom.

Remedial action - The actual construction or implementation phase of a contaminated site cleanup following remedial design.

Remediation - Cleanup technology or approach used to remove or contain contamination.

Remediation goal - Goals define what the remedial actions are intended to achieve or accomplish. Goals can be general for the overall remediation system (or treatment train), or they may be specific to one of the technologies in the treatment train (see "Treatment Goals" below). Under CERCLA goals are often numeric levels. For example, during the Feasibility Study process under CERCLA, preliminary remedial goals (PRGs) are the concentrations used to define the area to be remediated and to what level. PRGs become remedial goals (RGs) once a ROD specifies the selected remedy and modifies the PRGs.

Remediation objective - Remediation objectives can be established to state the purpose for which remediation is intended. They often tend to be high-level outcomes that are desired but, in and of themselves, are often not directly measurable. For example, an objective might be stated as: remediation of a contaminated site to reduce the risk to human health for unrestricted current and future land use. RAOs under the CERCLA represent an example of an objective.

Reynolds number (Re) - A dimensionless quantity that expresses the relative importance of inertial forces compared to viscous forces in a flow system. A small Reynolds number is associated with laminar flow; a large Reynolds number is associated with turbulent flow.

Residence time (retention time) - The average amount of time that a particle spends in a particular system.

Residual NAPL - Saturation level below which NAPL will no longer freely drain.

Residual saturation - Saturation level below which water will no longer freely drain.

Retardation - Slowing of the movement of substances in an aquifer relative to the groundwater velocity. For example, a contaminant plume exhibiting a retardation factor of 5 moves one-fifth as fast as the water itself or a non-reactive tracer such as chloride, which has a retardation factor near 1.0.

Reverse osmosis - A treatment process used in water systems by adding pressure to force water through a semi-permeable membrane. Removes most drinking water contaminants. Also used in wastewater treatment.

Salinity - Percentage of salt in water.

Saturated zone - Part of the subsurface that is beneath the water table and in which the pores are filled with water.

Saturation - Refers to the fraction of porous media pore space that contains fluid (for example, water or NAPL). If no fluid is specified, it is generally taken to refer to water saturation.

Scavenger - Refers to a substance that can react with a free radical to inhibit the free radical from participating in oxidation reactions with COCs. Scavengers include organic compounds like formate and ethanol and inorganic compounds like bicarbonate and carbonate.

Second-order reaction - A chemical reaction with a rate proportional to the concentration of the square of a single reactant or the product of the concentration of two reactants: rate $= k[A][B]$ or $k[A]^2$.

Sediments - Soil, sand, and minerals carried from land into water bodies.

Seepage velocity - The average pore water velocity, also known as average linear velocity. Since groundwater flow actually occurs only through interconnected pores and not through the entire subsurface volume – used in calculating the specific discharge or Darcy velocity (q) – the seepage velocity (v) is equal to the Darcy velocity divided by the porosity (n), or $v = q/n$.

Semi-passive treatment - *In situ* remediation approach in which amendments are added to the subsurface intermittently (at intervals of a few weeks to a few months).

Sheer stress - That component of stress which acts tangential to a plane through any given point in a body.

Site characterization - The collection of environmental data that are used to describe the conditions at a property and delineate the nature and extent of a site's contamination.

Slug test - A particular type of aquifer test where water is quickly added to or removed from a groundwater well, and the change in hydraulic head is monitored through time, to determine the near-well aquifer characteristics. It is a method used by hydrogeologists and civil engineers to determine the transmissivity and storativity of the subsurface material surrounding the well.

Sodium permanganate ($NaMnO_4$) - A chemical oxidant used for ISCO.

Sodium persulfate ($Na_2S_2O_8$) - A chemical oxidant commonly used for ISCO.

Soil mixing - An approach used to deliver and distribute chemical oxidants (or other remedial amendments) to contaminated soil.

Soil organic matter (SOM) - Organic constituents in the soil, including undecayed plant and animal tissues, their partial decomposition products, and the soil biomass. SOM includes high–molecular-weight organic materials (such as polysaccharides and proteins), simpler substances (such as sugars, amino acids, and other small molecules), and humic substances.

Soil oxidant demand (SOD) - Refers to one or more chemical reactions that can occur between an oxidant and the soil or porous media in the subsurface. The oxidant consumed during these reactions is unavailable for reaction with the target COCs. NOD is the preferred terminology for the same set of nonproductive reactions and NOD is used in this volume (see NOD).

Soil vapor extraction (SVE, soil venting) - An established technology for the *in situ* remediation of VOCs in the vadose (unsaturated) zone. The process removes soil vapor contaminated with VOCs and enhances the mass transfer of VOCs from the soil pores to the vapor phase by applying a vacuum to extract soil contaminants and gases.

Solubility - Ability of a substance to dissolve (or solubilize). The solubility of a specific solute is its maximum concentration in a given solvent at a reference temperature.

Solute - A substance dissolved in another substance. A relevant example is an oxidant dissolved in groundwater: oxidant is the solute and groundwater is the solvent.

Solvent - A substance, usually a liquid, capable of dissolving another substance.

Sorb - To take up and hold by either adsorption or absorption.

Sorption - Collection of a substance on or within a solid and held by physical or chemical attraction. Can refer to either absorption (in which one substance permeates another) or adsorption (surface retention of solid, liquid, or gas molecules, atoms, or ions).

Sorption isotherm - Describes the sorption of a material onto or within a surface at constant temperature. Determined by comparing the sorbed concentration of a compound to its concentration in solution. Describes the ability of a dissolved contaminant to adsorb or absorb onto or into the solid particles (soil or particulates).

Source strength - The mass discharge from a source zone. Represents the mass loading to a plume per unit time (e.g., grams TCE released per day).

Source zone - A subsurface zone that serves as a reservoir of contaminants that sustains a plume of dissolved contaminants in groundwater. Includes the subsurface material that is or has been in contact with the contaminants originally released into the subsurface (e.g., DNAPLs for chlorinated solvents); the source zone mass includes the sorbed and aqueous phase contaminants, as well as any residual NAPL.

Sparge - Injection of gases into water. As used for *in situ* remediation, air (or another gas) is sparged to strip dissolved VOCs and/or oxygenate groundwater to facilitate aerobic biodegradation of organic compounds. Ozone in air is sparged to oxidize organic compounds.

Specific conductance (electrical conductivity) - Rapid method of estimating the dissolved solid content (total dissolved solids) of a water by testing its capacity to carry an electrical current.

Specific discharge - The flow rate through a cross-section of porous medium divided by the total area of that cross section. Units of length per time [LT^{-1}].

Specific storage - The amount of water that a portion of an aquifer releases from storage per unit mass or volume of aquifer, per unit change in hydraulic head, while remaining fully saturated.

Specific yield - The volumetric fraction of the bulk aquifer volume that a given aquifer will yield when all the water is allowed to drain out of it under the forces of gravity.

Spreading - The stretching and deformation of a (contaminant) plume.

Stabilization/solidification - Remediation technique in which contaminants are physically bound or enclosed within a stabilized mass (solidification) or their mobility is reduced due to chemical reactions induced between a stabilizing agent and the contaminants (stabilization).

Stagnation point - A point in a flow field where the local velocity of the fluid is zero.

Stabilizer - Term used to describe a substance that can reduce the rate of reaction of a chemical oxidant during transport in the subsurface.

Stakeholder - A person (other than regulators, owners, or technical personnel) who has a legitimate interest in a contaminated site.

Steady-state - A condition of a physical system or device that does not change over time or in which any one change is continually balanced by another, such as the stable condition of a system in equilibrium.

Steam enhanced remediation (SER) - An *in situ* thermal treatment technology involving steam injection and aggressive vapor and liquid extraction to mobilize and remove organic contaminants from a source zone.

Steric effects - The influence of the structural configuration of reacting substances upon the rate, nature, and extent of reaction.

Sterilization - The removal or destruction of all microorganisms, including pathogenic and other bacteria, vegetative forms, and spores.

Stoichiometry - The quantitative (measurable) relationships between the reactants and products in a balanced chemical equation.

Storativity - The volume of water an aquifer releases from or takes into storage per unit surface area of the aquifer per unit change in head. It is equal to the product of specific storage and aquifer thickness. In an unconfined aquifer, the storativity is equivalent to the specific yield. Also called storage coefficient.

Streamline (flow line) - A line that is everywhere tangent to the flow velocity vector. This shows the direction a fluid element will travel in at any point in time.

Streamtube - Channels between streamlines.

Stratum (strata) - A layer of subsurface media with internally consistent characteristics that distinguishes it from contiguous layers. Each layer is generally one of a number of parallel layers that lie one upon another, laid down by natural forces. Typically seen as ands of different colored or differently structured material exposed in cliffs, road cuts, quarries, and river banks.

Substrate - The reactant which is consumed during a catalytic or enzymatic reaction.

Sulfate radical - A radical which can be produced during ISCO by activation of sodium persulfate.

Sulfate-reducing bacteria (SRB, sulfate reducer) - Bacteria that convert sulfate to hydrogen sulfide. Often play important roles in the oxygen-limited subsurface.

Superoxide radical anion - An anion with the chemical formula O_2-. It is important as the product of the one-electron reduction of dioxygen O_2, which occurs widely in nature. With one unpaired electron, the superoxide ion is a free radical and, like dioxygen, is paramagnetic. Superoxide is biologically quite toxic and is deployed by the immune system to kill invading microorganisms. Because superoxide is toxic, nearly all organisms living in the presence of oxygen contain isoforms of the superoxide scavenging enzyme, superoxide dismutase, which is an extremely efficient enzyme; it catalyzes the neutralization of superoxide nearly as fast as the two can diffuse together spontaneously in solution.

Surfactant - A material that can greatly reduce the surface tension of water when used in very low concentrations. Primary ingredient of many soaps and detergents.

Surfactant flushing (surfactant enhanced aquifer remediation [SEAR]) - Remediation technology involving injection of a solution of surfactants into a subsurface containing NAPLs. Surfactants increase the effective aqueous solubility of the NAPL constituents, greatly enhancing NAPL removal during flushing.

Target treatment zone (TTZ) - The portion of the subsurface that the remediation technology or approach is intended to treat.

Thermodynamics - The study of the conversion of energy into work and heat and its relation to macroscopic variables, such as temperature and pressure.

Tortuosity - A parameter used in some models. It represents the actual length of a fluid flow path in porous media, which is sinuous in form, divided by the straight-line distance between the ends of the flow path.

Total concentration - Linear combination of a species and the stoichiometric proportion of its reaction products such that the total concentration does not change upon reaction (it is conservative).

Total dissolved solids (TDS) - Combined content of all inorganic and organic substances in a liquid that are present in a molecular, ionized or micro-granular (colloidal sol) suspended as well as dissolved forms.

Total organic carbon (TOC) - A measure of the mass of carbon bound in organic compounds in a substance (e.g., soils, sediments, and water). Often used as a nonspecific indicator of water quality.

Toxicity - The degree to which a substance or mixture of substances can cause harm to organisms. Acute toxicity involves harmful effects in an organism through a single or short-term exposure. Chronic toxicity is the ability of a substance or mixture of substances to cause harmful effects over an extended period.

Tracer test - Used to "trace" the path of a migrating fluid. For groundwater applications, tracer tests are commonly conducted by dissolving a tracer chemical into groundwater at concentrations that do not significantly change the aqueous density. Tracer chemicals must behave conservatively, meaning that no mass is lost through reaction or partitioning into differing phases. Bromide ion is often employed as a solute tracer (added to groundwater as potassium or sodium bromide salt).

Transmissivity - Rate at which water of a prevailing density and viscosity is transmitted through a unit width of an aquifer or confining bed under a unit hydraulic gradient (units of area/time, e.g., ft^2/day). A function of properties of the liquid, the porous media, and the thickness of the porous media. For homogeneous aquifer, equal to hydraulic conductivity (K) times aquifer thickness.

Transverse dispersion - Dispersion in a direction perpendicular to the bulk flow (see also "dispersivity").

Transport - The processes of moving solutes (in fluid), namely, advection, diffusion, and dispersion.

Treatability test - A means of evaluating the suitability of treatment technologies or processes prior to their implementation. Treatability tests are commonly carried out under laboratory conditions.

Treatment goal - Treatment goals are specific criteria by which the successful completion of an activity can be determined.

Treatment performance monitoring - Monitoring to obtain data concerning the effectiveness of a technology or approach and achievement of treatment goals.

Transient - A condition of a physical system or device that is time-dependent.

Trichloroethane (TCA) - An industrial solvent (CH_3CCl_3) also called methyl chloroform. Occurs in two isomers: 1,1,1-TCA and 1,1,2-TCA.

Trichloroethene (TCE, trichloroethylene) - A stable, low boiling point colorless liquid ($CH_3Cl=CHCl_2$). Used as a solvent or metal-degreasing agent and in other industrial applications. Toxic if inhaled and a suspected carcinogen.

Turbulence (turbulent flow) - A flow regime characterized by chaotic property changes. This includes low momentum diffusion, high momentum convection, and rapid variation of pressure and velocity in space and time.

Unconfined aquifer - An aquifer that has no overlying confining impermeably layer.

Unsaturated zone - The region of the subsurface above the groundwater table where media pores are not fully saturated, although some water may be present. Also called the vadose zone.

Upscaling - The process of averaging over the local scale in order to determine parameters at the larger scale of interest that are consistent the important processes at the local scale. Mathematical upscaling methods include volume averaging, homogenization theory, and moment methods.

Vadose zone - The region of the subsurface above the groundwater table where pores are partially or largely filled with air. Also called the unsaturated zone.

Vapor intrusion - Migration of volatile chemicals from the subsurface into overlying buildings.

Vapor pressure - A measure of a substance's propensity to evaporate. The force per unit area exerted by vapor in an equilibrium state with surroundings at a given pressure. Increases exponentially with an increase in temperature. A relative measure of chemical volatility, vapor pressure is used to calculate water partition coefficients and in the determination of volatilization rate.

Vaporization - Conversion of a substance from the liquid or solid phase to the gaseous (vapor) phase.

Vinyl chloride (VC) - A chemical compound ($CH_2{=}CHCl$) that is highly toxic and known to be carcinogenic. A colorless compound and an important industrial chemical chiefly used to produce the polymer PVC.

Viscosity - The molecular friction within a fluid that produces flow resistance. It describes the resistance of a fluid which is being deformed by shear stress.

Volatile - Evaporates readily at normal temperatures and pressures.

Volatile organic compound (VOC) - Any organic compound that has a high enough vapor pressure under normal conditions to significantly vaporize and transfer from a liquid to a gas phase.

Volatilization - Transfer of a chemical from the liquid to the gas phase (as in evaporation).

Water solubility - The maximum amount of the chemical that will dissolve in pure water at a specified temperature.

Water table - The top of an unconfined aquifer at which the pressure is equal to that of the atmosphere. It is the surface between the vadose zone and the groundwater. Indicates the level below which subsurface porous media are saturated with water.

Wellhead - The assembly of fittings, valves, and controls located at the land surface and connected to the flow lines, tubing, and casing of the well so as to control the flow from a groundwater zone.

Wettability - The relative degree to which a fluid will spread into or coat a solid land surface in the presence of other immiscible fluids.

Zero-order reaction - Chemical reaction in which the rate is independent of the concentrations of the reactants.

INDEX

D

E

F

G

T